Electrochemistry of Immobilized Particles and Droplets

Fritz Scholz • Uwe Schröder • Rubin Gulaboski •
Antonio Doménech-Carbó

Electrochemistry of Immobilized Particles and Droplets

Experiments with Three-Phase Electrodes

2nd Edition

Springer

Fritz Scholz
Institute of Biochemistry
University of Greifswald
Greifswald, Germany

Uwe Schröder
Inst. of Environm. and Sustain. Chemistry
Technical University Braunschweig
Braunschweig, Germany

Rubin Gulaboski
Faculty of Medical Sciences
Goce Delcev University
Stip, Macedonia

Antonio Doménech-Carbó
Dept. of Analytical Chemistry
University of Valencia
Burjassot, Spain

ISBN 978-3-319-33043-3 ISBN 978-3-319-10843-8 (eBook)
DOI 10.1007/978-3-319-10843-8
Springer Cham Heidelberg New York Dordrecht London

Printed on acid-free paper

Springer is part of Springer Science+Business Media (www.springer.com)

Foreword to the First Edition

Fundamental Electrode Kinetics is represented by treatments of reactions such as $O_2 + 4H^+ + 4e^- \leftrightarrows 2H_2O$, still important in the theory of electrocatalysis. As for the cutting edge in electrochemistry is concerned, that is now far away into areas of complexity. In this fascinating book, Professor Scholz and his colleagues have written academically about a group of phenomena bordering on practical reality.

The material in the book develops out from a vast change which came to electrochemistry in the 1960s when it was realized that metals were not the only materials which could act as electrodes: germanium and silicon were the first to be examined, but the leading principle which allowed such developments was the many orders of magnitude change in conductivity which is brought about by impurities naturally present or added.

Professor Scholz' book takes a certain group of metallic and nonmetallic materials—immobilized particles and droplets—and presents their electrochemistry. Much of this detailed work involves three-phase boundary and many fuel cell-oriented electrochemists will be familiar with something similar arising in the theory of some types of porous electrodes.

The treatments in this book are certainly modern—to make sure of this, the authors have a preliminary chapter in which they compress a description of work before about 1990, dealing, e.g., with colloids and sandwich electrodes. Then comes Chapter 2 in which three-phase electrodes are met. After that comes a solid chapter on the experimental methods of the field: for example, how does one immobilize small particles for measurement or attach droplets? Chapter 4 is called Hyphenated Techniques and it brings out the applications to the field of spectroelectrochemistry. Chapter 5 concerns immobilized particles and here we realize the full complexity of the systems treated, e.g., voltammograms are shown of systems containing several phases; and relevant data on the electrochemistry of minerals is described. The theory of such systems is presented with impressive clarity and detail.

The last chapter (6) is about droplets and the electrochemistry of compounds dissolved therein and includes a properly modern discussion of the electrochemistry of liquid–liquid interfaces.

Treatments of the thermodynamics of these systems is fully described, but the kinetic treatments are less than full because of the interplay of resistance effects, concentration overpotential, and the basic electron transfer kinetics which is position dependent. There is good opportunity for impedance spectroscopy in these systems. Descriptions of many electrochemical events seldom met elsewhere is given, e.g., ion transfer from water to nitrobenzene, atom-force microscopy to study immobilized silver nanocrystals and their oxidation, which proceeds via an oversaturated silver halide solution.

The book gives strong attention to quartz crystal microbalance, able to record changes of nanograms. Impressive sensitivity of 10^{-3} K is exhibited by equipment used to determine entropy changes in hexacyanoferrates. The description given in Chapter 4 of the thermodynamics of minerals is impressive—and will find application, e.g., in rechargeable MnO_2 cells. Descriptions of the penetration of particles by redox reactions through microcrystals typify some of the material.

I read this book with increasing pleasure. It is certainly specialized, but in my opinion its significance is greatly increased by the fact that it clearly represents an attempt by these German colleagues to gather descriptions of work closely approaching that of real systems—and for that they should be greatly thanked, and their book widely distributed.

Gainesville, FL John O'M. Bockris

Preface to the Second Edition

The First edition of this book has been published in 2005. Now, 9 years later, the number of publications in the field of immobilized particles and droplets has increased by more than one third, and the authors felt the need to present these new developments. The authors of the previous edition are very glad that Professor Antonio Doménech-Carbó (University of Valencia, Spain) was so kind to join as coauthor. He has very extensively contributed to the further development of electrochemical techniques based on solid particles immobilized on electrode surfaces and has indeed opened new horizons. Most of the *new* text in the second edition has been written by him, so that finally it can be said that all 4 coauthors have equally contributed to the second edition.

The entire second edition has been thoroughly revised, updated, and corrected. The book is now organized in such way that each chapter is self-consistent with respect to the references. The Backmatter again contains a complete bibliography of all papers which have been published in this topic. Of course, only those papers have been included where an immobilization of particles was made with the deliberate goal to study the properties of the immobilized solid particles.

Greifswald, Germany — Fritz Scholz
September 2014

Preface to the First Edition

Electrochemistry provides a range of powerful techniques, which can be used to study liquid electrolyte solutions, solid electrolytes, metal and semiconductor electrodes, as well as mixed ion and electron conductors. The most important question in deciding on the applicability of electrochemical techniques has always been on the *how* of making an electrode or a solution of a compound to be studied. This has set the most serious limitations to general applications. An important step forward in the applicability of electrochemistry has been achieved by showing that a mechanical immobilization of *particles* onto the surface of inert electrodes allows electrochemical measurements on any compound or material, provided it does not dissolve in the electrolyte solution. Organic liquids and solutions immiscible with aqueous electrolyte solutions can be immobilized on electrode surfaces in the form of *droplets*. The electrochemistry presented in this monograph is very diverse while there is one common feature: all electrodes with immobilized electroactive particles and droplets are characterized by *electron* transfer and an *ion* transfer occurring simultaneously at one and the same electrode. As a result, the model of a *three-phase electrode* will be the guiding principle throughout the entire book.

This monograph summarizes the content of more than 300 publications listed at the end of the book in a bibliography. They are referenced as, e.g., "B 235." Additional references that do not belong to the bibliography are cited as footnotes on the respective pages.

The book introduces the experimental basis and theoretical foundations of the electrochemistry of immobilized particles and droplets. The authors are aware that various results and theoretical considerations from other fields of electrochemistry would deserve to be mentioned as well; however, the scope of this book limits such detail.

The mechanical immobilization of particles and droplets on electrode surfaces has proved to be useful for solving both fundamental and applied issues: quantitative analysis of alloys, qualitative detection of alloy phases, study of the electrochemical corrosion of metal particles, identification of mineral phases, quantitative analysis of minerals and inorganic compounds, determination of thermodynamic data of solid compounds and of the rate constants of electrochemically driven dissolution of solid compounds, qualitative and quantitative determination of the composition of solid solutions, elucidation of electrochemically driven isomerizations and other reactions of solid compounds, study of insertion electrochemical systems, elucidation of solid-to-solid electrochemical reactions and of structure-reactivity relations of solid compounds, understanding of the thermodynamics of solid-state electrochemical transformations, determination of the Gibbs energies of ion transfer between an aqueous electrolyte and an immiscible organic solvent, study of the geometric reaction pathways in solid particles and droplets, etc. The growing number of applications of the new technique in several laboratories around the world may lead to much broader usage in the future, once the results are critically compiled and presented in a comprehensive treatise. One of us (F. Sch.) has initiated this project, the other two coauthors having applied and developed it further within the past few years. We hope to provide a firsthand survey, to help others to make successful use of the new technique. Electrochemical measurements on immobilized particles and droplets are an attractive addition to spectroscopic, diffraction, and microscopic techniques, as used by inorganic, organic, physical, analytical, and material chemists, who face the task of characterizing solid compounds and materials as well as immiscible liquids and solutions.

Here I would like to acknowledge the contributions of many students, coworkers, cooperation partners, guest scientists, and friends. I cannot list them all; their names can be found in the bibliography at the end of the book. However, I would like to name just a few here: Lutz Nitschke was the student who performed the first experiments with immobilized particles; Birgit Meyer intensively studied minerals and contributed significantly to the general methodology; Ales Dostal from Prague started the work with metal hexacyanometalates that was continued by Uwe Schröder, Heike Kahlert, Michael Hermes, and Antje Widmann; Milivoj Lovrić and his wife Šebojka Komorsky-Lovrić (Croatia) contributed extensively to the theoretical and experimental studies with immobilized particles and droplets and became very good friends of my group in Berlin at the Humboldt University and later at the University of Greifswald. Maximiliano Bárcena-Soto (Mexico) developed the in situ calorimetry, and Ulrich Hasse successfully uses in situ AFM. Thanks also go to Valentin Mirčeski (Macedonia) and to Zbigniew Stojek (Poland). Uwe Schröder established the microscopic diffuse reflectance spectroscopy and was an active member in the hexacyanometalate project in my group and in the ionic liquid project in Richard Compton's group at Oxford University. Rubin Gulaboski (Macedonia), who performed most of the measurements in liquid–liquid ion transfer at Greifswald, and Uwe Schröder are thanked for agreeing to coauthoring this monograph. I am especially indebted to Alan Bond (Australia) who invited me to his lab and who became so much interested in this new field of electrochemistry that dozens of fascinating landmark papers have been published by his group. Keith Oldham (Canada) has been an inspiring cooperator and discussions with him are a pleasure to acknowledge. Frank Marken (UK) and Tomaš Grygar (Czech Rep.) are acknowledged for frequent exchanges of ideas and manuscripts. Mikhail Vorotyntsev (France) kindly discussed Chapter 2 with me. Finally I want to thank my family, Gudrun, Christiane, and Thilo for their patience.

Contents

Chapter 1
Earlier Developed Techniques

The first electrochemical experiments were performed with solid materials, esp. metals. However, these experiments, conducted in the eighteenth and nineteenth centuries, were directed toward the elucidation of the basic features of the electrical action of chemical substances and the chemical action of electricity. Initially, metals played the major role; only later it became obvious that many chemical compounds possess metallic or semiconducting properties that can be utilized in electrochemical cells. Parallel to the studies of new electrode materials, solid electrolytes were discovered and entire solid galvanic cells could be constructed. In this book, we will entirely neglect pure solid electrolytes because this is a field in its own and the subject of many thorough treatises.[1,2,3,4,5,6,7,8,9,10,11,12,13,14,15,16,17,18,19,20,21,22,23,24,25,26,27,28,29,30,31,32,33,34,35,36,37,38,39,40,41,42,43,44,45]

---

[1] Kharton V (ed) (2009) Handbook of solid state electrochemistry. Wiley-VCH, Weinheim

[2] Gellings PJ, Bouwmeester HJM (eds) (1996) The CRC handbook of solid state electrochemistry. CRC, Boca Raton, FL

[3] Bruce PG (ed) (1995) Solid state electrochemistry. Cambridge University Press, Cambridge

[4] Rickert H (1985) Solid state electrochemistry. An introduction. Springer, Berlin

[5] Maier J (2000) Festkörper – Fehler und Funktion, Prinzipien der Physikalischen Festkörperchemie. Teubner, Stuttgart

[6] Doménech-Carbó A, Labuda J, Scholz F (2013) Pure Appl Chem 85:609–631

[7] Kolb D (2001) Angew Chem 113:1198–1220

[8] Fischer H (1954) Elektrokristallisation von Metallen. Springer, Berlin

[9] Budevski E, Staikov G, Lorenz WJ (1996) Electrochemical phase formation and growth. VCH, Weinheim

[10] Slepushkin VV (1980) Zh Anal Khim 35:249–253

[11] Slepushkin VV, Mykovnina GS (1985) Zashcita Metallov 21:280–282

[12] Slepushkin VV (1987) Zh Anal Khim 42:606–616

[13] Slepushkin VV, Stifatov BM, Neiman EYa (1994) Zh Anal Khim 49:911–919

[14] Slepushkin VV, Stifatov BM (1996) Zh Anal Khim 51:553–511

F. Scholz et al., *Electrochemistry of Immobilized Particles and Droplets*,
DOI 10.1007/978-3-319-10843-8_1

Focusing on solid electrodes, two directions have been followed: one being the study of the properties of solid electrodes, esp. with respect to possible application, be it in electrochemical refining of metals, galvanic coating, primary and secondary batteries, technical electrolysis, or others and the second being the application of

---

[15] Canturija VA, Vigdergauz VE (1993) Elektrokhimija sulfidov, teorija i praktika flotacii. Nauka, Moskva

[16] Holliday RI, Richmond WR (1990) J Electroanal Chem 288:83–98

[17] Jirkovský R (1934) Mikrochemie (N.F. 9) 15:331–342

[18] Laitinen HA, Kolthoff IM (1941) J Phys Chem 45:1079–1093

[19] Laitinen HA, Jennings WP, Parks TD (1946) Ind Eng Chem Anal Ed 18:355–358

[20] Laitinen HA, Jennings WP, Parks TD (1946) Ind Eng Chem Anal Ed 18:358–359

[21] Kolthoff IM, Kuroda PK (1951) Anal Chem 23:1306–1309

[22] Kolthoff IM, Stock JT (1955) Analyst 80:860–870

[23] Micka K (1956) Coll Czech Chem Commun 21:647–651

[24] Micka K (1957) Coll Czech Chem Commun 22:1400–1410

[25] Micka K (1960) Depolarisation of the dropping mercury electrode by suspensions of insoluble substances. In: Longmuir IS (ed) Advances in polarography, Proceedings of the second international congress, vol 3. Pergamon, Oxford, pp 1182–1190

[26] Micka K (1965) Coll Czech Chem Commun 30:235–245

[27] Micka K, Kadlec O (1966) Coll Czech Chem Commun 31:3837–3844

[28] Micka K (1968) Fresenius' Z Analyt Chem 234:119–121

[29] Dausheva MR, Songina OA (1973) Uspekhi Khim 42:323–342 (English edition: Dausheva MR, Songina OA (1973) Russ Chem Rev 42:136–146)

[30] Scholz F (2013) J Solid State Electrochem 17:1493–1504

[31] Songina OA, Dausheva MR (1965) Elektrokhim 1:1464–1468

[32] Franklin TC, Nnodimele R, Adeniyi WK, Hunt D (1987) J Electrochem Soc 134:2150–2153

[33] Franklin TC, Nnodimele R, Adeniyi WK, Hunt D (1988) J Electrochem Soc 135:1944–946

[34] Franklin TC, Adeniyi WK (1988) Anal Chim Acta 207:311–317

[35] Franklin TC, Darlington J, Nnodimele R, Duty RC (1992) Heterogeneous catalysts for use in anodic electrosyntheses and electrodestruction of organic compounds in aqueous surfactant systems. In: Mackay RA, Texter J (eds) Electrochemistry in colloids and dispersions. VCH, New York, pp 319–329

[36] Majer V (1943) Chem listy 37:202–204

[37] Heyrovký J (1941) Polarographie. Theoretische Grundlagen, praktische Ausführung und Anwendungen der Elektrolyse mit der tropfenden Quecksilberelektrode. Springer, Wien, p 273

[38] Pauli W, Valkó E (1929) Elektrochemie der Kolloide. Springer, Wien, p 296

[39] Mackay RA, Texter J (eds) (1992) Electrochemistry in colloids and dispersions. VCH, New York

[40] Heyrovský M, Jirkovský J, Müller B (1995) Langmuir 11:4293–4299

[41] Heyrovský M, Jirkovský J, Štruplová-Bartáčková M (1995) Langmuir 11:4300–4308

[42] Heyrovský M, Jirkovský J, Štruplová-Bartáčková M (1995) Langmuir 11:4309–4312

[43] Mulvaney P (1998) Zeta potential and colloid reaction kinetics. In: Fendler JH (ed) Nanoparticles and nanostructured films. Wiley-VCH, Weinheim, pp 275–306

[44] Morehouse CK, Glicksman R (1956) J Electrochem Soc 103:94–98

[45] Glicksman R, Morehouse CK (1958) J Electrochem Soc 105:299–306

electrochemical measurements to acquire information on the composition or structure of solids. The latter direction of research has developed to what is now known as Solid-State Electroanalytical Chemistry (SSEAC) (see footnote 6). Parallel to these attempts, fundamental research on the electrochemical properties of solid electrodes has been performed.

In the first section (1.1), we want to discuss the methods developed in the past which have enabled electrochemical studies of solid materials. The second section (1.2) is devoted to techniques for studying electrochemical phenomena of immiscible liquids. To obtain information on the electrochemistry of liquids or solutions that are immiscible with water, several approaches have been developed so far. A brief survey of the available techniques is given to prepare the reader for the chapter on immobilized droplets.

1.1 Solid Materials

1.1.1 Compact Electrodes

When the electrochemical properties of metals or alloys are to be studied, the simplest way is to fabricate electrodes, e.g., as disks, cylinders, etc., from the target metal. These electrodes are generally polycrystalline and the energetic inhomogeneity of the electrode surface has to be born in mind. The use of well-characterized surfaces of single crystals has tremendously helped in understanding the electrochemistry of metals (see footnote 7). Since the electrochemical literature abounds in detailed descriptions of electrode fabrications, we omit this subject area along with that on metal electrodeposition (see footnotes 8 and 9) in this volume.

In practice, the fabrication of compact electrodes is comparably inefficient and time consuming. The development of *pressed cells* sought to overcome this problem. These are cells with a counter electrode and a reference electrode, but with a missing bottom (see footnotes 10–14). Instead of the bottom, there is only a rubber ring so that the cell can be pressed onto a metal specimen with a flat surface. Electrolyte solution can be filled in, and the electrochemical measurements can be performed with the metal at the bottom serving as the working electrode.

The preceding does not imply that in the case of metals and alloys, there is no reason for applying the technique of immobilized particles. There are cases where the metal to be studied cannot be destroyed, e.g., antique objects, etc., or cases where the fabrication of electrodes would be too expensive with respect to the obtainable information. In Sect. 5.1, we will give examples of such applications.

Compact electrodes made of minerals have been extensively studied in cases where the minerals exert sufficient conductivity and manage to be machined as electrodes (see footnotes 15 and 16).

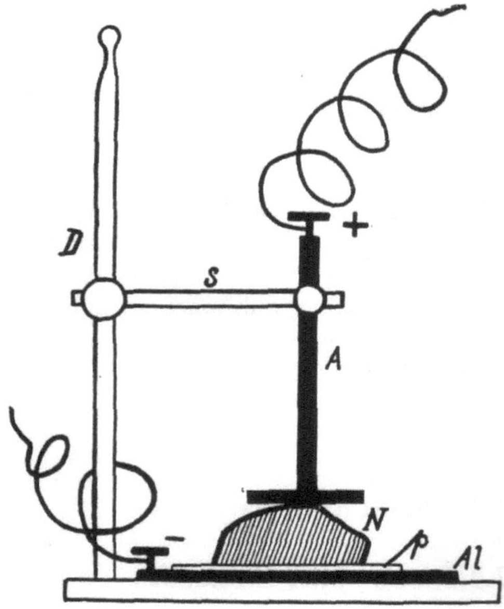

Fig. 1.1 Experimental arrangement in electrography: A anode, AL cathode, N mineral, p paper soaked with a reagent solution (see footnote 15)

Electrography

Electrography, introduced independently by A. Glazunov and H. Fritz, is an obsolete technique; however, it deserves reference, as it was a first direct electrochemical analysis of solid materials (see footnote 17). The principle is that a solid specimen is pressed on a paper which is soaked with an electrolyte solution. By anodic oxidation of the surface of the solid specimen, the reaction products (e.g., nickel(II) ions) react with a reagent in the paper (e.g., dimethylglyoxime) to give colored reaction product (red in case of nickel(II) and dimethylglyoxime). This produces a print that clearly shows the distribution of the reactive element (nickel, in our example) on the surface of the specimen. Figure 1.1 shows the experimental arrangement and Fig. 1.2 an electrographic print depicting the distribution of nickel in an ore.

1.1.2 Electrochemistry of Suspensions and Colloids

Noncolloidal Suspensions

It is very appealing to obtain information on the electrochemistry of solid materials by studying suspensions of solid particles in electrolyte solutions. Laitinen and Kolthoff have shown that suspended silver halides exhibit electroactivity

Fig. 1.2 Electrographic print of a nickel ore. The *black parts* represent the originally red spots of nickel dimethylglyoxime complex that have been formed where nickel ions were released into the reagent paper by oxidation of the nickel mineral (see footnote 15)

(see footnotes 18–22). Micka has performed a series of systematic studies on the behavior of particles suspended in electrolyte solutions (see footnotes 23–28). He has used the dropping mercury electrode in quiet and in stirred solutions. Most of the suspended compounds (insoluble salts and oxides) gave peak-shaped polarograms with a maximum at or near the point of zero charge (*pzc*) of the mercury electrode. In many cases, there was an additional reduction current at very negative potentials. Figure 1.3 shows the polarogram of a charcoal suspension exhibiting the typical peak-shaped signal. The signals were usually very noisy. Whereas Micka has explained the current peak at the point of zero charge by adsorption of the particles, Dausheva and Songina (see footnotes 29 and 30) have discussed the stability of the electrolyte film at the electrode in dependence on the electrode potential. They have assumed that only around the *pzc* this film is fragile enough to allow impinging particles penetrating it and contacting the metal surface allowing electron transfer. The same authors have experimentally shown that mercury(I) iodide particles strongly adhere at a mercury electrode only in the vicinity of the *pzc* (see footnote 31). Certainly, van der Waals forces and the surface charge of the particles have also to be taken into account to understand the interaction of suspended particles with a charged electrode surface.

Franklin et al. have pioneered work using electrodes covered with a polymer film in situ formed in suspensions of mineral particles (see footnotes 32–35). The authors have used cationic surfactant–styrene–aqueous sodium hydroxide emulsions and have observed specific signals of the suspended mineral particles. However, no detailed analyses of the electrode surface, of the state of the suspended particles, and of the mechanism of electrode reactions have been published.

The electrochemistry of slurries and suspensions has also attracted attention from a technical point of view.

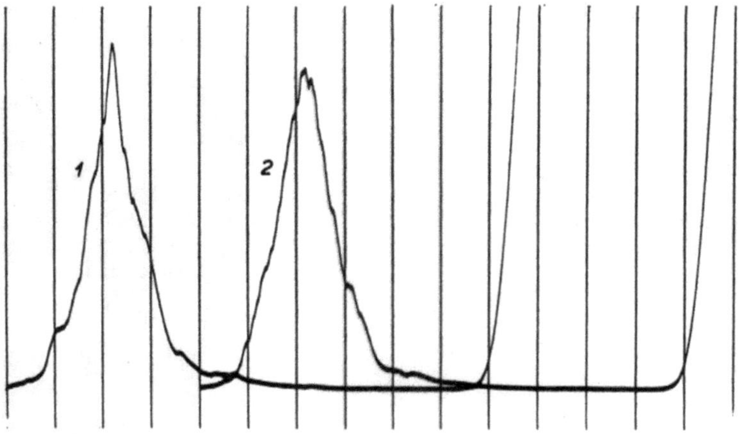

Fig. 1.3 Polarogram of a suspension of 40 mg charcoal in 10 mL 0.1 M KCl solution. Recording of the polarograms was started at 0 V in a Heyrovský cell so that the potential refers to that of the bottom mercury (see footnote 23). Copyright ©1956 with permission from publisher

Colloidal Solutions

Majer has been the first to observe a signal in colloidal iron(III) hydroxide solutions (see footnote 36) when magnesium ions were present. Since that first observation, for a long time no further studies have been published. All the classic books on polarography of the twentieth century do not refer to any work in this field. Perhaps the paradigmatic statement of J. Heyrovský that only "really dissolved compounds" give polarograms (see footnote 37) is responsible for this. Probably attempts have been made in vain to obtain voltammograms of colloidal solutions, and this experience did not help either in developing the subject. Pauli and Valkó have remarked in their monograph on the electrochemistry of colloids that "the simple case of a primary deposition by electron acceptance or delivery seems not to occur in case of colloids" (see footnote 38). However, they have given some examples of hydrogen and oxygen evolution in the course of electrolysis of colloid solutions. Sometimes, precipitations are caused by pH changes in the electrolysis. Within the last 30 years, the electrochemistry of colloidal solutions and that of colloidal particles on electrode surfaces has made advancements (see footnote 39); however, publications on the voltammetry of colloidal solutions reporting characteristic signals of colloidal substances are still rare. Interesting examples like tin, titanium, and mixed titanium–iron oxides have been published by M. Heyrovský et al. (see footnotes 40–42). However, even in these cases, it is a primary reduction of protons that causes chemical reductions of the metal centers in the colloidal particles. The majority of publications concern either the heterogeneous electron transfer between colloidal semiconducting particles and dissolved species (see footnote 43) or even secondary effects of colloids on other electrochemical systems. It is important for the subject of this book *that the use of colloidal solutions cannot be regarded as a simple way to access the electrochemistry of solid substances.* However, it is noteworthy that Compton and coworkers have shown that it is possible to measure the attachment of nanoparticles on electrode surfaces.[46,47,48,49] In the case of dispersed solid particles (organic and inorganic), it has been shown that the particles can be converted by Faraday reactions. The phenomenon of attachment of dispersed nano- and microparticles has been discovered earlier by Scholz et al. in case of liposomes and cell organelles.[50,51,52,53,54,55] In that case, all information is derived from the capacitive signals which result when the soft-matter particles disintegrate on the electrode surface and form island of adsorbed molecules. This can be easily studied on mercury electrodes.

1.1.3 Paste and Composite Electrodes

Paste electrodes have been introduced for the study of battery materials already in 1956–1959 (see footnotes 44 and 45).[56,57] Adams introduced graphite paste electrodes for analytical studies,[58,59] and Kuwana and French[60] used paste electrodes with additions of solid organic compounds to access the electrochemistry of these added compounds. Songina and coworkers started the application of paste electrodes to study the electrochemistry of solid inorganic and mineral compounds as

[46] Cheng W, Zhou X-F, Compton RG (2013) Angew Chem Int Ed 52:12980–12982

[47] Stuart EJE, Rees NV, Cullen JT, Compton RG (2013) Nanoscale 5:174–177

[48] Zhou Y-G, Rees NV, Compton RG (2014) Phys Chem Chem Phys 15:761–763

[49] Tschulik, K, Batchelor-McAuley C, Toh H-S, Stuart EJE, Compton RG (2014) Phys Chem Chem Phys 16:616–623

[50] Hellberg D, Scholz F, Schauer F, Weitschies W (2002) Electrochem Commun 4:305–309

[51] Hellberg D, Scholz F, Schubert F, Lovrić M, Omanović D, Agmo Hernández V, Thede R (2005) J Phys Chem B 109:14715–14726

[52] Agmo Hernández V, Scholz F (2006) Langmuir 22:10723–10731

[53] Agmo Hernández V, Scholz F (2008) Israel J Chem 48:169–184

[54] Hermes M, Scholz F, Härdtner C, Walther R, Schild L, Wolke C, Lendeckel U (2011) Angew Chem Int Ed 50:6872–6875

[55] Agmo Hernández V, Lendeckel U, Scholz F (2013) Electrochemistry of adhesion and spreading of lipid vesicles on electrodes. In: Vayenas CG, White RE (eds) Modern aspects of electrochemistry, vol 56; Applications of electrochemistry in medicine. Schlesinger M (ed). Springer, New York, pp 189–247

[56] Morehouse CK, Glicksman R (1958) J Electrochem Soc 105:306–311

[57] Glicksman R, Morehouse CK (1959) J Electrochem Soc 106:741–745

[58] Adams RN (1958) Anal Chem 30:1576

[59] Olson CJ, Adams RN (1960) Anal Chim Acta 22:582–589

[60] Kuwana T, French WG (1964) Anal Chem 36:241

admixtures.[61,62,63,64,65] These scientists also noted that modified paste electrodes allow analytical determinations of solid phases with standard deviation down to 2–3 %, although 2–11 % are usual.[66] Songina et al.[67] also discussed the different possible mechanisms of electrode reactions of solid compounds dispersed in carbon paste electrodes; she clearly identified three possibilities: (1) the solid compound does not chemically react with the electrolyte solution and is converted electrochemically in a solid-state reaction; (2) the products of the direct electrochemical conversion of a solid compound react with the electrolyte to form a sparingly soluble product which is not electroactive; and (3) the solid compound undergoes a chemical reaction (dissolution) with the electrolyte solution, in which case the dissolved species may be (a) less electroactive or (b) more electroactive than the solid. To distinguish these different cases, she has formulated clear criteria which refer to the dependence of peak currents on the electrolyte concentration and on the time of contact of the paste electrode with the electrolyte solution.

The use of paste electrodes has been by far the most popular technique for studying the electrochemistry of solid materials.[68,69,70] The powdered sample material is mixed with a binder and with an "inert" conducting material, e.g., graphite powder. The liquid binder can be an organic liquid immiscible with the aqueous electrolyte solution used for the electrochemical measurements or it can be exactly the same aqueous electrolyte. Both techniques have their advantages and disadvantages. An organic liquid binder prevents ingress of the aqueous electrolyte into the paste and allows the paste to be stable toward the electrolyte solution. The paste can be thick enough to be kept in a holder even when the outlet is turned downward. However, organic binders, e.g., paraffin oil, may always spread on the surface of the particles of the graphite and the solid substance. Such films can influence the electrochemistry of solid compounds, and one can never be sure of how severe this effect will be. In this respect, the use of aqueous electrolytes as

---

[61] Barikov VG, Songina OA (1966) USSR patent No. 191209

[62] Barikov VG, Rozhdestvenkaya ZB, Songina OA (1969) Zavod lab 35:776–778 (in Russian); (1969) Ind Lab 35:928–930 (in English)

[63] Songina OA, Trushina IM, Rozhdestvenskaya ZB, Cherkasova NM (1973) Elektrokhimiya 9:1310–1312 (in Russian)

[64] Songina OA (1978) Talanta 25:116–118, and papers cited there

[65] Rozhdestvenskaya ZB, Sigitov VB, Songina OA (1979) Zhur analit khim 34:455–458 (in - Russian); (1979) J Anal Chem USSR 34:350–353 (in English)

[66] Songina OA, Rozhdestvenskaya ZB, Medvedeva EP (1976) Zavod lab 42:379–381 (in Russian); (1976) Ind Lab 42:505–507 (in English)

[67] Rozhdestvenskaya ZB, Medvedeva EP, Songina OA (1981) Izvest Vyssh Uchebn Zaved, Khim Khimicheskaya Tekhnol 24:1389–1393

[68] Brainina KhZ, Neyman EJa (1982) Tverdofaznye reakcii v elektroanaliticheskoy khimii. Khimija, Moskva

[69] Brainina KhZ, Neyman EJa, Slepushkin VV (1988) Inversionnye elektroanaliticheskie metody. Khimija, Moskva

[70] Brainina Kh, Neyman E (1993) Electroanalytical stripping methods. Wiley, New York

binders is preferable.[71] However, they can prompt chemical reactions with the solid, e.g., dissolutions. In some studies, such pre-dissolutions have been deliberately used for producing electrochemical signals of the "in situ" dissolved species. It has been shown that these signals depend to some extent on the starting material; however, the processes at such electrodes are rather complex and such electrodes should be used with caution.

When the binder is solid, the produced electrodes are usually called "composite electrodes." The solid binder can be paraffin, polyethylene, Teflon, acryl polymers, etc. Also silica gel formed by hydrolysis of appropriate esters can be used.

1.1.4 Sandwich Electrodes

Studying the electrochemistry of a solid material by sandwiching it between two solid electrodes requires the material to be negligibly electronically conductive, since, otherwise, it would short-circuit the cell. If the sandwiched compound is an ion conductor, such experiments can yield information on its conductivity and on reactions of the solid ion conductor with the solid electrodes or with a gaseous phase. Such experiments have been performed in classic solid state electrochemistry, esp. at elevated temperatures (see footnotes 1–3). However, there is a class of solid materials that lends itself to room temperature experiments because the compounds possess electroactive redox centers and additionally exert ion conductivity. No electronic current can flow through these materials, unless an ionic current accompanies the electrons. Prussian blue and its analogues and many solid heteropoly salts and acids belong to this class. Various studies of the electrochemical behavior in such sandwich cells have been published so far.[72] Usually they have been performed in two-electrode arrangements, however, sometimes also under potentiostatic control using a reference electrode. Because of the more complex experimental design and the restrictions with respect to materials, this approach has found only limited application.

1.2 Immiscible Liquids and Solutions

Compounds, esp. organics that are too sparingly soluble in water, may be dissolved in organic solvents for electrochemical measurements. This is a classic approach; however, in several cases, it is of interest to study the electrochemical reactions in

---

[71] Alonso Sedano A, Tascón García LM, Vázquez Barbado DM, Sánchez Batanero P (2003) J Solid State Electrochem 7:301–308

[72] Kulesza P, Malik A (1999) Solid-state voltammetry. In: Wieckowski A (ed) Interfacial electrochemistry, theory, experiment, and applications. Marcel Dekker, New York, pp 673–688

the presence of an aqueous electrolyte solution. For this purpose, the idea of studying emulsions of the nonaqueous phase in an aqueous phase suggests itself. However, a price has to be paid, as many complicating additional phenomena are observed (see footnote 36).

At the interface of two immiscible liquid phases, ions partition and thus a galvanic potential difference is established. Hence, by applying a potential difference across the interface, it is possible to transfer ions. The ion transfer between two immiscible liquid phases has been established as a research field in its own.[73,74,75,76] These studies require the use of four-electrode potentiostats. In addition, supporting electrolytes must be present in the aqueous and the nonaqueous phases. Static liquid–liquid interfaces are most frequently used, although also dropping electrolyte electrodes have been developed and applied. In Chap. 6 of this monograph, the electrochemistry of immobilized droplets will be discussed. In this case, ion transfer reactions occur across the liquid–liquid interface accompanied by electron transfer reactions across the liquid (nonaqueous)–solid interface.

[73] Girault HH, Schiffrin DJ (1989) Electrochemistry of liquid-liquid interfaces. In: Bard AJ (ed) Electroanalytical chemistry, vol 15. Marcel Dekker, New York

[74] Vanysek P (1996) Modern techniques in electroanalysis. Wiley, New York

[75] Volkov AG (ed) (2001) Liquid interfaces in chemical, biological, and pharmaceutical applications. Marcel Dekker, New York

[76] Scholz F (2006) Annu Rep Progr Chem C 102:43–70

Chapter 2
Electrodes with Immobilized Particles and Droplets: Three-Phase Electrodes

It is a common feature of electrodes with immobilized particles and droplets that three phases are in close contact with each other, i.e., each phase having an interface with the two other phases. This situation exists also in most of the so-called surface-modified or film electrodes, many battery and fuel cell electrodes, electrodes of the second kind, etc. In fact, the majority of surface-modified electrodes consist of arrays of particles that partially cover the electrode surface. It would be far beyond the scope of this book to include all chemical and electrochemical techniques to deposit films on electrodes. Here we shall deal only with electrodes where the particles or droplets have been mechanically attached with the aim of studying their electrochemistry. Before going into the details in Chaps. 5 and 6, we now like to outline the specificity of *three-phase electrodes*.

Figure 2.1 depicts the situation at a three-phase electrode, phase II being a particle or a droplet. Since the particle or droplet contains neutral molecules and ions with equal amounts of positive and negative charges, any electron transfer between phase I and phase II must be accompanied by an ion transfer between the phases II and III. The ion transfer is an indispensable reaction to maintain the electroneutrality of phase II, provided that phase II does not only pass the electrons on from phase I to a redox species in solution. The latter may take place in systems where the particles are used as electrocatalysts. It is a most important feature of the three-phase electrode with immobilized electroactive compounds that the *electron* and *ion* transfer must take place simultaneously at one electrode (cf. Fig. 2.2). The electron transfer between phase I and II can be disregarded only in the case of simple metal oxidations.[1]

[1] When a metal particle, e.g., Ag, attached to a gold electrode is anodically oxidized, the ion transfer is the transfer of Ag^{+} ions from the metal to the solution and the electron transfer occurs between gold and silver. Such kind of electron transfer between electronically conducting phases will, of course, always occur in electrochemistry, because any electrode needs another conductor at its terminal. It should be remembered that the interfacial potentials that build up between the

F. Scholz et al., *Electrochemistry of Immobilized Particles and Droplets*,
DOI 10.1007/978-3-319-10843-8_2

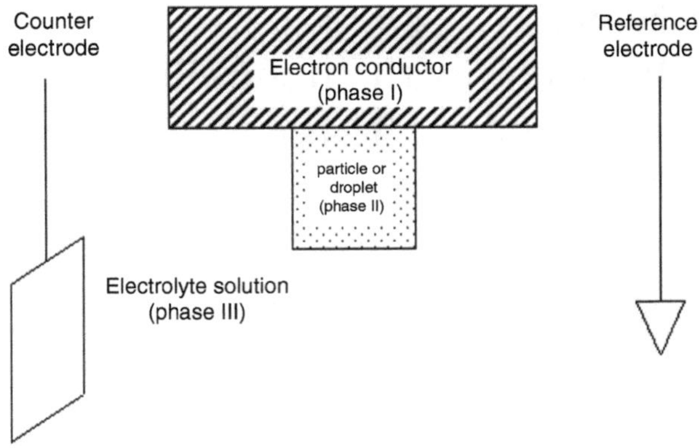

Fig. 2.1 Schematic drawing of a three-electrode cell with a working electrode on the surface of which a particle or a droplet is attached

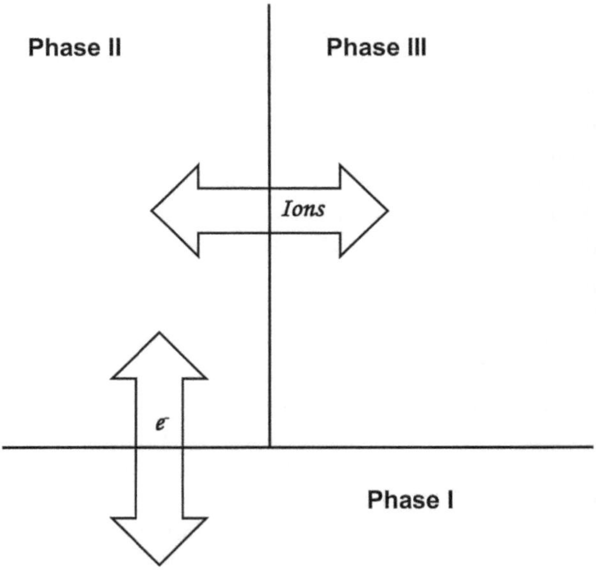

Fig. 2.2 Scheme of the simultaneous electron and ion transfer at a three-phase electrode

Thermodynamically speaking, the three-phase electrode comprising a phase II that has redox centers and ion conductivity is a *double electrode* (the German expression is *zweifache Elektrode*[2]). The formal thermodynamic analysis of a three-phase electrode is rather simple; however, there are general problems involved that have not been solved until today: Formally, one can split the overall equilibrium equation

electrodes and its terminal conducting connectors are responsible for the inaccessibility of single electrode potentials.

[2] Lange E, Göhr H (1962) Thermodynamische Elektrochemie. Hüthig Verlag, Heidelberg

$$\mathrm{Ox}^{x+}_{\mathrm{phaseII}} + ne^{-}_{\mathrm{phaseI}} + n\mathrm{C}^{+}_{\mathrm{phaseIII}} \leftrightarrows \mathrm{Red}^{(x-n)+}_{\mathrm{phaseII}} + n\mathrm{C}^{+}_{\mathrm{phaseII}} \tag{I}$$

with the Nernst equation

$$E = E^{\ominus}_{\mathrm{Ox/Red/C}} + \frac{RT}{nF}\ln\frac{a_{\mathrm{Ox}^{x+}_{\mathrm{phaseII}}}\,a^{n}_{\mathrm{C}^{+}_{\mathrm{phaseIII}}}}{a_{\mathrm{Red}^{(x-n)+}_{\mathrm{phaseII}}}\,a^{n}_{\mathrm{C}^{+}_{\mathrm{phaseII}}}} \tag{2.1}$$

into two equilibria, one involving the transfer of electrons

$$\mathrm{Ox}^{x+}_{\mathrm{phaseII}} + ne^{-}_{\mathrm{phaseI}} \leftrightarrows \mathrm{Red}^{(x-n)+}_{\mathrm{phaseII}} \tag{Ia}$$

and one involving the transfer of ions

$$\mathrm{C}^{+}_{\mathrm{phaseIII}} \leftrightarrows \mathrm{C}^{+}_{\mathrm{phaseII}} \tag{Ib}$$

with the following Nernst equations

$$E_{\mathrm{I/II}} = E^{\ominus}_{\mathrm{Ox/Red}} + \frac{RT}{nF}\ln\frac{a_{\mathrm{Ox}^{x+}_{\mathrm{phaseII}}}}{a_{\mathrm{Red}^{(x-n)+}_{\mathrm{phaseII}}}} \tag{2.2}$$

$$E_{\mathrm{II/III}} = E^{\ominus}_{\mathrm{C}} + \frac{RT}{F}\ln\frac{a_{\mathrm{C}^{+}_{\mathrm{phaseIII}}}}{a_{\mathrm{C}^{+}_{\mathrm{phaseII}}}} \tag{2.3}$$

The standard potentials are interrelated by the simple relation:

$$E^{\ominus}_{\mathrm{Ox/Red/C}} = E^{\ominus}_{\mathrm{Ox/Red}} + E^{\ominus}_{\mathrm{C}} \tag{2.4}$$

Both equilibria Ia and Ib are of *electrochemical* nature since a transfer of charged species between two phases takes place. When phase II is a solid, it is not yet clear how the activities of the species Ox, Red, and Cat^{+} in the solid have to be defined and how they could be determined. Further, no experiments are known that would lead to a separation of the free energies of the equilibria Ia and Ib in the case of solids. When phase II is a solution phase, the activities of Ox, Red, and Cat^{+} are in principle accessible; however, it remains the problem that an extrathermodynamic assumption is necessary for quantifying the free energy of ion transfer between the liquid phases II and III. Despite these inherent and still unresolved problems, progress has been made by rather simple models, e.g., assuming for solids that the activities of Ox and Red can be approximated by their molar ratios (Chap. 5) and by assuming that ions like tetraphenylarsonium cations and tetraphenylborate anions have the same free energy of ion transfer in systems like water–nitrobenzene (see Chap. 6).

The kinetic description of a three-phase electrode is much more complex since several aspects have to be taken into account: (1) the kinetics of the electron transfer, (2) the kinetics of the ion transfer, (3) the electron and ion conductivity of the immobilized phase, and (4) the size and shape of the immobilized phase. Thus, it is not surprising that compared to the thermodynamics, the kinetics of three-phase electrodes with immobilized particles and droplets have been much less studied. Of course, most systems are even more complex than illustrated in Fig. 2.2. For example, the reactions at the three-phase electrode may lead to the dissolution of phase II or to the formation of a new solid (or liquid) phase IV. Such phase transformation may proceed via the solution phase (see Sect. 4.2: the oxidation of Ag to AgI, where the oxidation of Ag nanocrystals results first in the formation of an oversaturated AgI solution from which by a nucleation–growth mechanism, the AgI crystals are formed in a follow-up process), or it can be a solid-to-solid phase transformation. An example of such a mechanism is the reduction of PbO to Pb as discussed in Sect. 4.2 [B184]. The model depicted in Fig. 2.3 was derived from the AFM results that show a reaction front proceeding through the PbO crystal. It was postulated that a reaction zone forms as an inter*phase* between the PbO and Pb phases.

Further mechanistic variations may result from the ion conductivity of phase II: if this phase is a good ion conductor, the electron transfer can start and proceed at the entire interface between the phases I and II. Parallel to this electron transfer, the ion transfer may take place across the interface between the phases II and III. Examples are the insertion electrochemical reactions of metal hexacyanoferrates (see examples in Sects. 5.3, 5.5, 5.6) and many other insertion electrochemical reactions, e.g., the reduction of "MnO_2." Even these reactions can be complicated, for example, when the reaction product is a poor ion conductor.

Of course, phase II may also be an insulator for both ions and electrons, like white phosphorus [B157]. In this case, the observed currents are much too high to be explained by a reaction that is confined to the three-phase junction line. It has been concluded from the electrochemical behavior of white phosphorus that the

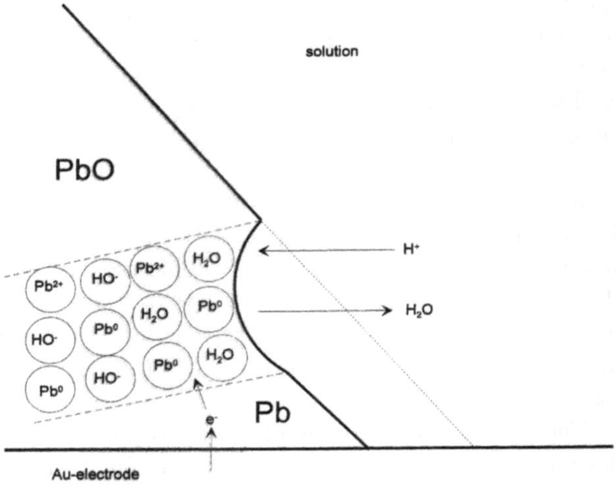

Fig. 2.3 Schematic model for the processes at a three-phase electrode gold–PbO solution when PbO is reduced to Pb as a fourth phase [B184]. Copyright ©2001 with permission from Springer

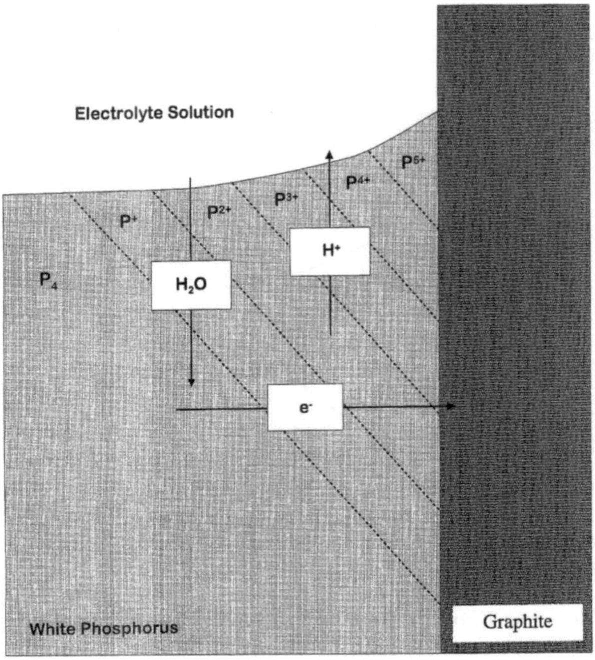

Fig. 2.4 Model for the electrochemical oxidation of white phosphorus at the three-phase junction electrolyte solution–phosphorus–graphite. The surface of the phosphorus will bend up due to the ingress of water and the expanding volume of the reaction products. The boarders between the different phosphorus oxides, symbolized by the oxidation state of phosphorus, are certainly diffused [B157]. Copyright ©2001 with permission from Elsevier

reaction proceeds via the formation of an ion-conducting phase IV, as depicted in Fig. 2.4.

Another possibility is that phase II attains ionic conductivity in the course of the electrochemical reaction, partly because ions are created from atoms or molecules in II, partly because ions are transferred from phase III to phase II. Such a reaction mechanism leads to an activation of the interface between phases I and II with respect to the electron transfer. Plenty of examples are given in Chap. 6, where the electrochemistry of immobilized droplets is discussed. Figure 2.5 illustrates such a situation, when an electroactive compound dissolved in a droplet of an organic solvent is oxidized, accompanied by a simultaneous ion transfer between the aqueous environments of the droplet and the organic phase. In Sect. 6.2, a detailed description of this situation will be given.

These examples of three-phase electrodes are given here to prepare the reader that this monograph deals with the special situation that arises when particles or droplets are immobilized on electrode surfaces. However, electrodes with immobilized particles or droplets must still be considered as simplified models for much more complex electrode situations, as they are frequently encountered in battery systems, technical electrolysis, etc. When carbon powder, manganese dioxide, and an alkaline electrolyte are mixed to make the cathode of a Leclanché battery, the latter is nothing else but an extended three-phase electrode.

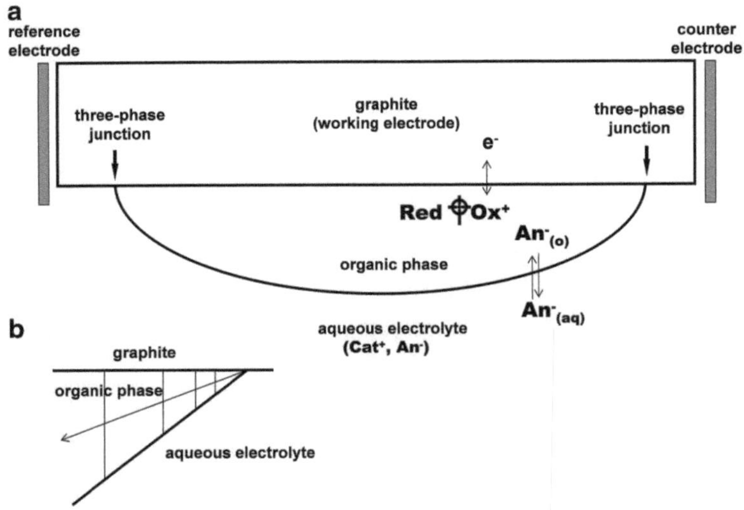

Fig. 2.5 (**a**) Schematic presentation of the situation when an organic phase droplet contains an electroactive form Red that is oxidized to Ox^+ accompanied by a transfer of the anions An^- from an aqueous electrolyte solution to the organic phase. (**b**) Shows how the reaction products Ox^+ and An^- spread from the three-phase junction into the droplet creating an ionic conductivity in the organic phase

Experiments of the kind described in this book may widen our understanding of such complex electrodes. In Fig. 2.6, an attempt has been made to give a general scheme for reactions starting at a three-phase junction line of a three-phase electrode. The properties of phase II determine the conductivity for electrons and ions in this phase. The properties of the two interfaces I/II and II/III determine the rate of electron transfer (I/II) and ion transfer (II/III). The transfer rates across the interfaces and the transport rates in phase II and phase III of the electrons and ions determine the size and direction of the fluxes of ions and electrons. The two fluxes determine the resulting flux of the products and the direction in which the reaction front spreads. The ion flux can be also directed into phase III, the solution, in which case phase II will dissolve, an example being the electrochemical dissolution of metal oxides as discussed in Sect. 5.7. Usually, one will observe that the advancement of the reaction into phase II or III will lead to an electron transfer spreading along the interface I/II and a spreading of ion transfer along the interface II/III. This way, the reaction that started at the three-phase junction line, i.e., a reaction that proceeds in that starting moment "around" a line, becomes a reaction across the two interfaces. There is only one example known so far, where the reaction remains confined to the three-phase junction line, because the transferred electrons and protons are used for a reduction of a chemical compound (see Sect. 6.1).

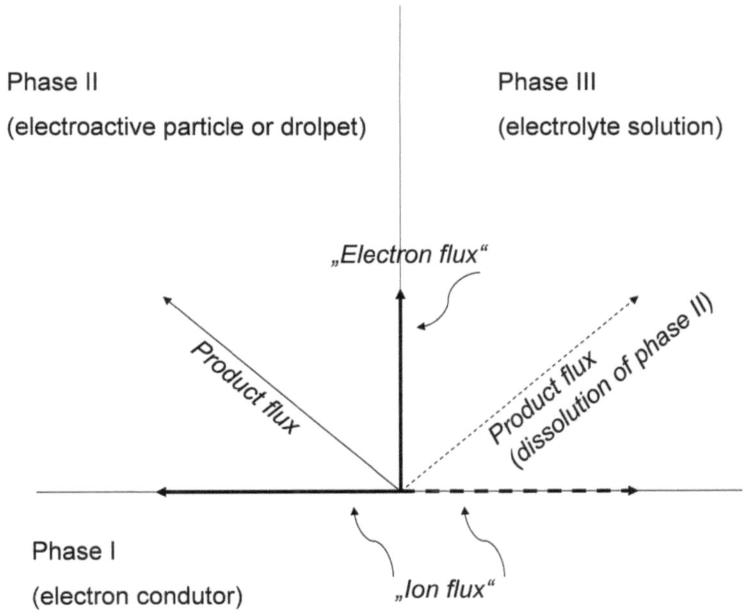

Fig. 2.6 Schematic view of a three-phase electrode. Phase I is an electron conductor (a metal or graphite), phase II is the immobilized solid or liquid electroactive phase, and phase III is the electrolyte solution. The electron flux shows the direction in which electrons can be transferred across the interface I/II and transported within phase II. The ion flux shows the direction in which ions can be transferred across interface II/III and within phase II (*full arrow*) or within phase III (*dashed arrow*). The product fluxes show in what direction the products may go

References

[B157] Hermes M, Scholz F (2000) The electrochemical oxidation of white phosphorus at a three-phase junction. Electrochem Commun 2:845–850

[B184] Hasse U, Scholz F (2001) In situ atomic force microscopy of the reduction of lead oxide nanocrystals immobilized on an electrode surface. Electrochem Commun 3:429–434

Chapter 3
The Experiment

3.1 Electrodes, Electrolytes, and Electrochemical Cells

3.1.1 Electrodes

Generally, all kinds of solid electrode materials can be used and have successfully been used to perform voltammetric measurements of immobilized particles and droplets. However, the suitability of an electrode material to a great deal depends on the electrode preparation, i.e., the immobilization procedure. Especially the mechanical immobilization of solid microparticles (see Sect. 3.2) requires a careful choice of electrode material.

Graphite has become most widely used as an electrode base material. It combines high chemical inertness with a wide potential window, low price, the ease of processing, and the necessary softness, enabling the mechanical "sticking" of powder particles of the sample on an electrode surface. One of the greatest advantages of graphite-based electrodes is the ease of cleaning of the electrode from attached sample particles which is achieved by removing the upper electrode surface layer by abrasion or cutting (see Sect. 3.2).

Paraffin-Impregnated Graphite Electrodes Among the graphite-based electrodes, the paraffin-impregnated graphite electrodes (PIGEs) are most successful. These electrodes consist of high-purity graphite rods whose pores are filled with solid paraffin in order to prevent high background currents and the spoiling of the electrode by an ingressing solution [B85]. To prepare PIGEs paraffin (melting point between 56 and 70 °C or higher) is melted in a closed vessel in a water bath. The graphite rods are added to the melt and the vessel is carefully evacuated (a water pump vacuum is sufficient). The melt remains under vacuum until no more gas bubbles evolve from the rods. Establishing ambient pressure then leads to a complete penetration of the rod by the paraffin. The rods are removed from the melt before the paraffin solidifies and are placed onto filter paper in order to remove

F. Scholz et al., *Electrochemistry of Immobilized Particles and Droplets*,
DOI 10.1007/978-3-319-10843-8_3

excess paraffin. There is no need to isolate the electrode shaft for later use (see Sect. 3.1.2).

Composite Electrodes Composite electrodes based on graphite powder as the electrode base material usually require polymer-resin-based binders, e.g., polyester [B195, B217] or wax ("sticky carbon" [B223]) in order to seal the pores and give the electrodes mechanical stability. The choice of a suitable binder is crucial as the chemical and physical inertness against the electrolyte solution, the thermal stability, and the mechanical properties are dominantly ruled by the binder. The preparation procedure of composite electrodes is generally straightforward. It consists of a thorough mixing of the graphite powder and the binder, the mass ratio depending on the used binder, the desired mechanical stability of the electrode, and the process of solidification. The latter may be accomplished by curing of the polymer resins [B195] or by resolidification of molten wax or paraffin.

Basal Plane Pyrolytic Carbon Basal plane pyrolytic carbon [B28, B29] is very well suited for the voltammetry of particles and droplets. Since basal plane pyrolytic carbon is not penetrated by the electrolyte solution, these electrodes do not require sealing with organic binder. This makes them equally well suited for measurements in aqueous and organic environments.

Highly Oriented Pyrolytic Graphite (HOPG) This material possesses a highly fragile layer structure that bans the mechanical immobilization of particles as that could severely damage the material surface. However, HOPG is a favorite electrode material for scanning electrochemical microscopy (SECM) and atomic force microscopy (AFM) and has successfully been used for the microscopic investigation of solid microparticles (see, e.g., [B147]).

Pencil Lead Electrodes Pencil lead has been proposed for use as electrode material [B48], and a comparison with PIGEs showed comparable results. Its major advantage is its low price and universal availability.

Boron-Doped Diamond Electrodes Recently, boron-doped diamond electrodes have received much attention [B123, B168, B224] as their extreme hardness allows the direct sampling from compact and hard materials like metals and ceramics. Desirable properties like a high chemical inertness connected with a wide potential window as well as a low interfacial capacity in aqueous electrolytes and hence low background currents add to the value of these electrodes. However, the cleaning of the electrodes after a mechanical (abrasive) sample immobilization is rather tedious.

Marken et al. [B168] have demonstrated that in boron-doped diamond electrodes, a facile oxidation and reduction of immobilized microdroplets of tetraalkylphenylenediamines can be achieved, which the authors attributed to a good adhesion of the organic material to the electrode surface.

Glassy Carbon Electrodes Glassy carbon is less suited for a mechanical immobilization as particles do not easily adhere to it. Exceptions are soft organic or organometallic compounds which are able to stick to the glassy carbon surface

[B138]. At polished glassy carbon electrodes, the deposition of microdroplets by evaporation from organic solvents is also rather difficult as the poor wetting properties and the therefore rather bad adherence of the droplets at the carbon surface often lead to the formation of larger droplets rather than the desired formation of an array of microdroplets. A possible way out is to roughen the surface of a glassy carbon electrode which fixes the position of the microdroplets at the surface. This procedure has also been used for the mechanical immobilization of solid inorganic compounds at glassy carbon [B60]. In this communication, it is demonstrated that the adherence of microparticles at GC electrodes is much weaker than at graphite electrodes, where particles are literally embedded into the electrode surface. However, glassy carbon electrodes have been deliberately chosen for particle immobilization to have an electrolyte layer sandwiched between the carbon and the particles to study the electron transfer catalysis of ions and compounds adsorbing at the particle surface [B60].

Metal Electrodes Due to their hardness and similar to glassy carbon, metal electrodes are generally less suited for the voltammetry of mechanically immobilized particles. However, metal electrodes have been successfully used for the investigation of soft samples like organic and organometallic solids. As demonstrated (see, for instance, [B138]), with gold, glassy carbon, and platinum electrodes, very similar results are obtained. Applications like the electrochemical quartz crystal microbalance (EQCM) (Sect. 4.4) require the use of gold electrodes [B54, B65]. See also AFM/STM (Sect. 4.2, [B184, B278, B279]) and in situ calorimetry (Sect. 4.6, [B219, B220]).

3.1.2 Electrolytes and Electrochemical Cells

The equipment to perform electrochemical studies of immobilized microparticles and microdroplets is the same as that used for conventional electrochemical measurements. Generally, a three-electrode arrangement consisting of a working, an auxiliary, and a reference electrode is utilized, the electrode control being achieved via a potentiostat. The cell design depends on the utilized working electrode and the purpose of the experiment and thus ranges from conventional glass cells to specially adapted constructions like for flow-through experiments or for spectroelectrochemistry (Sect. 4.1).

The choice of electrolyte depends on the nature of the studied compound and processes. In the vast majority of cases, aqueous electrolyte solutions are used. However, organic solvents like acetonitrile can be used as well, provided their dielectric constant allows electrolyte salts to be dissolved in order to achieve sufficient conductivity. The major prerequisite for the choice of electrolyte solvent is the insolubility or at least a very low solubility of the studied compound in the solvent.

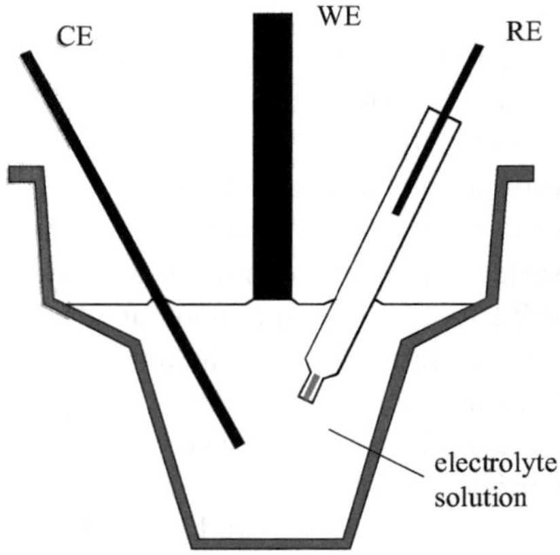

Fig. 3.1 Configuration of electrodes in an electrochemical cell for the voltammetry of immobilized microparticles and microdroplets

After immobilization of the sample particles or droplets (see Sects. 3.2 and 3.3), the electrode is ready for measurement.

Figure 3.1 depicts a most appropriate configuration of the working electrode in the electrochemical cell. The working electrode (e.g., a PIGE) is dipped into the electrolyte solution and then slightly raised in order to make the solution adhere to the lower circular electrode surface. This ensures a reproducible active surface area without the necessity of insulating the shaft of the electrode. However, since the electrode shaft is not contaminated with sample particles or droplets, even an immersion of the electrode into the solution for some millimeters would only affect the background current.

3.2 Immobilization of Particles

When the sample is a powder with grain size in the μm range, the easiest way to immobilize the particles on a paraffin-impregnated electrode is to place a few milligrams on a glass plate or a glazed tile and gently rub the electrode rod with its lower circular surface on the sample spot (see Fig. 3.2). Usually, this results in the attachment of several particles on the electrode. Most of them will be partially embedded in the soft graphite surface. Some of them will keep attached by adhesion, and they may adhere so weakly that they can be washed off with water. Normally, no water remains at the hydrophobic electrode, so that this washing cannot dilute the electrolyte solution. In fact, washing off the loose particles is even unnecessary because they would fall off from the electrode upon introduction to the electrolyte solution. A second way of immobilizing particles is to place a few milligrams of the powder on filter paper and rub the electrode on the sample spot. This has the advantage that a more even distribution is achieved and the loose particles are kept on the paper at the same time. Some preliminary experiments will give the necessary experience on how much sample should be used, what is the

Fig. 3.2 Transfer of microparticles from a glass plate onto the surface of a paraffin-impregnated graphite electrode (PIGE)

right pressure when rubbing the electrode on the sample spot, and what are the best movements when rubbing. Occasionally, it may be adequate just to press the electrode on the sample spot without circular movements of the electrode. When metals and alloys are studied, it may suffice to gently rub the electrode on the surface of the metals and alloys, which is perhaps, but not necessarily, cleaned before. When the metals and alloys are very hard, some abrasive powder (e.g., high-purity Al_2O_3 or diamond powder) may be placed between the electrode and the compact sample. Because of its inertness, the abrasive powder does not interfere in the electrochemistry. Alternatively, very hard electrodes, e.g., boron-doped diamond electrodes, can be used. In the case of a very rare or very small sample, the sampling can also be performed with a thin glassy carbon pin. In special cases, for example, in AFM studies, immobilization can be achieved by slow evaporation of a very dilute suspension of the powder (or even colloid) on the electrode surface. Care must be taken to avoid the formation of large droplets in suspension that could leave patches of adhering particles, which are unsuitable for AFM/STM. When the sample is a rather coarse powder, it is best to crush a small amount in a mortar to obtain fine particles.

After having performed the electrochemical measurements, it is very important to clean the electrode surface for the next measurements. The best way to clean a PIGE is to rub and polish its surface on filter paper. This rubbing must be repeatedly interrupted so that sample residues cannot be smeared from one end to the other. This is easily achieved by producing traces as depicted in Fig. 3.3. The electrode must be polished by keeping it perpendicular to the paper surface as to avoid rounding of the edges. The electrode surface should be kept flat, as this helps very much in follow-up cleaning and immobilization operations. Sometimes it may be advisable to fold a piece of clean filter paper around the shaft of the electrode and to wipe off possible remains of the sample that may have been transferred to that side during the polishing. After that, it is suggested to polish the lower circular surface again. The success of cleaning must be checked by recording blank voltammograms. Without this, one can never be sure that the

Fig. 3.3 Cleaning the surface of a PIGE by interrupted polishing on paper

measured response of a sample is caused by the fresh sample only and not resulting from previous measurements. In rare cases, when the particles are very hard and when they are not dissolved in the electrochemical measurements (this can happen with some rather coarse steel particles or any particles which form solid surface layers as a result of electrochemical reactions), the leftover particles may be even pressed further into the electrode during polishing, and it is not possible to get rid of them by this cleaning procedure. In such cases, one has to remove a thin layer of the electrode on a very fine abrasive paper, followed by polishing the electrode surface on paper. The surface of the electrode can also be cleaned with the help of a razor blade. The best movement to polish an electrode surface is to write the number "8" on the paper, however, keeping the electrode as upright as possible. Other points to remember in immobilizing samples are discussed in connection with the electrodes in Sect. 3.1.

Several reports describe different procedures for attaching single particles to electrodes using microscope-aided protocols: painting the crystals with silver paste [B447] and encapsulation into epoxy resin [B448, B449].

3.3 Attachment of Droplets

Generally speaking, the major advantage of the voltammetry of droplets over other conventional techniques of probing liquid–liquid interfaces lies in the formation of *three-phase electrodes* comprising the *triple-phase junction* electrolyte–electrode–droplet, i.e., the circumference of the droplets. By allowing charge compensation along and across this boundary, it is possible to study electrochemical reactions of ionically nonconducting liquids and nonpolarizable liquids without the addition of supporting electrolyte to the organic liquid.

Consequently, the electrode preparation procedure must always ensure the presence of three-phase boundaries, and a complete coverage of the electrode surface should be prevented. A full coverage with ionically nonconducting liquids can ultimately block the electrode, although, in some cases, a quick partition of ions between the electrolyte solution and the film may happen. This obviously happened in experiments of Shi and Anson.[1,2]

One can distinguish between two approaches—the attachment of single droplets and the attachment of an array of microdroplets. In the following sections, the different strategies for the droplet immobilization will be listed and discussed.

Independent of the applied approach, like in the case of other techniques that probe liquid–liquid interfaces, the voltammetry of droplets requires the studied liquids to be sufficiently immiscible with the chosen electrolyte solution. This requirement is crucial since due to the ratio of the volumes of the immobilized droplets and of the electrolyte solution, the small droplets would very quickly dissolve in the surrounding electrolyte solution.

However, the solubility of many organic solvents in water cannot be neglected (e.g., the solubility of nitrobenzene in water is 0.2 mass% at 20 °C). Thus, when working with solvent droplets, the electrolyte solution should always be saturated with the respective droplet solvent in order to prevent their dissolution and to ensure a long lifetime of the deposited droplet.

3.3.1 Attachment of Single Droplets

The attachment of a single droplet on the surface of an electrode can be achieved in many ways. The probably most sophisticated method is the laser trapping technique (see, e.g., Nakatani et al.[3]). In the following sections, however, only the approaches that are available without disproportional technical expenses will be considered.

Generally, the droplet attachment is straightforward. In order to transfer a defined, small volume of the liquid sample onto a large electrode surface, either μL syringes or pipettes can been used. As an example, droplets of polar and nonpolar aprotic solvents containing dissolved electroactive species have been deposited in order to study transfer processes across the solvent–water interface facilitated by electrochemical reactions of electroactive species dissolved in the droplet [B132]. For that, the authors have deposited 1–2 μL of nitrobenzene, containing 1–100 mmol L^{-1} of the electroactive species, on the surface of a paraffin-impregnated graphite electrode, e.g., [B132, B188] or a glassy carbon electrode [B242] (5-mm diameter). This procedure forms a droplet with a diameter of approximately 0.5–2 mm [B188, B242].

More complicated is the attachment of single droplets at microelectrodes. This usually requires the use of an optical microscope and a micromanipulation system.

---

[1] Shi Ch, Anson FC (1998) J Phys Chem B 102:9850–9854

[2] Shi Ch, Anson FC (1999) J Phys Chem B 103:6283–6289

[3] Nakatani K, Chikama K, Kitamura N (1999) Laser trapping–spectroscopy–electrochemistry of individual microdroplets in solution. In: Advances in photochemistry, vol 25. Wiley-Interscience, New York, pp 173–223

Terui et al., for instance, have used a glass fiber with a tip radius of 5 μm in order to deposit single droplets of nitrobenzene of a diameter between 25 μm and 100 μm in gold microelectrodes [B153]. The attachment of single droplets at microelectrodes usually leads to conditions in which the droplet size is larger than the diameter of the electrode. Due to the absence of a three-phase junction water–electrode–droplet, this setup requires to add supporting electrolytes to the droplets.

As the surface tension of water is much higher than that of most organic liquids, it is comparably easier to study the voltammetry of immobilized droplets and films of organic liquids immersed in aqueous solution than vice versa. Immersing an electrode with attached droplets of an aqueous solution into an organic electrolyte solution can easily lead to the detachment of the droplets from the electrode. This is the reason why this experimental setup is only rarely used. An example is the study of Ulmeanu et al. [B180] who deposited single drops of a volume of 10 μL of the appropriate aqueous solution containing supporting electrolyte and electroactive species on a freshly polished platinum electrode (surface area 0.06 cm^2) mounted in a Teflon holder and immersed the electrode in 1,2-dichloroethane.

3.3.2 Attachment of Arrays of Microdroplets

In contrast to the above technique, the deposition of an array of microdroplets leads to the formation of extremely small droplets, down to femtoliter volumes [B285]. The resulting drawback of a small electrochemical signal of the droplets compared to the large capacitive background current of the electrode surface is sidestepped by depositing a great number of droplets, distributed over the electrode surface.

Apart from chemical means (which are not going to be considered here), the most widely used technique for the deposition of arrays of microdroplets is the evaporation from a volatile solvent. In the first step, a defined amount of the sample liquid is dissolved in a volatile solvent (e.g., acetonitrile, dichloromethane). It has to be taken into account that growing concentrations of the compound in the solvent will eventually lead to an increased surface coverage and, eventually, to the formation of a continuous film rather than an array of microdroplets.

In the next step, a defined volume of this solution is transferred to the surface of an appropriate electrode. Graphite electrodes like paraffin-impregnated electrodes and basal plane pyrolytic carbon electrodes seem most suitable, as the hydrophobic surface allows a good wetting and thus adherence of the organic solution (see also Sect. 3.1). Then, the solvent is evaporated in air or, if necessary, under inert gas atmosphere.

Naturally, this technique can be applied only for nonvolatile compounds. Examples are redox liquids like tetraalkyl-substituted phenylenediamines (TRPDs) which have been deposited at basal plane pyrolytic carbon electrodes (see Sect. 6.1). As an example, Marken and coworkers transferred 1–50 μL of freshly prepared solution of the compound dissolved in acetonitrile (0.1–1 mmol L^{-1}) onto the electrode surface and evaporated the acetonitrile in air (e.g., [B74]).

An important question to be addressed is, in what form and size are the microdroplets distributed at electrode surfaces? In a recent work, Wadhawan et al. [B230] answered this question in an elegant manner. The authors examined the deposition pattern of *para-N,N,N′,N′*-tetrahexylphenylenediamine (THPD) at basal plane pyrolytic carbon electrodes by taking advantage of the well-known abilities of the related *para*-phenylenediamine as a photographic developer. After the deposition of microdroplets of THPD, the modified electrode was dipped for a few seconds into a 0.1 M aqueous solution of silver perchlorate allowing the following reaction to take place:

$$\mathrm{THPD_{(oil)}} + \mathrm{Ag^{+}_{(aq)}} + \mathrm{ClO^{-}_{4(aq)}} \leftrightarrows \left[\mathrm{THPD^{+\bullet}ClO_4^{-}}\right]_{(oil)} + \mathrm{Ag_{(s)}} \qquad \text{(I)}$$

The chemical reaction takes place at the oil–water interface, covering the THPD droplets with a silver shell. These silver-plated droplets could now be imaged with the help of scanning electron microscopy (SEM) as shown in Fig. 3.4. For both depicted cases, the droplets exhibit a very narrow size distribution, with an average size of 4.2 μm (for 4-nmol THPD deposited) and 20.6 μm (for 40-nmol THPD

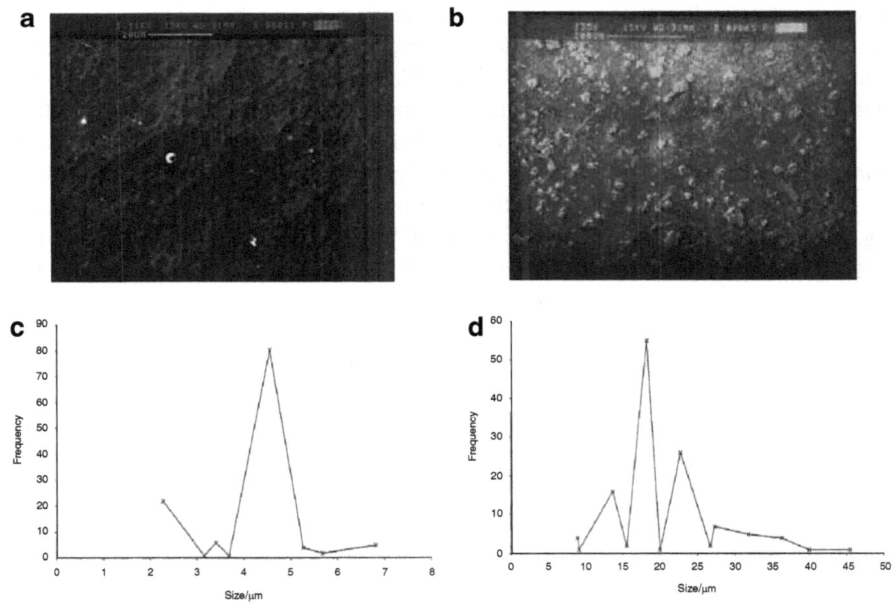

Fig. 3.4 (**a** and **b**) SEM image of (**a**) 4-nmol and (**b**) 40-nmol THPD deposited on a basal plane pyrolytic carbon electrode and immersed in $AgClO_4$ for 10 s (at open circuit potential). (**c** and **d**) Measured droplet distribution for the two coverages indicated in (**a**) and (**b**) [B230]. Copyright ©2001 with permission from Amer. Chem. Soc

deposited). The experiments also supported earlier assumptions that the formation of the microdroplets takes place mainly within the troughs of the graphite surface.

At very smooth electrode surfaces, the decreasing size of the solvent droplet during evaporation can lead to the unwanted concentration of the redox liquid in the remaining solvent and thus to the formation of few bigger droplets in the center of the electrode. In order to yield a more even distribution of the droplets across the electrode surface, it is possible to roughen the electrode surface mechanically prior to the droplet deposition. By doing this, spots with high surface energy are formed at which the droplet formation takes place, inhibiting a coalescence of the droplets during the process of solvent evaporation.

In order to achieve a sufficiently even distribution of the droplets at transparent indium tin oxide (ITO) electrodes, Marken [B74] siliconized the ITO electrodes prior to the deposition of microdroplets of THPD redox liquid, which makes the electrode surface hydrophobic and thus allows a better adherence of the organic droplets at the electrode surface.

3.4 The Electrochemical Measurement

In principle, there is no limitation with respect to the electrochemical measuring technique when immobilized particles and droplets are studied. The entire arsenal of electrochemical techniques and also hyphenated (in situ) techniques is applicable.[4] It can be recommended to perform preliminary studies with cyclic voltammetry or linear scan voltammetry. Care must be taken in choosing the starting potential so that the compound to be studied is not undergoing any reaction at that potential. Generally, the scan rates most appropriate for studies of immobilized particles and droplets are similar to those used in solution studies, i.e., between 10 and 500 mV s^{-1}. If the voltammograms exhibit very broad signals, e.g., when they possess a half-peak width of 100 mV or more, it may be that too large amounts of the compound were immobilized on the electrode surface. In such a case, some trials are necessary with deliberately chosen smaller amounts. Usually the signals will become narrower with decreasing amounts. If that is not the case, the broad signals may be an inherent property of the compound. When the electrode reaction is confined to the surface or the three-phase junction region, i.e., when no diffusion is involved, scan rates as high as 5 V s^{-1} can be used without any remarkable peak broadening [B83]. With electrode diameters in the range of 0.5–5 mm and a particle size of about 10 μm, one observes peak currents between 1 and 500 μA in cyclic voltammetric experiments, at voltammetric scan rates between 1 and 100 mV s^{-1}.

[4] Scholz F (ed) (2010) Electroanalytical methods. Guide to experiments and applications, 2nd edn. Springer, Berlin

In the case of completely chemically reversible systems, e.g., in the case of Prussian blue and its analogues, it can be useful quantifying the amount of deposited material with the help of chronocoulometry. Of course, chronocoulometry will yield only the amount of electrochemically active particles. Particles that are adhering to the surface so loosely that an electron transfer is not possible cannot contribute to the measured charge. Another problem could arise when the electrochemical reaction prompts the formation of an insulating layer on the surface of the particles or in contact with the electrode. Although no such examples have been reported so far, one must be aware of this possibility.

AC and pulse techniques can be suggested due to their high resolution power and simple peak evaluation. In case that metals or alloys are electrochemically dissolved by oxidation or some oxides are dissolved by reduction, one must bear in mind that the scanning time with these techniques may be too small for a complete stripping. In that case some of the material will remain on the electrode surface and must be carefully removed before the next measurement.

Electrochemical impedance spectroscopy (EIS) is a technique widely used in the study of corrosion phenomena, redox polymers, etc. This technique has been applied to the study of electrochemical doping in platinum phthalocyanine microcrystals in contact with nonaqueous electrolytes [B167] and the growth of microcrystalline films of metal oxides on metal surfaces [B442, B443] as well as for dating [B332] and monitorizing [B356] corrosion microparticulate deposits on archaeological metal.

Data processing can involve background subtraction, smoothing, convolution/deconvolution and derivation/integration, all accessible to conventional electrochemical equipments. Advance methodologies for data processing can also be applied in the electrochemistry of immobilized particles and droplets. The elimination voltammetry, developed by Trnkova et al. [B434, B435, B436], consists of the elimination of one or two selected particular voltammetric currents to be eliminated. The eliminated currents are expressed as a product of potential function $\Psi(E)$ and scan rate function $W(v)$. For digitally obtained data at a given potential, this product can be written as $I = (\text{constant})v^x$ so that diffusion currents ($x = ½$), charging currents ($x = 1$), and kinetic currents ($x = 0$) can be selectively eliminated using a mathematical transformation achieved by the elimination function obtained by a linear combination of total currents measured at different potential scan rates. The application of this methodology to the electrodissolution of iron nanoparticles in chloride-containing acetate buffers provides increased sensitivity and evidences the involvement of adsorbed species in the process [B437]. A combination of alternating current relaxation techniques, namely, electrical charge, mass, and color impedance spectroscopy have been developed and applied to the study of the electrochemistry of Prussian blue films [B438, B439]. The definition of simultaneous electrochemical impedance, mass transfer, and absorbance functions provides insight into electrochemical mechanisms with application to redox polymers [B440] and Prussian blue films [B441].

References

[B28] Bond AM, Colton R, Daniels F, Fernando DR, Marken F, Nagaosa Y, Van Steveninck RFM, Walter JN (1993) Voltammetry, electron microscopy, and X-ray electron probe microanalysis at the electrode-aqueous electrolyte interface of solid microcrystalline *cis*- and *trans*-$Cr(CO)_2(dpe)_2$ and *trans*-$[Cr(CO)_2(dpe)_2]^+$ complexes (dpe = $Ph_2PCH_2CH_2PDh_2$) mechanically attached to carbon electrodes. J Am Chem Soc 115:9556–9562

[B29] Bond AM, Marken F (1994) Mechanistic aspects of the electron and ion transport processes across the electrode|solid solvent (electrolyte) interface of microcrystalline decamethylferrocene attached mechanically to a graphite electrode. J Electroanal Chem 372:125–135

[B48] Blum D, Leyffer W, Holze R (1996) Pencil-leads as new electrodes for abrasive stripping voltammetry. Electroanalysis 8:296–297

[B54] Shaw SJ, Marken F, Bond AM (1996) Detection of new features associated with the oxidation of microcrystalline tetrathiafulvalene attached to gold electrodes by the simultaneous application of electrochemical and quartz crystal microbalance techniques. Electroanalysis 8:732–741

[B60] Lange B, Lovrić M, Scholz F (1996) The catalytic action of adsorbed thiocyanate ions and thiourea in the electron transfer from glassy carbon to solid copper(I) selenide and copper (I) sulfide particles. J Electroanal Chem 418:21–28

[B65] Bond AM, Colton R, Mahon PJ, Tan WT (1997) Tetrabutylammonium cation expulsion versus perchlorate electrolyte anion uptake in the electrochemical oxidation of microcrystals of $[(C_4H_9)_4N][Cr(CO)_5I]$ mechanically attached to a gold electrode: a voltammetric and quartz crystal microbalance study. J Solid State Electrochem 1:53–61

[B74] Marken F, Webster RD, Bull SD, Davies SG (1997) Redox processes in microproplets studied by voltammetry, microscopy, and ESR spectroscopy: oxidation of *N,N,N′,N′*-tetrahexylphenylene diamine deposited on solid electrode surfaces and immersed in aqueous electrolyte solution. J Electroanal Chem 437:209–218

[B83] Friedrich A, Hefele H, Mickler W, Mönner A, Uhlemann E, Scholz F (1998) Voltammetric and potentiometric studies of the stability of vanadium(IV) complexes. A comparison of solution phase voltammetry with the voltammetry of the microcrystalline solid compounds. Electroanalysis 10:244–248

[B85] Scholz F, Meyer B (1998) Voltammetry of solid microparticles immobilized on electrode surfaces. In: Bard AJ, Rubinstein I (eds) Electroanalytical chemistry, a series of advances, vol 20. Marcel Dekker, New York, NY, pp 1–86

[B123] Manivannan A, Tryk DA, Fujishima A (1999) Abrasive stripping voltammetry at polycrystalline diamond electrodes. Chem Lett 851–852

[B132] Scholz F, Komorsky-Lovrić Š, Lovrić M (2000) A new access to Gibbs free energies of transfer of ions across liquid-liquid interfaces and a new method to study electrochemical processes at well-defined three-phase junctions. Electrochem Commun 2:112–118

[B138] Jiang J, Kucernak A (2000) The electrochemistry of platinum phthalocyanine microcrystals. I. Electrochemical behaviour in acetonitrile electrolytes. Electrochim Acta 45:2227–2239

[B147] Kucernak AR, Chowdhury PB, Wilde CP, Kelsall GH, Zhu YY, Williams DE (2000) Scanning electrochemical microscopy of a fuel-cell electrocatalyst deposit onto highly oriented pyrolytic graphite. Electrochim Acta 45:4483–4491

[B153] Terui N, Nakatani K, Kitamura N (2000) Kinetic analysis of electrochemically induced ion transfer across a single microdroplet/water interface. J Electroanal Chem 494:41–46

[B167] Jiang J, Kucernak A (2001) The electrochemistry of platinum phthalocynine microcrystals. II. A microelectrode observation of nucleation-growth controlled solid-solid phase transformation in non-aqueous solvent. Electrochim Acta 46:1223–1231

[B168] Marken F, Compton RG, Goeting CH, Foord JS, Bull SD, Davies SG (2001) Fast electrochemical triple-interface processes at boron-doped diamond electrodes. J Solid State Electrochem 5:88–93

[B180] Ulmeanu S, Lee HJ, Fermin DJ, Girault HH, Shao Y (2001) Voltammetry at a liquid-liquid interface supported on a metallic electrode. Electrochem Commun 3:219–223

[B184] Hasse U, Scholz F (2001) In situ atomic force microscopy of the reduction of lead oxide nanocrystals immobilized on an electrode surface. Electrochem Commun 3:429–434

[B188] Komorsky-Lovrić Š, Lovrić M, Scholz F (2001) Cyclic voltammetry of decamethylferrocene at the three-phase junction organic liquid|aqueous solution|graphite. J Electroanal Chem 508:129–137

[B195] Doménech-Carbó A, Doménech-Carbó MT, Gimeno-Adelantado JV, Bosch-Reig F, Saurí-Peris MC, Sánchez-Ramos S (2001) Electrochemistry of iron oxide pigments (earths) from pictorial microsamples attached to graphite-polyester composite electrodes. Analyst 126:1764–1772

[B217] Doménech-Carbó A, Doménech-Carbó MT, Osete-Cortina L, Gimeno-Adelantado JV, Bosch-Reig F, Mateo-Castro R (2002) Electrochemical identification of metal ions in archaeological ceramic glazes by stripping voltammetry at graphite/polyester composite electrodes. Talanta 56:161–174

[B219] Bárcena Soto M, Kubsch G, Scholz F (2002) Cyclic voltammetry of immobilized microparticles with *in situ* calorimetry. Part I: The thermistor electrode. J Electroanal Chem 528:18–26

[B220] Bárcena Soto M, Scholz F (2002) Cyclic voltammetry of immobilized microparticles with *in situ* calorimetry. Part II: Application of a thermistor electrode for in situ calorimetric studies of the electrochemistry of solid metal hexacyanoferrates. J Electroanal Chem 528:27–32

[B223] Long JW, Ayers KE, Rolison DR (2002) Electrochemical characterization of high-surface-area catalysts and other nanoscale electroactive materials at sticky-carbon electrodes. J Electroanal Chem 522:58–65

[B224] Chatterjee A, Wiltshire R, Holt KB, Compton RG, Foord JS, Marken F (2002) Abrasive stripping voltammetry of silver and tin at boron-doped diamond electrodes. Diamond Relat Mater 11:646–650

[B230] Wadhawan JD, Evans RG, Banks CE, Wilkins SJ, France RR, Oldham NJ, Fairbanks AJ, Wood B, Walton DJ, Schröder U, Compton RG (2002) Voltammetry of electroactive oil droplets: electrochemically-induced ion insertion, expulsion and reaction processes at microdroplets of *N,N,N′,N′*-tetraalkyl-para-phenylenediamines (TRPD, R = *n*-butyl, *n*-hexyl, *n*-heptyl and n-nonyl). J Phys Chem B 106:9619–9632

[B242] Tasakorn P, Chen J, Aoki K (2002) Voltammetry of a single oil droplet on a large electrode. J Electroanal Chem 533:119–126

[B278] Hasse U, Wagner K, Scholz F (2004) Nucleation at three-phase junction lines: *in situ* atomic force microscopy of the electrochemical reduction of sub micrometer size silver and mercury(I) halide crystals immobilized on gold electrodes. J Solid State Electrochem 8:842–853

[B279] Hasse U, Scholz F (2004) In situ AFM evidence of the involvement of an oversaturated solution in the course of oxidation of silver nanocrystals to silver iodide crystals. Electrochem Commun 6:409–412

[B285] Banks CE, Davies TJ, Evans RG, Hignett G, Wain AJ, Lawrence NS, Wadhawan JD, Marken F, Compton RG (2003) Electrochemistry of immobilised redox droplets: concepts and applications. Phys Chem Chem Phys 5:4053–4069

[B332] Doménech-Carbó A, Doménech-Carbó MT, Peiró MA, Martínez I, Barrio J (2012) Application of the voltammetry of microparticles for dating archaeological lead using polarization curves and electrochemical impedance spectroscopy. J Solid State Electrochem 16:2349–2356

[B356] Doménech-Carbó A, Lastras M, Rodríguez F, Cano E, Piquero J, Osete-Cortina L (2014) Monitoring stabilizing procedures of archaeological iron using electrochemical impedance spectroscopy. J Solid State Electrochem 18:399–409
[B434] Dracka O (1996) Theory of current elimination in linear scan voltammetry. J Electroanal Chem 402:19–28
[B435] Trnkova L, Dracka O (1996) Elimination voltammetry. Experimental verification and extension of theoretical result. J Electroanal Chem 413:123–129
[B436] Trnkova L (2005) Identification of current nature by elimination voltammetry with linear scan. J Electroanal Chem 582:258–266
[B437] Pikna L, Kovacova Z, Hezelova M, Trnkova L (2013) Anodic dissolution of iron nanoparticles in chloride-containing acetate buffer electrolytes by elimination voltammetry. J Solid State Electrochem 17:2687–2696
[B438] Agrisuelas J, García-Jareño JJ, Giménez-Romero D, Vicente F (2009) Innovative combination of three alternating current relaxation techniques: electrical charge, mass, and color impedance spectroscopy. Part I: The tool. J Phys Chem C 113:8430–8437
[B439] Agrisuelas J, García-Jareño JJ, Giménez-Romero D, Vicente F (2009) Combination of three alternating current relaxation techniques: electrical charge, mass, and color impedance spectroscopy. Part II: Prussian blue $\leftrightarrows$ Everitt's salt process. J Phys Chem C 113:8438–8446
[B440] Agrisuelas J, Gabrielli C, García-Jareño JJ, Perrot H, Vicente F (2012) Kinetic and mechanistic aspects of poly(o-toluidine)-modified electrode. 1. Simultaneous cyclic spectroelectrochemistry and electrogravimetric studies in H_2SO_4 solutions. J Phys Chem C 116:15620–15629
[B441] Agrisuelas J, García-Jareño JJ, Vicente F (2012) Identification of processes associated with different iron sites in Prussian blue structure by in situ electrochemical, gravimetric, and spectroscopic techniques in the dc and ac regimes. J Phys Chem C 116:1935–1947
[B442] Li C, Li Y, Delaunay J-J (2014) A novel method to synthesize highly photoactive Cu_2O microcrystalline films for use in photoelectrochemical cells. Appl Mater Interfaces 6:480–486
[B443] Palacios-Pedrós A, Caballero-Briones F, Díez-Pérez I, Sanz F (2013) Tin passivation in alkaline media: formation of SnO microcrystals as hydroxyl etching product. Electrochim Acta 111:837–845
[B447] Brahimi R, Trari M, Bouguelia A, Bessekhouand Y (2010) Electrochemical intercalation of O^{2-} in $CuAlO_2$ single crystal and photoelectrochemical properties. J Solid State Electrochem 14:1333–1338
[B448] Ketir W, Saadi S, Trari M (2012) Physical and photoelectrochemical characterization of $CuCrO_2$ single crystal. J Solid State Electrochem 16:213–218
[B449] Eghbalnia M, Dixon DG (2013) In situ electrochemical characterization of natural pyrite as a galvanic catalyst using single-particle microelectrode technique in ferric sulfate solutions. J Solid State Electrochem 17:235–267

Chapter 4
Hyphenated Techniques

The depth of information obtainable from a single electroanalytical technique like the voltammetry of immobilized particles and droplets is naturally limited. Consequently, the combinations of electroanalytical techniques with non-electrochemical techniques become important and attractive. In particular, in situ combinations are powerful tools for investigating electrode processes. They are based on a simultaneous recording of electrochemical and non-electrochemical signals.

A number of restrictions apply to in situ combinations, mainly connected with the presence of the aqueous electrolyte solution. Thus, the use of vacuum spectrometric techniques employing electrons, ions, or atoms for excitation or detection is restricted to be applied ex situ only. Most important are the spectroelectrochemical techniques, i.e., the combination of electroanalytical and spectroscopic techniques. The combination of electron spin resonance spectroscopy[1,2] with Raman,[3,4,5] infrared,[6] and UV–VIS spectroscopy[7] proved to be very powerful instruments for investigating electrode processes.

The adaptation of hyphenated techniques to the voltammetry of immobilized particles and droplets is not always straightforward. Some methods, like the electrochemical quartz crystal microbalance, EQCM, can be utilized literally instantaneously. Other, especially spectroelectrochemical techniques generally require the

---

[1] Goldberg IB, McKinney TM (1984) Principles and techniques of electrochemical electron spin resonance experiments. In: Kissinger PT, Heinemann WR (eds) Laboratory techniques in electroanalytical chemistry. Dekker, New York Basel, pp 675–699

[2] Bagchi RN, Bond AM, Scholz F (1989) Electroanalysis 1:1–11

[3] Birke RL, Lu T, Lombardi JR, (1990) Surface enhanced Raman spectroscopy. In: Varma R, Selman JR (eds) Techniques for characterization of electrodes and electrochemical processes. Wiley, New York, pp 211–277

[4] Bard AJ, Faulkner LR (2001) Electrochemical methods, 2nd edn. Wiley, New York, pp 704–709

[5] Arsov LD, Plieth W, Koßmehl G (1998) J Solid State Electrochem 2:355–368

[6] Kulesza PJ, Malik MA, Denca A, Strojek J (1996) Anal Chem 68:2442–2446

[7] Mortimer RJ, Rosseinski DR (1984) JCS Dalton Trans 2059–2062

F. Scholz et al., *Electrochemistry of Immobilized Particles and Droplets*,
DOI 10.1007/978-3-319-10843-8_4

building of adapted electrochemical cells in order to meet the special conditions that arise from the specific geometry of the electrodes.

4.1 Optical Techniques and Spectroelectrochemistry

4.1.1 In Situ Spectroscopy

Optical spectroscopy in the UV–VIS range lends itself very well for the in situ observation of electrochemical reactions. For studying electrochemical reactions of species that are confined to an electrode surface, like of immobilized films, particles, and droplets, a number of spectroelectrochemical techniques have been developed. They are based on two major setups:

- Measuring the transmission of a redox species and the spectral changes during an electrochemical reaction
- Measuring the reflectance of a deposited compound and the changes in the reflectance spectra during the electrochemical experiment

Basically, the simplest spectroelectrochemical experiment is to measure the change of the transmission of a light beam that is directed through an immobilized sample. Obviously, this requires the electrode to be optically transparent. The probably most used transparent electrode material for spectroelectrochemical applications is the semiconducting indium tin oxide (ITO), which is usually vacuum deposited on a quartz or glass substrate. ITO electrodes have been used, for example, for the spectroelectrochemical investigation of the electrochromic properties of inorganic and organic polymer films deposited electrochemically on the ITO electrodes.[8,9,10] However, since the vast majority of electrode materials are optically opaque and many of the materials are highly reflecting, the utilization of reflectance measurements, e.g., at polished metal sheet electrodes, is widespread in electrochemistry.[11]

Generally, both transmission and reflectance techniques have been used based on a preliminary, mostly electrochemical, deposition of the target compound onto the electrode. However, they can also be exploited for studying mechanically immobilized particles and droplets. Due to the nature of the voltammetry of immobilized microparticles and microdroplets, however, they require some modification in order to make use of their full potential. One of the features of the

[8] Kulesza PJ, Zamponi S, Malik MA, Miecznikowski K, Berrettoni M, Marassi R (1997) J Solid State Electrochem 1:88–93

[9] Malinauskas A, Holze R (1998) Electrochim Acta 43:2563–2575

[10] Monk PMS, Mortimer RJ, Rosseinsky DR (1995) Electrochromism: fundamentals and applications. VCH, Weinheim

[11] Plieth W, Wilson GS, Gutuiérrez de la Fe C (1998) Pure Appl Chem 70:1395–1414

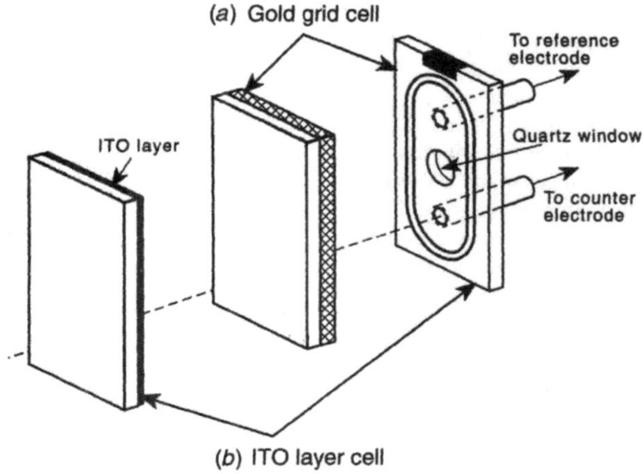

Fig. 4.1 Thin layer transparent electrochemical cells with a gold grid electrode for solution experiments and an ITO electrode coated on glass for electrochemical experiments using solids [B72]. Copyright ©1997 with permission from Royal Society of Chemistry

voltammetry of immobilized microparticles and droplets is the generally low and partially uneven surface coverage of the compound at the electrode. Generally, it is not advisable to use thicker particle layers, since, apart from the generally poor adherence, the growing resistance between the particles often does not allow the entire layer to react through. In the case of microdroplets, a growing surface coverage generally leads to an increasing droplet size connected with the relative decrease of the length of the triple-phase junction and eventually to the formation of an insulating liquid layer. As it will be demonstrated in Sect. 4.1.2, the problem of applying optical observation and spectroscopic analysis to the voltammetry of immobilized microparticles and microdroplets has been successfully solved using optical microscopy.

An important example for the exploitation of transmission spectroelectrochemical measurements of immobilized microparticles is the study of indigo by Bond et al. [B72]. Bond and coworkers have utilized UV–VIS spectroelectrochemistry in order to unravel the complex redox chemistry of indigo. They compared the spectroelectrochemical behavior of microparticles of indigo mechanically attached onto the surface of an ITO electrode and that of the compound dissolved in DMSO. For their investigation, the authors have used a spectroelectrochemical cell that allowed the study of both the dissolved and the microcrystalline compound (Fig. 4.1). For measurements of indigo microparticles, the compound has been transferred onto the ITO electrode surface by rubbing a thin layer with a cotton bud onto the electrode surface. This immobilization procedure must be performed very carefully as a scratching of the thin indium tin oxide layer would compromise the function of the electrode. For the study of the dissolved compound, the authors have used a thin layer cell with a gold micro-grid electrode as the working electrode.

Figure 4.2 shows the comparison of the changes of the absorption spectra of indigo during an electrochemical reduction of (b) the immobilized solid and (c) the dissolved compound. From the comparison of these spectra, which indicate remarkable differences in the reductive conversion of indigo, the authors have proposed a reaction scheme for the conversion of the immobilized compound that, without the

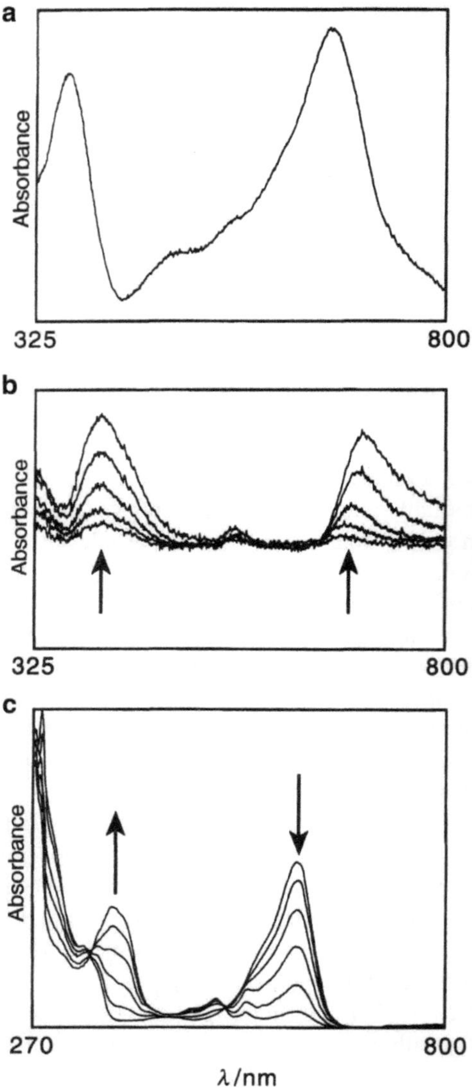

Fig. 4.2 (**a**) UV–VIS absorption of indigo mechanically attached to an ITO electrode and immersed in 1.28 mol L^{-1} NaCl electrolyte solution. (**b**) Changes in the UV–VIS absorption of indigo on ITO (spectrum **a** was used as the background and only the background corrected absorbance was recorded) when a potential of –1.0 V versus SCE is applied (curves correspond to 5 s, 30 s, 2 min, 10 min, and 30 min electrolysis). (**c**) In situ UV–VIS spectroelectrochemical monitoring of the reduction of 2.4×10^{-4} mol L^{-1} indigo in DMF (0.1 mol L^{-1} $NBut_4PF_6$, 10 nmol L^{-1} benzoic acid) (applied potential –1.0 V vs. SCE) [B72]. Copyright ©1997 with permission from Royal Society of Chemistry

spectroelectrochemical study, would not have been possible to derive (for further details, see Sect. 5.4).

4.1.2 In Situ Light Microscopy and Microscopic Spectroelectrochemistry

The low surface coverage in combination with the uneven distribution of the microparticles and droplets at a given electrode surface favors light microscopy for the observation of electrochemical reactions of these redox systems. As we will show in the following sections, the accessible information can range from the purely qualitative, visual observation of electrochemical processes to the quantitative evaluation with the help of coupled spectrometric equipment.

In situ light microscopy is a cheap and straightforward technique that enables the reaction to be followed with a spatial resolution down to micrometer scale. Compared to atomic force microscopy (AFM), this resolution appears to be rather low, but considering the low investment and operational costs of the equipment, the possible real-time observation of the electrode surface and additional information like color changes, and the interrogation of the spatial progression of a reaction, the in situ light microscopy has a great potential. In general, again two techniques can be utilized—incident light microscopy and transmission light microscopy. Due to the different optical setup and therewith different requirements toward electrode material and sample nature, both techniques have their application domains.

In Situ Incident Light Microscopy

For the in situ observation of electrochemical reactions at nontransparent electrodes, conventional light microscopes equipped with an incident light unit can be used. Microscopes of this kind are typically used for ore microscopy, material, and life science. The setup requires an adapted electrochemical cell that accommodates the working electrode, a reference, and a counter electrode, equipped with a glass or quartz window. At very low cost, such cells can conveniently be produced from acryl. The geometry of the setup, i.e., the $_0R_0$ geometry of the optical path, usually requires the specular reflectance of the electrode surface to be minimized. This can be realized with the help of two polarizing filters, which are generally available at incident light microscopes, and which have to be arranged in a crossed (90°) position.

Two basic setups are possible, which are described in the following sections. As an example, an electrochemical cell has been presented for the in situ microscopic observation of electrochemical reactions of microcrystalline samples mechanically attached to the surface of paraffin-impregnated graphite electrodes [B58]. The authors have assembled an electrochemical cell with an ore microscope (Leitz Laborlux 12 POL S, Leica Mikroskopie und Systeme GmbH, Germany) in an upright position. This setup, however, requires the electrode to be fitted through the cell bottom, making the exchange of electrodes rather tedious and time-consuming. Turning the microscope upside down (or using an inverse microscope in the first place) offers the chance to construct a cell in which the electrode can be fitted from the top of the cell. An example for such a setup is presented in [B67]. The authors have brought the microscope into the inverse position with the help of a scaffold. Now, a cell arrangement (see Fig. 4.3) could be used that allowed a rod electrode (paraffin-impregnated graphite electrode, PIGE) to be fitted through the lid of the cell. This setup massively simplifies the electrode handling.

The use of a 20-fold magnifying objective lens in combination with 10-fold oculars resulted in a 200-fold magnification which allowed visualizing objects of a size of approximately 10 μm (see Fig. 4.4). Higher spatial resolutions are principally possible; however, the shrinking electrolyte layer thickness between the

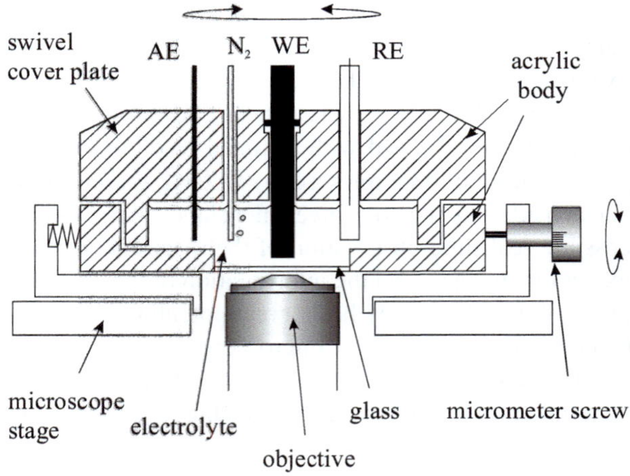

Fig. 4.3 Cross section of the electrochemical cell for in situ microscopic spectroelectrochemistry [B67]. Copyright ©1997 with permission from Springer

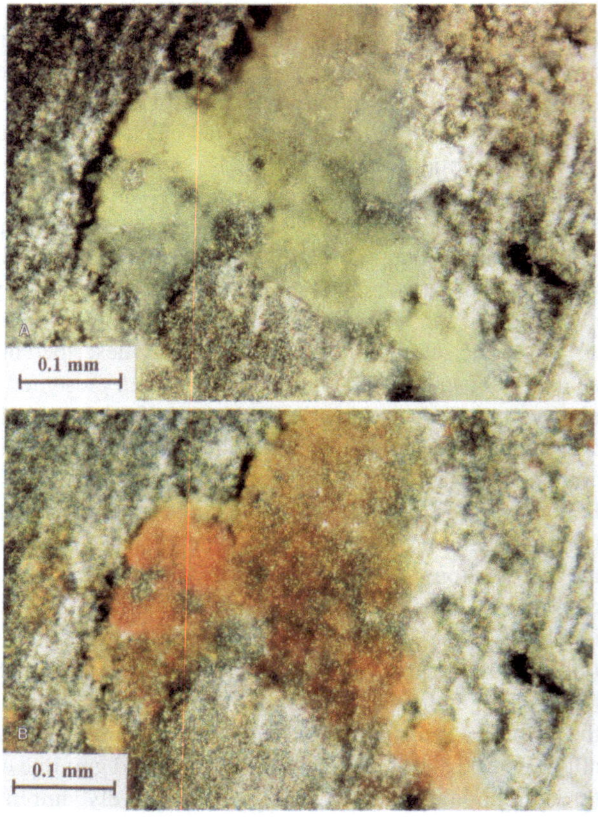

Fig. 4.4 (*Top*) Photographic image of silver octacyanomolybdate (IV) immobilized on a PIGE and immersed into 0.1 mol L^{-1} $AgNO_3$. (*Bottom*) The same particles after electrochemical oxidation at 0.75 V versus Ag/AgCl to silver octacyanomolybdate (V) [B58]. Copyright ©1996 with permission from Springer

working electrode and the glass window goes along with a restriction of the diffusion from or to the electrode surface and thus with a strong deterioration of the voltammetric signals. The use of long-range object lenses might help in overcoming this problem. For the study of immobilized microparticles, the electrode preparation is straightforward and follows the usual procedure of mechanical immobilization (Sect. 3.2).

A similar technique has been used by Bond et al. [B56, B101] who employed a video capturing system in combination with a Nikon Epiphot inverted metallurgical microscope fitted with a long-working-distance objective lens and a CCD TV camera for a subsequent analysis of the optical data. From the captured images, the authors could deduce that the electrochemical reactivity of microparticles of 7,7,8,8-tetracyanoquinodimethane (TCNQ) immobilized on polished gold electrodes strongly depends on the size of the particles. The authors found that the voltammetric responses of TCNQ at a scan rate of 100 mV s^{-1} were principally associated with the conversion of nano-size particles (<1 μm), whereas the conversion rate of larger crystals did not suffice to contribute to the voltammetric signal.

The purely qualitative analysis of the visual observation, however, leaves a lot of information concealed, and it is only a logic step to combine the microscopic observation with the in situ recording of the spectral information of the studied compound. This step can be done by taking advantage of the property of immobilized microparticles to scatter incident light. The resulting technique, the *microscopic in situ diffuse reflectance spectroelectrochemistry* [B67], not only allows quantifying the electrochromic properties of the immobilized particles, it also gives access to mechanistic information on the studied electrochemical reaction. In reference [B67], the authors demonstrate that for microparticles of silver octacyanomolybdate (IV/V), immobilized at a PIGE, in situ recorded microscopic diffuse reflectance spectra can be analyzed with the help of the Kubelka–Munk function (Eq. (4.1) and Fig. 4.5). This function, which can be referred to as the analogue of the law of Bouguer–Lambert–Beer for powder samples, describes the ratio of absorbance and scattering of a given sample layer.[12]

$$F(R_\infty) = \frac{K}{S} = \frac{(1 - R_\infty)^2}{2R_\infty} \tag{4.1}$$

(K = absorption coefficient, S = scattering coefficient, R_∞ = reflectance of a sample of infinite thickness)

Although the Kubelka–Munk function, like the law of Bouguer–Lambert–Beer, is strictly speaking exclusively valid for highly diluted samples, the authors have demonstrated a good agreement of voltammetric and spectrometric data. For the electrochemical oxidation of immobilized silver octacyanomolybdate, they have shown validity of Eq. (4.2), a modification of the Kubelka–Munk function

[12] Kortüm G (1969) Reflexionsspektroskopie. Springer, Berlin

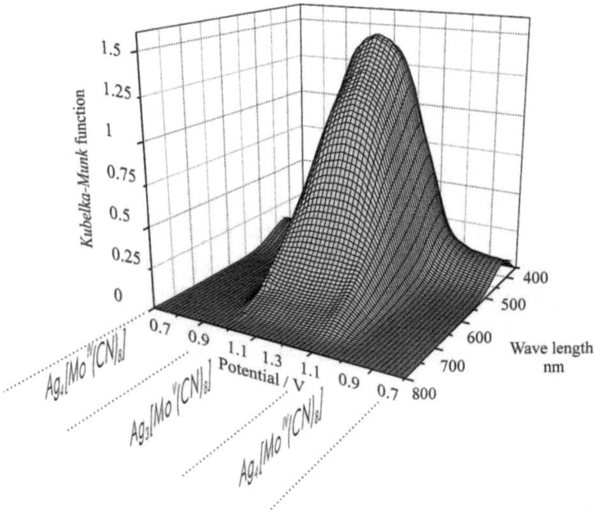

Fig. 4.5 Kubelka–Munk function of silver octacyanomolybdate (IV) immobilized on the surface of a paraffin-impregnated graphite electrode and immersed in a 0.1 M $AgNO_3$ solution. The Kubelka–Munk function was recorded in situ during the cyclic oxidation and re-reduction of the compound [B136]. Copyright ©2000 with permission from American Chemical Society

connecting the absorbance/reflectance date of the immobilized particles and their Red/Ox ratio.

$$\frac{c_{\mathrm{ox}}}{c_{\mathrm{red}}} = \frac{\frac{\left(F(R)-F'_{\mathrm{red}}\right)}{\left(F'_{\mathrm{ox}}-F'_{\mathrm{red}}\right)}}{\left(1-\frac{\left(F(R)-F'_{\mathrm{red}}\right)}{\left(F'_{\mathrm{ox}}-F'_{\mathrm{red}}\right)}\right)} \tag{4.2}$$

($F'_{\mathrm{ox/red}}$ are the values of the Kubelka–Munk function at 530 nm for $c_{\mathrm{ox}} = 1$ and $c_{\mathrm{red}} = 1$, respectively. $F(R)$ is the actual value of the Kubelka–Munk function for a certain Red/Ox ratio.)

The microscopic in situ diffuse reflectance spectroelectrochemistry can also be utilized to interrogate mechanisms of electrochemical reactions of immobilized particles or droplets. As an example, information on the geometric course of the electrochemical redox processes at the deposited silver octacyanomolybdate (Ag ocm) has been derived for the oxidation and reduction of immobilized microparticles of silver octacyanomolybdate (IV)/(V) [B67]. Based on a comparison of the change of the diffuse reflectance (d*R*/d*E*) of differently thick patches of silver octacyanomolybdate and the simultaneously recorded cyclic voltammograms (Fig. 4.6a, b), the authors have derived a model of the geometrical course of the electrochemical reaction. They have found that whereas for thin sample layers the voltammetric and the spectrometric curves are virtually identical, there is a strong retardation of the d*R*/d*E* curve on the potential scale for the case of a thick sample layer (Fig. 4.6a). In principle, one could suppose that the oxidation of solid silver octacyanomolybdate(IV), Ag ocm(IV), may start either at the graphite-sample interface or at the sample-electrolyte interface. The same applies for the reduction of Ag ocm(V). However, only in the case where the reaction advances from the

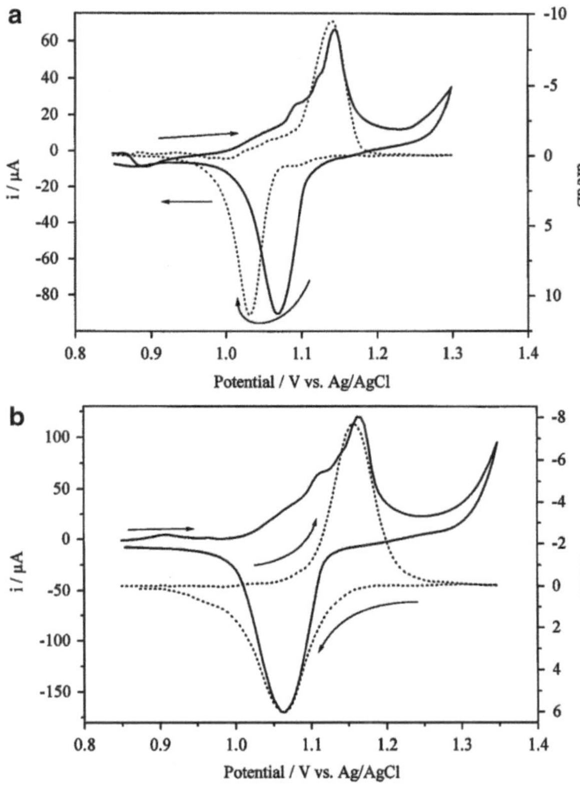

Fig. 4.6 Cyclic voltammogram (*solid line*) of silver octacyanomolybdate (IV/V) in 0.1 mol L^{-1} $AgNO_3$, scan rate 1 mV s^{-1}; first derivative of the reflectance over the d*R*/d*E* potential (*dotted line*) versus potential. (**a**) The reflectance was measured for a sample layer of about 10-µm thickness; (**b**) the reflectance was measured for a sample layer of about 1-µm thickness [B67]. Copyright ©1997 with permission from Springer

graphite-sample interface into the compound crystal one can explain the described optical behavior: Upon oxidation, a layer of higher optical density grows from the graphite octacyanomolybdate (ocm) interface, and the incident light beam can fully penetrate the upper layer of optically less dense Ag ocm(IV) (cf. Fig. 4.7). Thus, the incident light will be the less reflected the thicker the Ag ocm(V) layer becomes. If the opposite case was true, i.e., if the optically less trans-parent layer grew from the solution ocm interface downward, the reflectance would decrease to its minimum before the electrochemical reaction reaches the graphite interface. Obviously, this does not happen. But, upon reduction of the optically dense Ag ocm(V), especially in the case of a thick sample layer, the effect of electrochemical reduction is mirrored by the reflection only with strong retardation. This is the result of a growth of the optically less dense layer from the graphite interface in the direction of the solution interface.

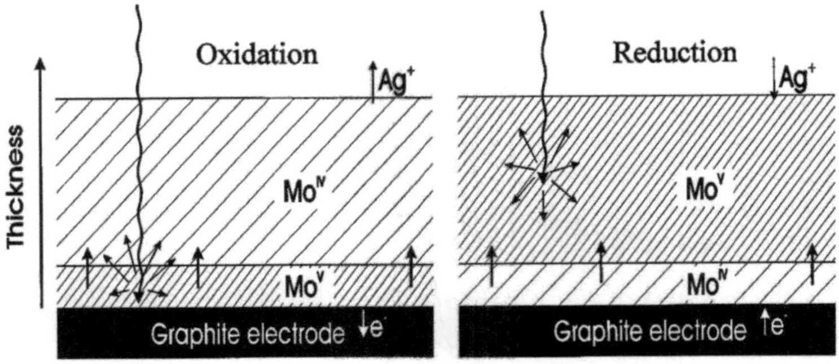

Fig. 4.7 Schematic drawing of a model for layer growing and light penetration for the processes of oxidation and reduction of silver octacyanomolybdate immobilized on an electrode surface. This scheme is based on the fact that the absorption coefficient of silver octacyanomolybdate (IV) is much smaller than that of silver octacyanomolybdate(V) [B67]. Copyright ©1997 with permission from Springer

In a later contribution, the same technique has been exploited to unravel the voltammetric responses of crystalline and amorphous sample material of the oxidation and reduction of silver octacyanomolybdate and tungstate [B136].

In Situ Transmission Light Microscopy

The coupling of transmission light microscopy and electrochemistry can be performed for (1) a purely qualitative evaluation and (2) for a quantitative analysis of electrochemical reactions of immobilized particles or droplets. Probably due to the delicate surface of ITO electrodes, however, so far it has only been utilized for the study of droplets. Of course, ITO electrodes have been used for decades for the study of electrochemically deposited films of electrochromic compounds; however, such films are not considered here.

Marken and coworkers [B74] have used the microscopy in order to study redox processes of microdroplets of immobilized microdroplets of *N,N,N',N'*-tetrahexylphenylenediamine (THPD). Marken took advantage of the possibility to visualize the spatial progress of the oxidation of the immobilized THPD droplets. The setup is, in principle, straightforward: an open cell with a counter and a reference electrode was used, and an ITO electrode was located in the optical path of the transmission microscope. A camera, mounted to the microscope, allowed taking images of the electrode surface during the voltammetric experiment. The THPD was deposited on the electrode via evaporation from acetonitrile (Sect. 3.3). The problem with commercial ITO electrodes is a relatively high hydrophilicity of the ITO, which can make the wetting and thus the adherence of droplets of organic liquids difficult. Marken circumvented the problem by siliconizing the ITO surface prior to droplet deposition (Sect. 3.3). The result of the experiment is presented in Fig. 4.8. The photographic images, taken at different stages of the electrochemical oxidation of THPD, gave support to the assumption of an initiation

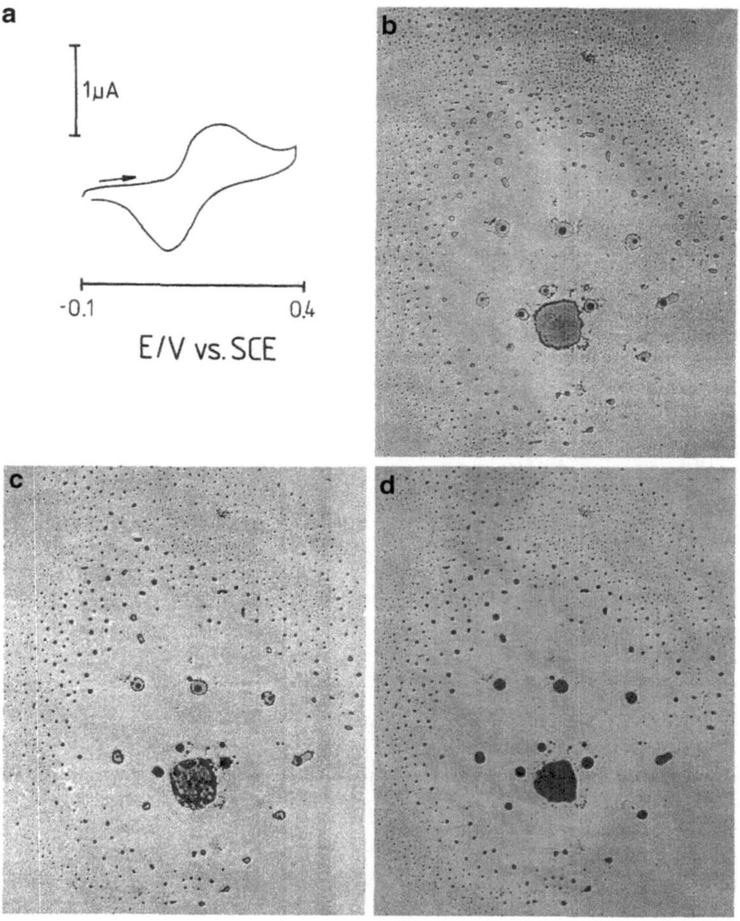

Fig. 4.8 (**a**) Cyclic voltammograms for the oxidation of 1.24 μg *N,N,N′,N′*-tetrahexylphenylenediamine (THPD) deposited on a siliconized ITO electrode immersed in aqueous 1 M $NaClO_4$ solution (22 °C, 1 mV s^{-1}). (**b–d**) Photographs of microscopic images of THPD deposited on a siliconized ITO electrode immersed in aqueous 1 M $NaClO_4$. The applied potential has been stepped from –0.1 to +0.4 V versus SCE for (**b**) 10 s, (**c**) 30 s, (**d**) 5 min (scale 780 × 920 μm) [B74]. Copyright ©1997 with permission from Elsevier

of the reaction at the three-phase boundary electrolyte–electrode droplet, from which the reaction proceeded into the bulk of the droplets. A closer look also revealed a separation of the oxidized and the reduced form of the compound during the reaction.

A quantitative analysis of a microscopic observation can be achieved with a spectroelectrochemical cell as presented in Fig. 4.9, coupled to a microscope and a diode array spectrometer [B280]. The cell body, made of acrylic, consists of an ITO plate (working electrode), sandwiched with a glass plate with attached platinum

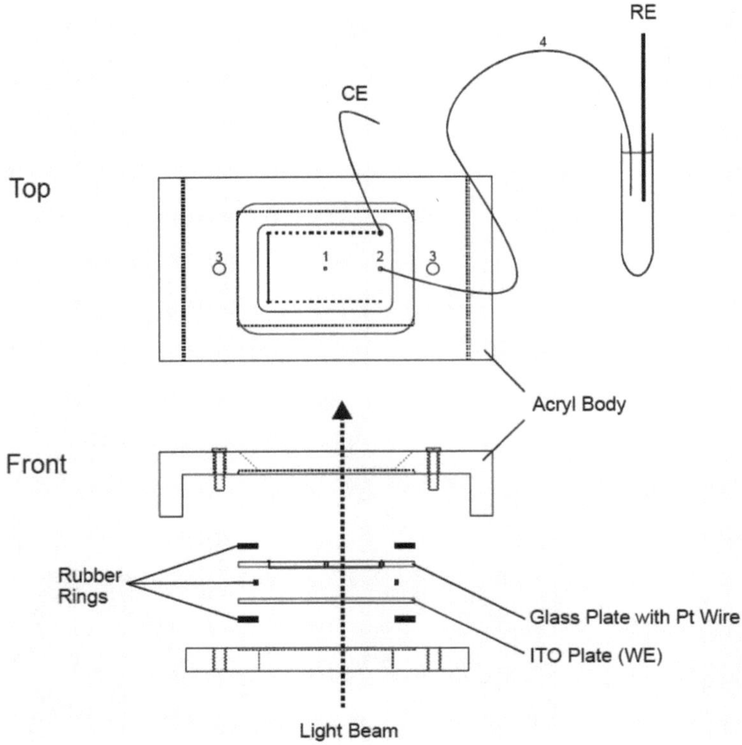

Fig. 4.9 Scheme of the spectroelectrochemical cell. CE is the counter electrode, RE is the reference electrode; (1) hole for inserting the nitrobenzene droplet, (2) hole for inserting the electrolyte solution, (3) threads and screws, and (4) tube filled with agar/KCl [B280]. Copyright ©2004 with permission from Elsevier

wire (counter electrode). The spacing between the plates is realized with rubber rings of a defined thickness.

Two holes enable the injection of the electrolyte solution and of a sample droplet. Komorsky-Lovrić and coworkers have used this setup for studying the ion transfer from water into nitrobenzene at the three-phase junction when decamethylferrocene, dissolved in the nitrobenzene droplet, becomes electrochemically oxidized. By spectrometrically recording the absorbance at different spots of the droplet (see Fig. 4.10), concentration-time plots can be derived which can help to unravel reaction at triple-phase electrodes (Fig. 4.11).

4.1.3 In Situ Electron Spin Resonance Spectroscopy

Many electrochemical reactions involve the formation of radical species, either as products or as intermediates. In situ electron spin resonance (ESR) measurements can help in identifying these species, in giving information on their chemical and

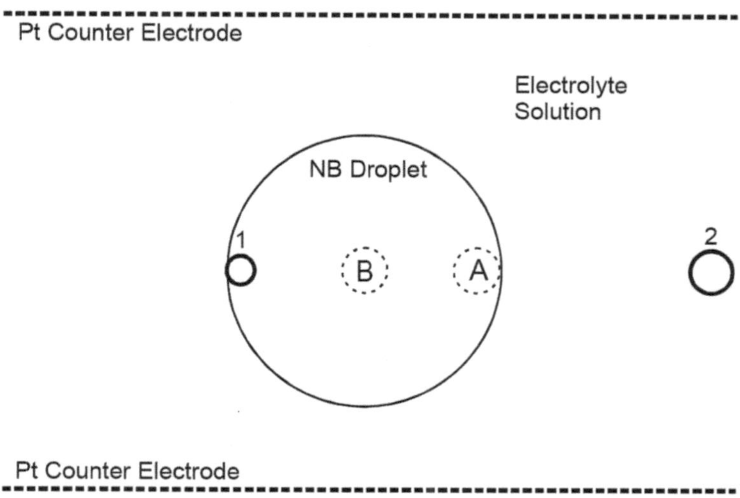

Fig. 4.10 Schematic drawing of the spectroelectrochemical cell with locations of the spectrometric measuring points. Spot (A) is located approximately 0.5 mm and spot (B) approximately 3 mm away from the junction line [B280]. Copyright ©2004 with permission from Elsevier

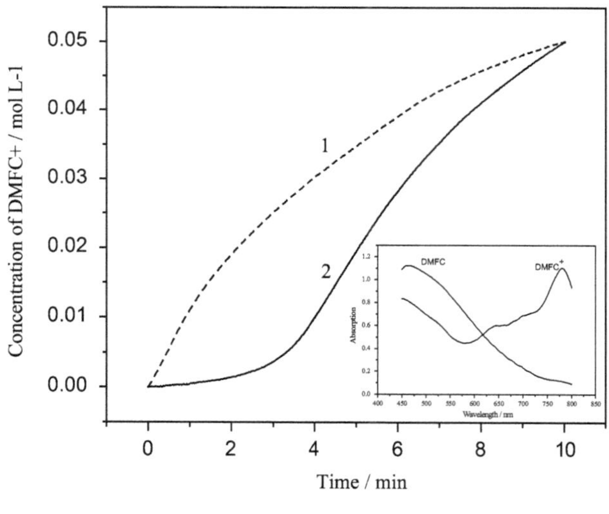

Fig. 4.11 The variation of the dmfc$^+$ concentration measured close to the three-phase junction line (curve 1) and in the middle of the droplet (curve 2). The concentration of dmfc in NB was 0.05 mol L^{-1}. The concentration of aqueous KNO_3 electrolyte solution was 0.1 mol L^{-1}. The ITO electrode was polarized at 0.4 V. The *inset* shows the spectra of the completely reduced solution (dmfc) and the completely oxidized solution (dmfc$^+$, ClO_4^-) [B280]. Copyright ©2004 with permission from Elsevier

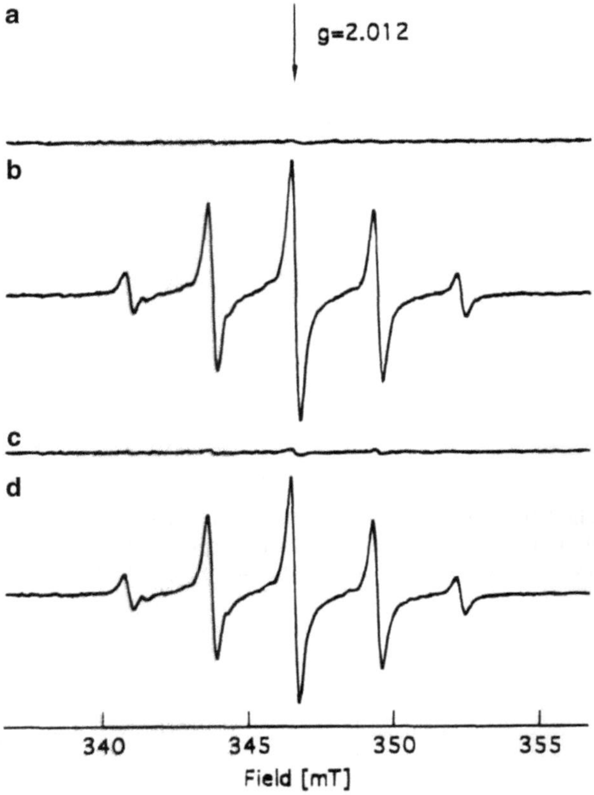

Fig. 4.12 In situ electrochemical generation at a platinum electrode of the solution-phase ESR spectrum of *trans*-[Cr$(CO)_2(dpe)_2]^+$ by oxidation of *trans*-Cr$(CO)_2(dpe)_2$ in a 1:1 mixture of acetonitrile and aqueous K^+-buffer. (**a**) Surface attached *trans*-Cr$(CO)_2(dpe)_2$ to oxidation; (**b**) oxidation of *trans*-Cr$(CO)_2(dpe)_2$ to *trans*-[Cr$(CO)_2(dpe)_2]^+$; (**c**) regeneration of solid *trans*-Cr$(CO)_2(dpe)_2$ by electrochemical reduction of *trans*-[Cr$(CO)_2(dpe)_2]^+$; (**d**) reoxidation of *trans*-Cr$(CO)_2(dpe)_2$ to *trans*-[Cr$(CO)_2(dpe)_2]^+$ after experiment **c** [B70]. Copyright ©1997 with permission from Elsevier

physical environment, and thus in elucidating the mechanism of electrode processes. As an example, in situ ESR can be utilized for distinguishing if an electrochemical oxidation or reduction process of an immobilized compound proceeds as a pure solid–solid conversion or if it proceeds via the solution phase. The first study of this kind was the reduction of solid 7,7,8,8-tetracyanoquinodimethane (TCNQ) and the oxidation of solid organometallic *trans*-Cr$(CO)_2(dpe)_2$ (see Fig. 4.12) by Bond and Fiedler [B70].

In situ ESR can also be applied to the study of immobilized microdroplets. Thus, Marken and coworkers have studied the voltammetry of THPD (tetrahexylphenylenediamine, see Sect. 6.1) immobilized on an electrode surface or dissolved in acetonitrile [B74]. The in situ recorded spectra of the electrochemically oxidized compound, $THPD^{+\cdot}$, are shown in Fig. 4.13a, b. The difference between the solution signal and that of the pure redox liquid is clearly visible.

As demonstrated in Fig. 4.13c, transients of the intensity of the ESR signal versus time can be recorded.

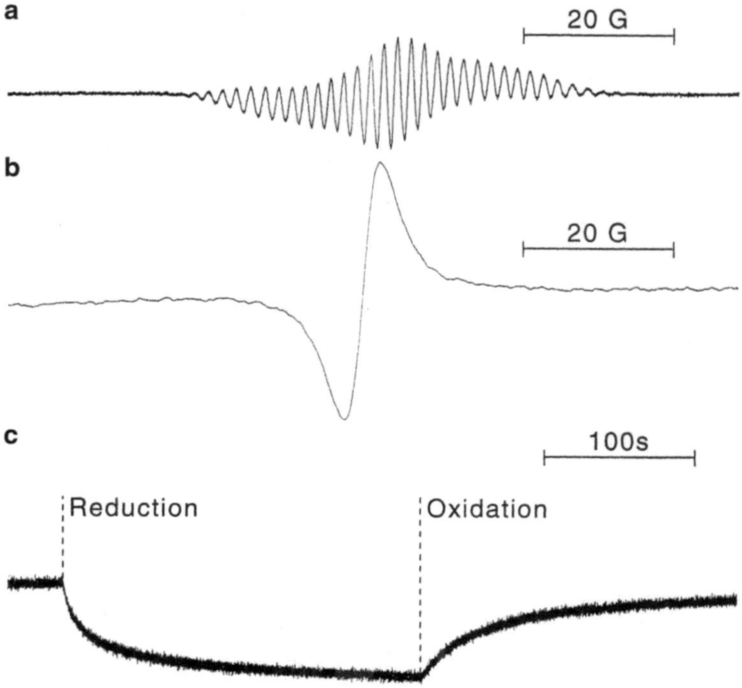

Fig. 4.13 (**a**) ESR signal for $THPD^+$ recorded during in situ electrolysis of 0.6 mmol L^{-1} THPD in acetonitrile (0.1 mol L^{-1} NBu_4PF_6) at $E = 0.4$ V versus SCE. (**b**) ESR signal for $THPD^+$ recorded during in situ electrolysis of 5 μg THPD deposited on a 1.2 cm^2 area gold electrode immersed in aqueous 0.1 M $NaClO_4$. (**c**) ESR signal intensity versus time transient for the ESR signal described in (**b**) when a double potential step between $E = 0.4$ V and 0 V versus SCE is applied [B74]. Copyright ©1997 with permission from Elsevier

For both studies, which represent examples for the application of in situ ESR for the study of the voltammetry of immobilized microparticles and microdroplets, conventional flat cells[13] have been used.

4.1.4 Ex Situ Spectroscopy

Due to the wavelength of the applied radiation, many spectroscopic techniques require the experiments to be performed under vacuum conditions. This makes them unsuitable for in situ electrochemical applications. Nevertheless, very often these techniques are of inestimable value for the study of electrochemical reactions.

---

[13] Piette LH, Ludwig P, Adams RN (1962) Anal Chem 34:916–921

Among other things, they, for example, allow the determination of the elemental composition of a sample after an electrochemical reaction.

Scanning Electron Microscopy and X-Ray Electron Probe Microanalysis

The scanning electron microscopy (SEM) and X-ray probe microanalysis are perfectly suited as a supporting tool for the voltammetry of immobilized microparticles. The value of this combination, i.e., the imaging of the electrode surface by SEM and the subsequent single particle elemental analysis, cannot be overestimated.

SEM measurements can be utilized to obtain information on the morphology of the electrode surface and the attached microparticles and their changes as a consequence of a voltammetric experiment. It can also be used to determine size and size distribution of the attached particles. The X-ray microprobe technique, which allows the qualitative and the quantitative surface analysis down to a spatial resolution of less than 1 μm—the usual diameter of the electron beam—is based on the identification and quantification of elements by means of their electron stimulated characteristic X-ray emission. The detection is usually performed with an energy dispersive X-ray detector (EDX). A major limitation of X-ray electron probe microanalysis is its inability to detect lightweight elements.

An example for the utilization of SEM and X-ray electron probe microanalysis is shown in Figs. 4.14 and 4.15. The first figure represents a surface map depicting the distribution of Prussian blue microparticles, mechanically immobilized at a paraffin-impregnated graphite electrode. The single particle analysis of immobilized and voltammetrically treated Prussian blue particles is shown in Fig. 4.15. The analysis clearly shows evidence for the insertion of potassium ions upon reduction of Prussian blue and for the expulsion of potassium upon oxidation.

4.2 Atomic Force Microscopy and Scanning Tunneling Microscopy

Atomic force and scanning tunneling microscopy are very useful techniques for direct observation of electrochemical conversions of solid particles immobilized on electrodes. Bond and Marken [B29] have been the first to apply scanning tunneling microscopy to make visible the deposit of decamethylferrocene on a basal plane pyrolytic graphite electrode before and after the electrochemical reaction. In a later study, they used AFM to detect submicrometer deposits of indigo on the same kind of electrodes [B72]. In case of solid deposits of C_{60} on the surface of a glassy carbon electrode, electrochemical studies with in situ recording of AFM images allowed showing that the morphology changes indicate nucleation–growth and redistribution processes as well as dissolutions of the solid material [B119]. The high

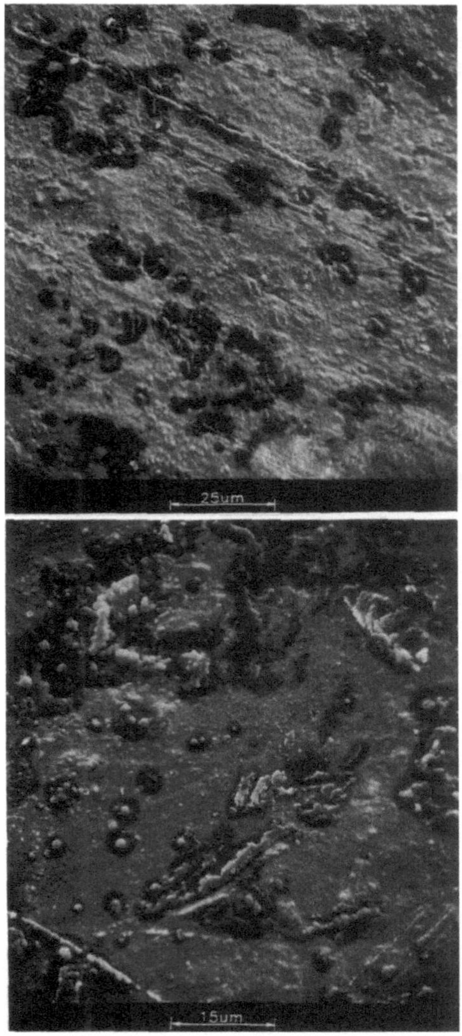

Fig. 4.14 SEM micrograph of a paraffin-impregnated electrode (*top*) and SEM micrograph of Prussian blue mechanically attached to a paraffin-impregnated electrode (*bottom*) [B35]. Copyright ©1995 with permission from American Chemical Society

potential of in situ AFM for the elucidation of electrode reactions of solid particles was also demonstrated in a study of TCNQ: the authors could prove that the electrochemical reduction of TCNQ is associated with the formation of well-formed microcrystals [B124].

The first case where the reaction front of a solid-to-solid electrochemical transformation could be detected was reported by Hasse and Scholz [B184]. This was possible when a submicrometer-size crystal of α-PbO was subject to a short reduction pulse that did not allow the entire crystal to be converted to Pb. Figure 4.16 shows a cut through such a partially reduced crystal where the reaction front appears as a rather steep step.

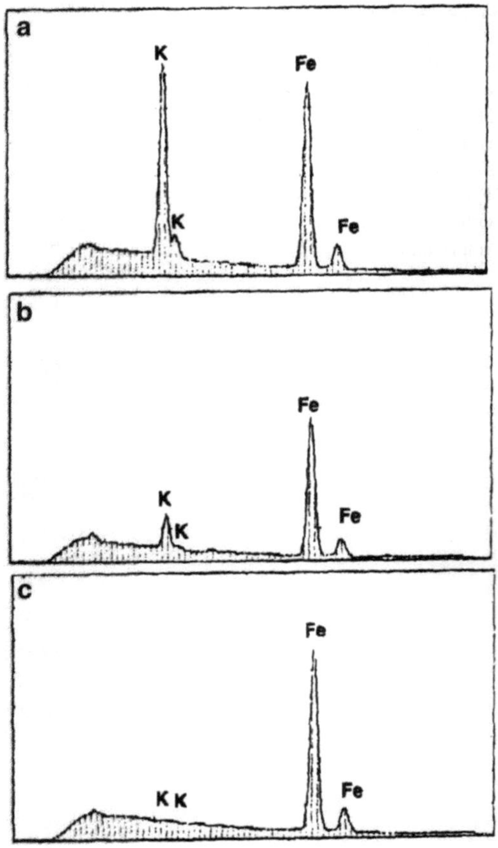

Fig. 4.15 Typical examples for single particle electron microprobe analysis of various Prussian blue samples mechanically attached to a paraffin impregnated graphite electrode. (**a**) Prussian blue electrochemically reduced in 0.1 M KNO_3 at −100 mV vs. Ag/AgCl for 5 min, (**b**) without treatment, (**c**) electrochemically oxidized in 0.1 M KNO_3 at +1,100 mV for 5 min [B35]. Copyright ©1995 with permission from American Chemical Society

The electrochemical reductive dissolution of goethite (α-FeOOH) has been also studied by in situ AFM [B276]. In this study, it could be shown that multidomain and twinned crystals dissolve at less negative potentials than single domain crystals. Further it was possible to follow the dissolution kinetics of single submicrometer-size crystals. Although the kinetics of different crystals varied considerably, the averaged dissolution kinetics agreed very well with what was found earlier in purely electrochemical studies of rather large assemblies of crystals. Figure 4.17 depicts goethite crystals immobilized on the surface of a gold electrode before and after reduction at −0.15 V versus Ag/AgCl, a potential where only multidomain and twinned crystals dissolve.

AFM studies can also provide direct evidence of the role of the three-phase boundary in an electrochemical reaction. When silver halide crystals, immobilized on a gold electrode, are reduced to silver, this reduction starts at the three-phase boundary gold–silver halide solution. When the silver halide crystal is only partially

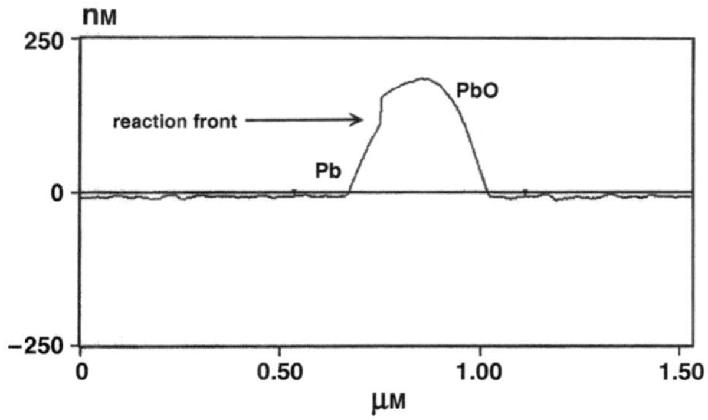

Fig. 4.16 Cut through the AFM image of a α-PbO crystal recorded after a 1 s reduction at 0.695 V versus Ag/AgCl. The crystal was deposited on a gold electrode and the electrolyte on a 1 mol L^{-1} KCl solution [B184]. Copyright ©2001 with permission from Elsevier

reduced during a short reduction pulse, and the remaining silver halide is dissolved with an ammonia solution, a ring of metallic silver is visible in the AFM image right at the place of the three-phase boundary (Fig. 4.18) [B278]. In this study, it has been shown that the reduction of silver halide crystals of submicrometer size proceeds considerably different than the reduction of much larger crystals: in the case of the small size crystals, the reduction converts a surface layer of the silver halide crystals to silver metal. The average thickness of this silver layer is remarkably constant: around 8 nm in case of AgCl, 8.8 nm in case of AgBr, 9.5 nm in case of AgI, and 9.14 nm in case of $AgBr_{0.95}I_{0.05}$. This corresponds in all cases to the formation of 16–21 atomic layers of silver. In no case, there has been a filamentous growth of silver for submicrometer-size silver halide crystals; however, this was observed in case of crystals of several micrometer sizes. Of course, the filamentous growth has been known since long to be the typical growth feature in silver halide photographic development. The authors of [B278] concluded that the "surface silvering" of the submicrometer-size crystals is the result of the specific mass transfer conditions: the surface reduction is obviously undisturbed by dislocations on the crystal surface. In the case of large crystals, the surface transformation will reach screw dislocations on the crystal surface where the mass transport of silver ions is very fast and the growth pattern changes from a surface layer formation to a whisker growth (see Fig. 4.19).

Very interesting results have also been obtained in studies of the electrochemical oxidation of nanocrystals of silver in an electrolyte solution that contained iodide ions [B279]. When the AFM image was focused on one single silver crystal, in many cases it disappeared from observation. In a few cases where the crystals did not disappear, their volume growth exceeded the theoretically expected value for a reaction in which Ag is oxidized to AgI by a factor of about 4! Figure 4.20 shows such a silver crystal and the resulting silver iodide crystal. The volume increased by a factor of 14. To understand the phenomenon, the AFM images were recorded in a range to cover many silver crystals. Figure 4.21a shows a gold electrode with an array of silver crystals. When they were oxidized to AgI, most of them disappeared and only a small number of AgI crystals grew (cf. Fig. 4.21b). The explanation of this phenomenon is that the oxidation of the silver nanocrystals first produces an

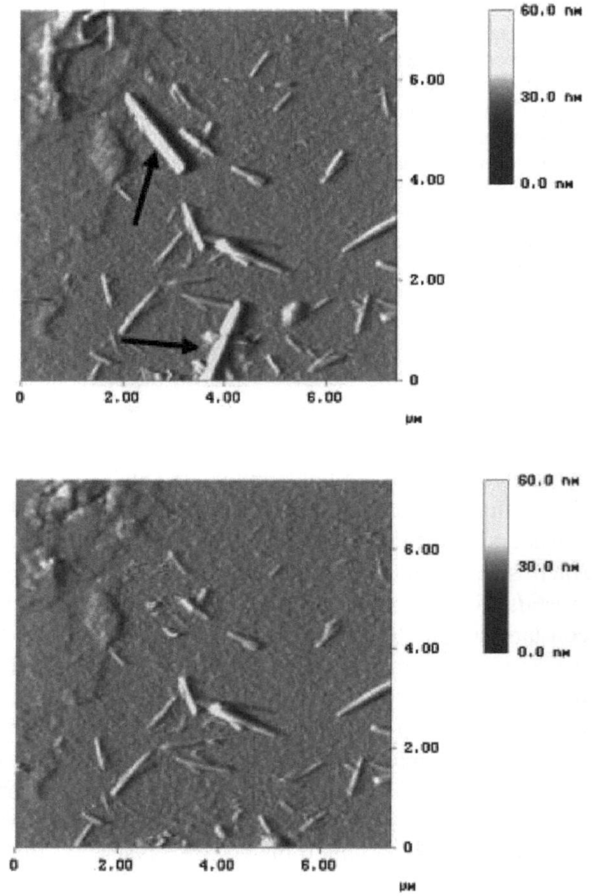

Fig. 4.17 AFM images of goethite crystals immobilized on a gold electrode before (*top*) and after (*bottom*) a reduction at –0.15 V versus Ag/AgCl. The *arrows* indicate multidomain and twinned crystals [B276]. Copyright ©2003 with permission from Elsevier

oversaturated AgI solution from which only a few but large AgI crystals grow. The number of Ag crystals in Fig. 4.21a was 2036 and the number of AgI crystals in Fig. 4.21b was only 27! However, when the volumes of all silver crystals and the volumes of all AgI crystals were calculated, it turned out that the overall volume increase was 3.8, which is very near to the theoretical value of 3.9. Figure 4.22 shows a similar experiment in which silver nanocrystals have been oxidized to silver bromide. Also in this case the number of AgBr crystals was very much smaller than the initial number of silver crystals, the ratio of the volumes being close to the theoretical value only when integration over all crystals of the image was performed. These in situ AFM studies proved that the oxidation of immobilized silver nanocrystals indeed proceeds via a detectable oversaturated AgI and AgBr solution, respectively.

This was the first clear observation of an electrochemical solid-to-solid transformation to proceed via the solution phase.

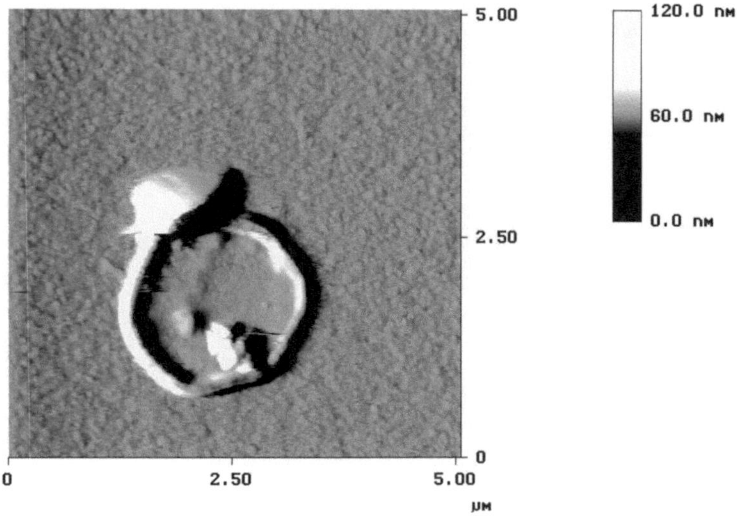

Fig. 4.18 Atomic force micrograph of a silver bromide crystal with 5 mol% iodide immobilized on a gold electrode after a reduction at –700 mV versus Ag/AgCl (3 mol L^{-1} KCl) for 20 s and dissolving the remaining silver halide with 5 mol L^{-1} ammonia [B278]. Copyright ©2004 with permission from Springer

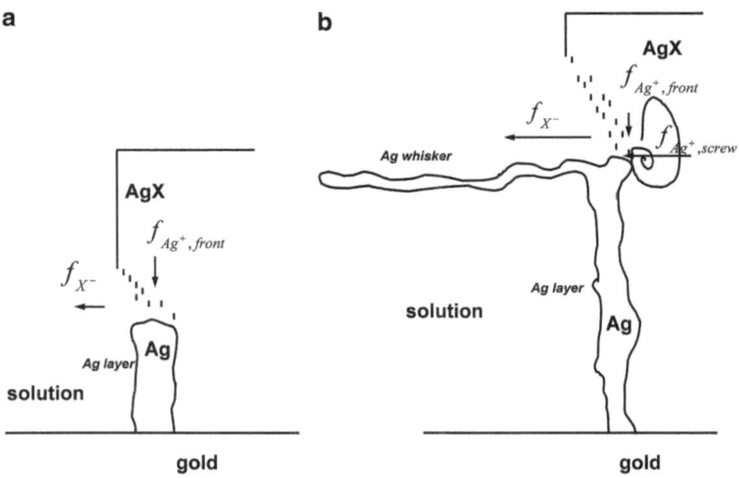

Fig. 4.19 Schematic drawing of the reduction of a silver halide crystal to silver: (**a**) Surface silvering of submicrometer-size crystals (initial stage). (**b**) Whisker growth following the initial formation of silver near the three-phase junction after prolonged electrolysis of a large crystal. The *arrows* indicate the transport of silver ions to the silver–silver halide interface [B278]. Copyright ©2004 with permission from Springer

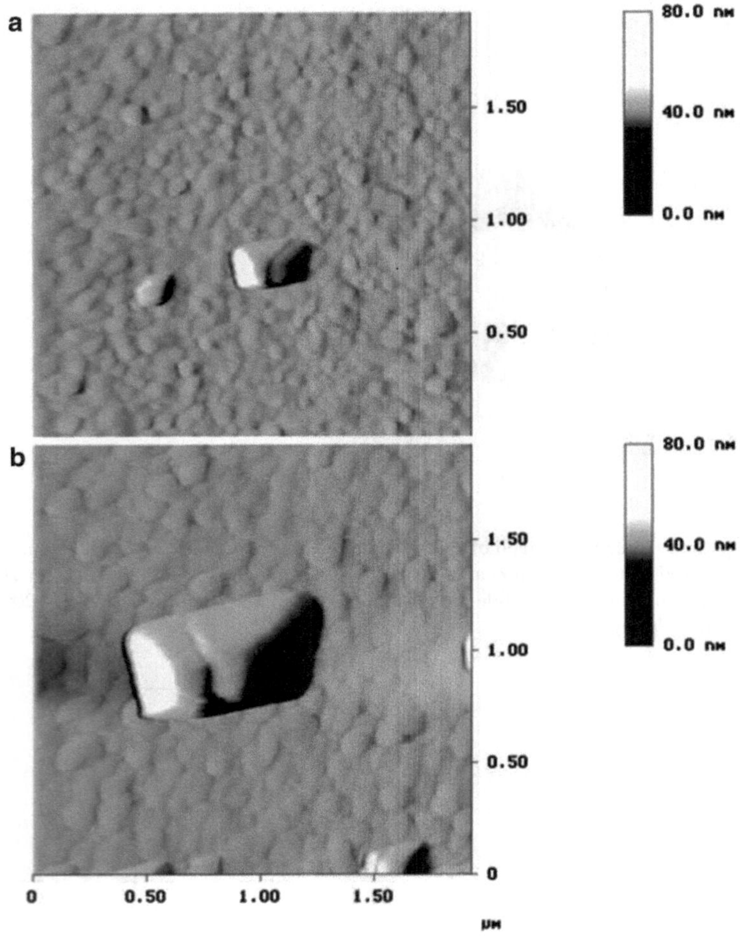

Fig. 4.20 (**a**) Silver nanocrystal immobilized on a gold electrode before oxidation. (**b**) The resulting silver iodide crystal formed during an oxidation at $E = 0.15$ V versus Ag/AgCl for 100 min. Electrolyte solution: 10^{-4} mol L^{-1} KI and 0.1 mol L^{-1} KNO_3 [B279]. Copyright ©2004 with permission from Elsevier

The above-discussed examples clearly show that AFM/STM can provide extremely important information on the course of solid-state electrochemical reactions of microparticles. The main problem of the experiments is certainly the procedure to immobilize the particles on suitable electrode surfaces. Many experiments are usually necessary to find the suitable immobilization technique for a special system. When the crystals are of a size in the range between 50 and 1,000 nm, very good results can be obtained with gold and platinum electrodes that are commercially manufactured by sputtering these metals on a chromium layer on quartz [B276, B278]. The optimum immobilization procedure was the evaporation from very dilute suspensions of the crystals in water or organic liquids

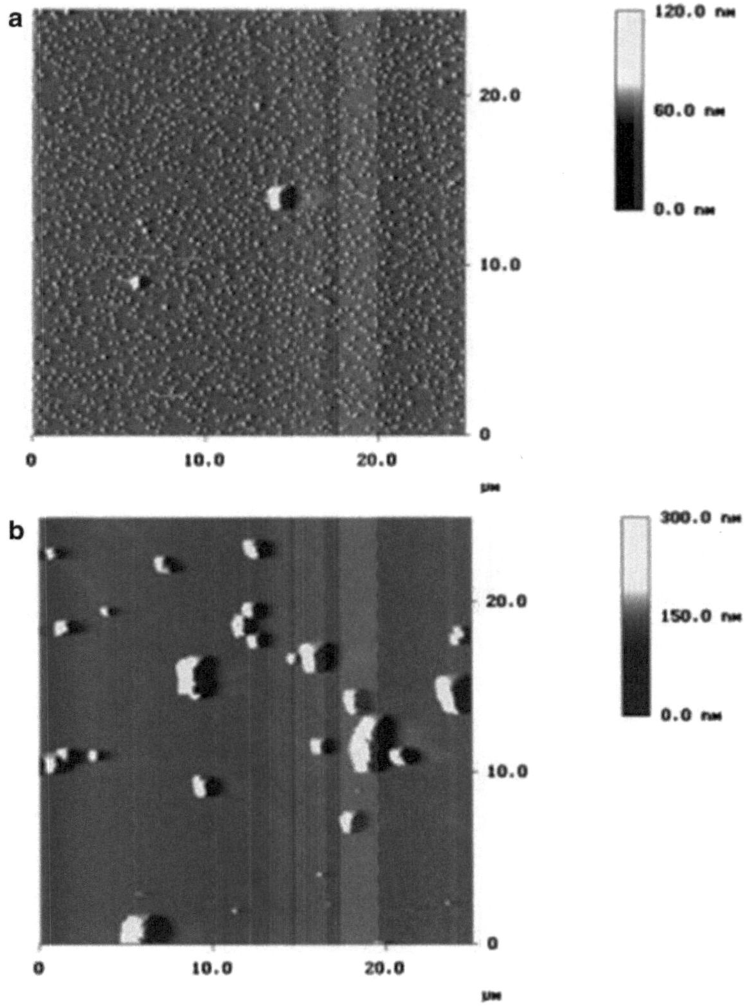

Fig. 4.21 (**a**) Array of immobilized silver nanocrystals before oxidation and (**b**) array of the resulting AgI crystals after oxidation of the silver at $E = 0.15$ V versus Ag/AgCl for 5 min. Electrolyte solution: 10^{-4} mol L^{-1} KI and 0.1 mol L^{-1} KNO_3 [B279]. Copyright ©2004 with permission from Elsevier

onto the electrodes. Mostly, the suspensions that did not settle after 5–10 min contain crystals of a suitable size.

AFM imaging of crystal aggregates of organic compounds provides evidence of changes in the size of the solid particles upon application of reductive and oxidative potentials. Figure 4.23 shows topographic maps of in situ AFM images from the upper face and sides of crystals of indigo attached to a graphite plane in contact with aqueous acetate buffer before (a, c) and after (b, d) application of a linear potential ramp from 0.0 to +0.75 V with a potential scan rate of 10 mV s^{-1}. Under these

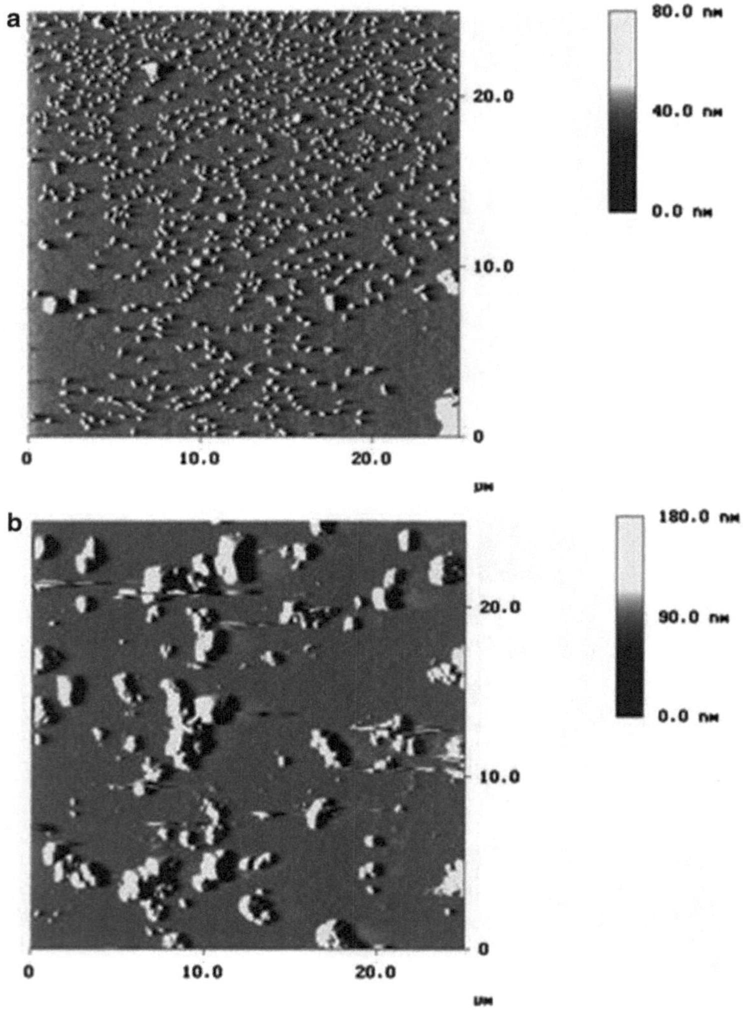

Fig. 4.22 (**a**) Array of immobilized silver nanocrystals before oxidation and (**b**) array of the resulting AgBr crystals after oxidation of the silver at $E = 0.20$ V versus Ag/AgCl for 5 min. Electrolyte solution: 10^{-4} mol L^{-1} KBr and 0.1 mol L^{-1} KNO_3 [unpublished results, Hasse U, Scholz F]

conditions, the dye is oxidized to dehydroindigo in a two-proton, two-electron process [B72, B319, B320], i.e., a process similar to those of anthraquinonic, naphtoquinonic, and flavonoid dyes [B321, B322] which have also been studied by AFM. The image shows an agglomerate of indigo crystals sized ca. 1×0.5 µm which, after application of the oxidation potential, exhibits an apparent erosion of several regions at the lateral sides, while other regions become apparently extended. The maximum extent of the distortions of crystals can be estimated as long as ca. 50–80 nm, but the progress of the electrochemical reaction was apparently

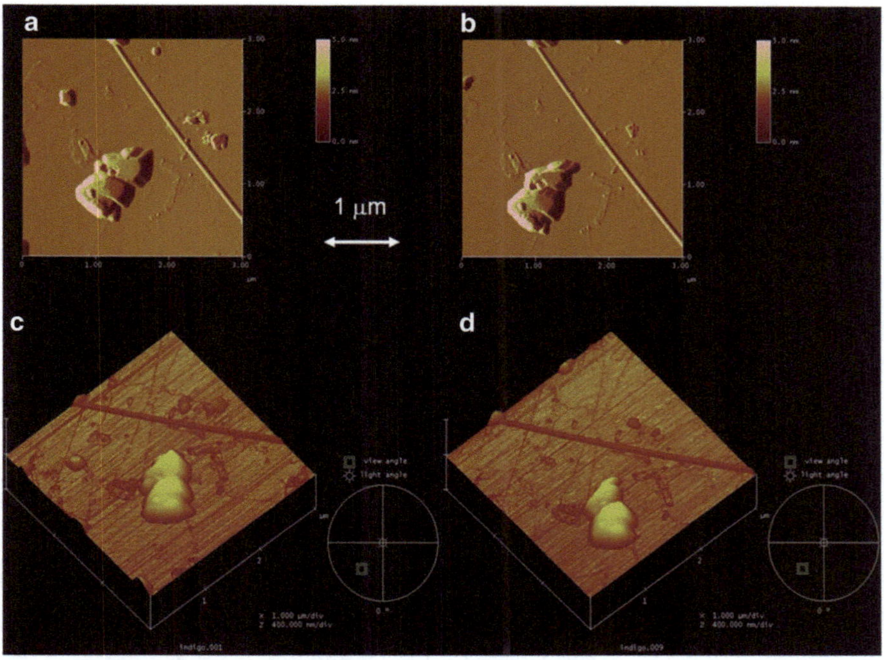

Fig. 4.23 AFM images (**a**, **b**) and topographic maps (**c**, **d**) of indigo crystals adhering to a graphite plate in contact with 0.50 mol L^{-1} aqueous acetate buffer before (**a**, **c**) and after (**b**, **d**) application of a linear potential scan from 0.00 to +0.75 V with a sweep rate of 10 mV s^{-1} [B320]. Copyright ©2008 with permission from Elsevier

lowered with increasing time, thus suggesting that no reductive/oxidative dissolution processes occur [B320]. The electrochemical process can be viewed as a topotactic formation of a layer of dehydroindigo at the lateral faces of the indigo crystals, the different density of the phases, resulting from the different intermolecular hydrogen bonding system, explaining the apparent size changes.

Comparable features were obtained by applying reductive steps on microparticulate deposits of organic compounds. Figure 4.24 shows the AFM images of a deposit of crystals of clonazepam, a widely used benzodiazepinic anxiolytic, on graphite plate in contact with aqueous acetate buffer before and after application of a reductive potential step between 0.00 and –0.65 V at 10 mV s^{-1}. Under these conditions, the drug displays a solid-state, proton-assisted cathodic process presumably involving the one-electron reduction of the $–NO_2$ group [B323, B324]. The image shows clonazepam crystals of ca. 1-µm size which shrink in several regions of the lateral sides. Such features, which were not observed in blank experiments performed at clonazepam deposits in contact with acetate buffer but without application of potential changes, can be attributed to a crystal contraction presumably associated with a topotactic solid-state transformation. Upon

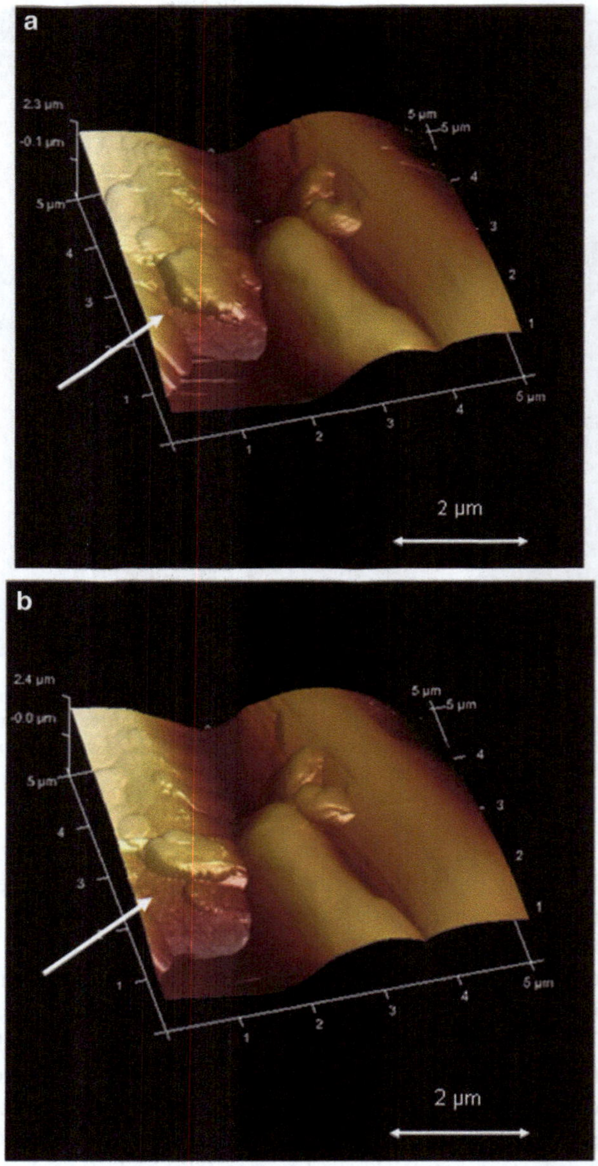

Fig. 4.24 AFM images of clonazepam crystals adhering to a graphite plate in contact with 0.50 mol L^{-1} aqueous acetate buffer (**a**) before and (**b**) after application of a linear potential scan between 0.00 and −0.65 V at a sweep rate of 10 mV s^{-1} [B323]. Copyright ©2013 with permission from Elsevier

application of a potential step from 0.0 to −1.0 V, however, the size of the crystals grows, associated with a lateral expansion, as can be seen in Fig. 4.25. At these potentials, the C=O and C=N units of clonazepam are reduced to –OH and –NH groups, respectively [B323]. This could explain the lateral expansion of the crystals as a result of the modification of the lattice parameters due to the formation of additional intermolecular hydrogen bonds.

4.3 Scanning Electrochemical Microscopy

Scanning electrochemical microscopy (SECM)[14] has also been applied for monitoring solid-state electrochemical processes. Thus, Bond et al. have recently monitored the solid–solid interconversion of TCNQ into CuTCNQ using the substrate generation tip collection mode [B325]. In the presence of aqueous Cu^{2+}-containing solutions, TCNQ yields solid CuTCNQ whose subsequent oxidation does not involve simple expulsion of copper ions into solution but a soluble complex attributed to $Cu^{2+}TCNQ^{-}(aq)$. A mechanistic understanding of the electrochemical conversion of CuTCNQ phase I into phase II by repetitive potential cycling and electrochemical formation of KTCNQ was established with the help of SECM. This technique has also been used to characterize praseodymium centers in praseodymia-doped zirconias [B326] and praseodymia-doped zircons [B327]. Here, differential electrocatalytic effects were monitored on the basis of the redox competition [B328] SECM mode. The current was recorded flowing through the tip electrode close to the substrate electrode where the solid redox probe was immobilized. Several authors have recently used this technique for mapping and visualizing the electrocatalytic activity of nanoparticle-laccase thin film [B329] and microstructured metal hexacyanoferrates [B330]. Figure 4.26 compares the images of Prussian blue spots deposited on a glassy carbon slide, in contact with H_2O_2 solutions in PBS at pH 6.0 upon application of different potentials to the substrate electrode. After background subtraction, clear differences were obtained in the catalytic effect depending on the applied potential, as indicated by variations of the tip current, also reflected in local variations of the same [B330].

If the Prussian blue spot is covered by a layer of enzyme glucose oxidase embedded in a polymer matrix and immersed in a glucose solution, the tip current profiles vary with the position over the spot as shown in Fig. 4.27. Under these conditions, the tip is sensitive to the H_2O_2 generated from the enzyme-assisted oxidation of glucose. Local variations of the tip current are associated to the corresponding variations of the enzyme distribution and become sensitive to the potential applied to the substrate electrode.

[14] Bard AJ, Mirkin MV (eds) (2001) Electrochemical microscopy. Taylor & Francis, Boca Raton, FL

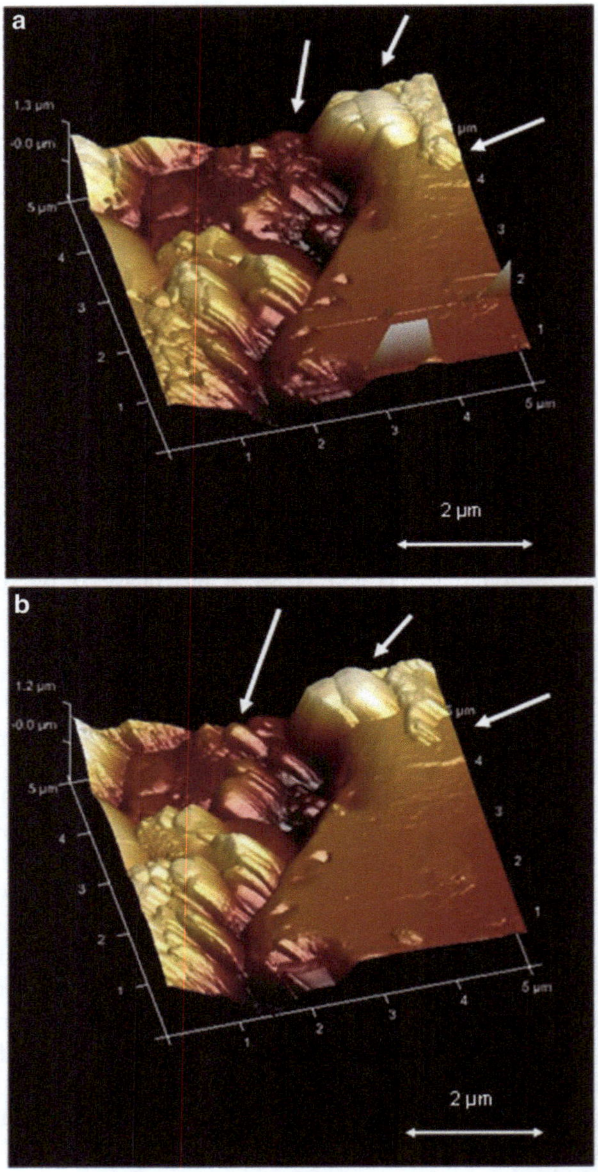

Fig. 4.25 AFM images of a microparticulate deposit of clonazepam adhered to a graphite plate in contact with 0.50 mol L^{-1} aqueous acetate buffer (**a**) before and (**b**) after application of a linear potential scan between 0.00 and –1.00 V at a sweep rate of 10 mV s^{-1} [B323]. Copyright ©2013 with permission from Elsevier

4.4 Electrochemical Quartz Crystal Microbalance

One of the most versatile and hence widespread hyphenated techniques for the interrogation of electrode processes is the electrochemical quartz crystal microbalance technique, EQCM. This technique allows to continuously monitor mass

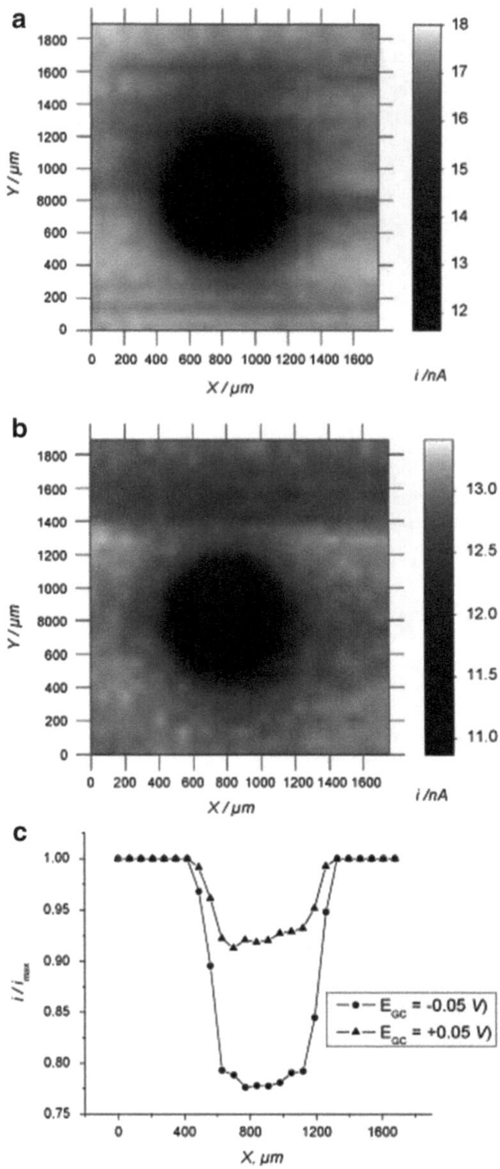

Fig. 4.26 Visualization of local activity of the Prussian blue spot deposited on glassy carbon slide. (**a**) SECM image after background correction. Sample potential −0.05 V. Solution: 0.1 M PBS (pH 6.0) with 1 mM H_2O_2. Distance increments (*x*, *y*): 70 μm. Scan rate (*x* direction): 35 μm s^{-1}. (**b**) SECM image after background correction. Sample potential +0.05 V. Solution: 0.1 M PBS (pH 6.0) with 1 mM H_2O_2. Distance increments (*x*, *y*): 70 μm. Scan rate (*x* direction): 35 μm s^{-1}. (**c**) Superimposed normalized current profiles across the spot under the experimental conditions employed for the scans shown in (**a**) and (**b**) [B330]. Copyright ©2009 with permission from Elsevier

changes at an electrode surface with a sensitivity of few nanograms by means of tracking the frequency of the resonator, a gold-coated quartz crystal, which at the same time serves as the working electrode in the electrochemical experiment. EQCM measurements have been widely used for studies of all kinds of monolayer and multilayer depositions and dissolutions, mass transport in polymer films on

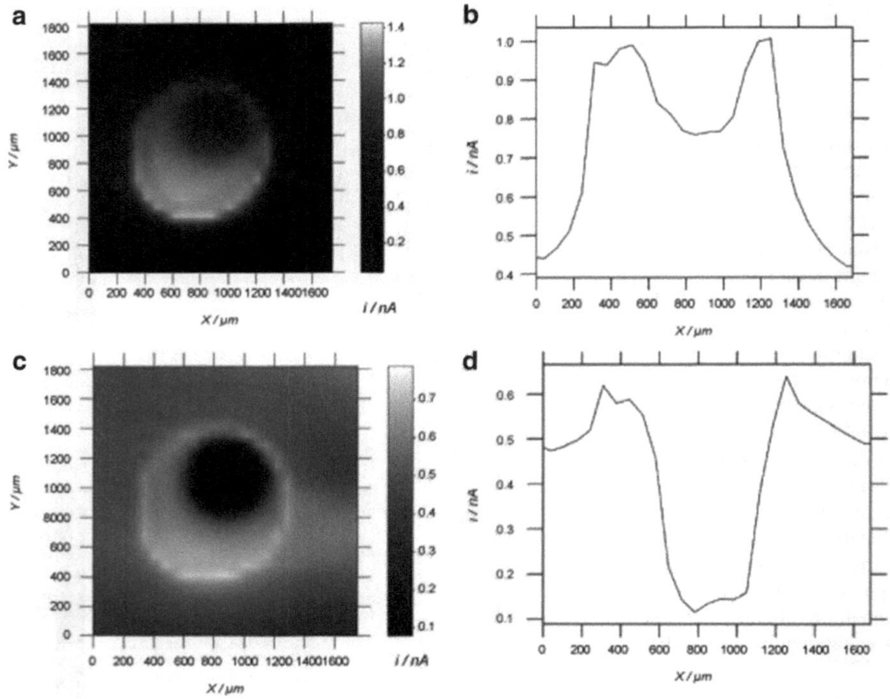

Fig. 4.27 Visualization of local activity of a polymer/glucose oxidase spot covering a Prussian blue spot deposited on glassy carbon electrode in contact with 100 mM glucose solution in PBS, pH 6.0. Potentials applied to the substrate electrode were (**a**, **b**) +0.275 V; (**c**, **d**) +0.175 V. Conditions such as in Fig. 4.26 [B330]. Copyright ©2009 with permission from Elsevier

electrodes, corrosion processes, etc.[15] Basically, EQCM measurements can easily be applied to the study of microparticles. The method is readily available and does not need special adaptation to the voltammetry immobilized microparticles. The simplest procedure to stick a powder sample onto a quartz crystal electrode is by using a cotton swab and gently rubbing a small amount of the solid onto the electrode. Alternatively, organic compound microcrystals can be attached via evaporation of the dissolved material from a volatile solvent (see Sect. 3.2).

One of the major domains of EQCM in the voltammetry of immobilized particles is the interrogation of electrochemically driven solid-to-solid transformations, which are generally accompanied by the charge-compensating ion transfers between the electrolyte phase and the solid. Here, EQCM helps in identifying the nature of the transferred ions. In the first study utilizing EQCM, Shaw and coworkers investigated the electrochemistry of nonconducting microcrystalline particles of *trans*-$Cr(CO)_2(dpe)_2$ and [*trans*-$Cr(CO)_2(dpe)_2]^+$ (dpe = $Ph_2PCH_2CH_2PPh_2$) attached to

[15] See, for instance, Buttry DA (1991) Applications of the quartz crystal microbalance to electrochemistry. In: Bard AJ (ed) Electroanalytical chemistry, vol 17. Dekker, New York

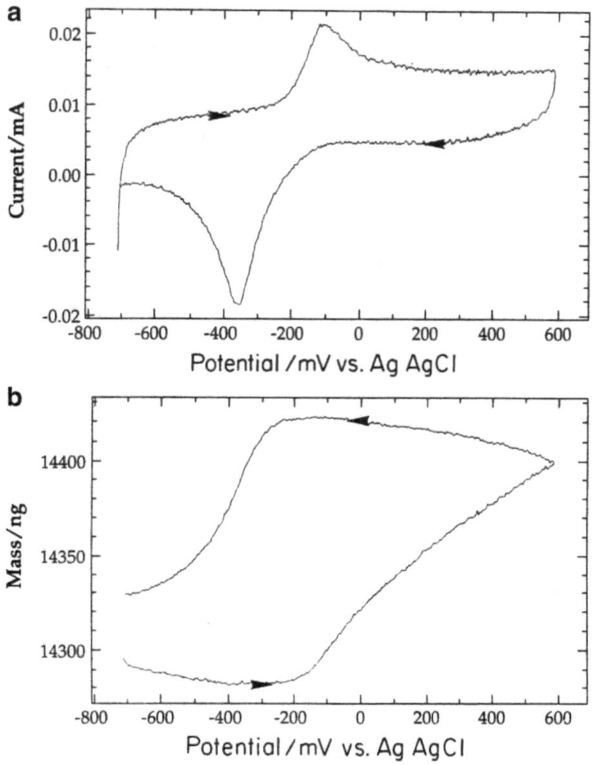

Fig. 4.28 (**a**) Cyclic voltammogram and (**b**) the corresponding mass change diagram for microcrystalline particles of *trans*-$[Cr(CO)_2(dpe)_2]Cl$ attached to a gold quartz crystal which has been placed in 0.1 M KCl aqueous electrolyte [B53]. Copyright ©1996 with permission from Elsevier

gold electrodes [B53]. From a combination of voltammetric measurements, EQCM experiments (see Fig. 4.28), and EDX data (see Sect. 4.1.3), the authors proved a reaction mechanism involving the incorporation of non-solvated anions into the solid upon oxidation and their expulsion upon reduction. EQCM also revealed a rapid ion exchange process when *trans*-$[Cr(CO)_2(dpe)_2]^+X^-$ salts were immersed into electrolyte solutions containing other anions than that contained in the attached solid.

Since the first application of the EQCM in the voltammetry of microparticles, a wealth of studies have been published, including the electrochemical interrogation of tetrathiafulvalene, TTF [B54], TCNQ [B101], of C-60 fullerenes [B154], iodopentacarbonyl chromium(0) [B65], of solid complex cyanides (e.g., [B33, B35, B96, B136]), and of photoactive ruthenium [B102] and manganese [B128] complexes. In the latter study, the authors demonstrate that EQCM measurements can be conveniently combined with photoelectrochemical experiments. For that, an adapted electrochemical cell has been presented as shown in Fig. 4.29.

The application of EQCM for studying immobilized microdroplets is less straightforward. First of all, the viscoelasticity of immobilized droplets does not allow applying the Sauerbrey equation and thus to easily quantify the mass changes

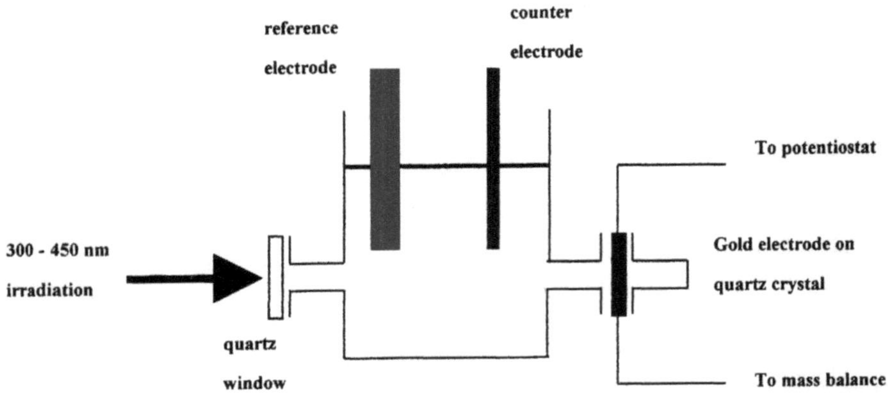

Fig. 4.29 Schematic diagram of an EQCM cell used in photochemical experiments [B128]. Copyright ©1999 with permission from American Chemical Society

at the electrode surface. Especially reactions at immobilized redox liquids—generally accompanied by ion expulsion or uptake—lead to considerable changes of the phase properties like the viscoelasticity of the liquid. Here it is very difficult to distinguish between frequency effects caused by mass changes and those caused by viscoelasticity changes.

4.5 In Situ X-Ray Diffraction

In situ X-ray diffraction is a well-established technique for studying the phase transformations in solid-state electrochemical reactions. However, due to the following reasons, the application to microparticles immobilized on electrode surfaces is not simple: (1) the overall amount of immobilized substance is usually very small, (2) the electrolyte solution adjacent to the electrode surface diminishes the X-ray intensity, and (3) for technical reasons, the X-ray diffraction can be measured only in reflection mode. Until now, there is only one publication in which an in situ X-ray diffraction study has been described with microparticles immobilized on an electrode surface [B36]. Figure 4.30 shows the cell construction.

The electrochemical cell was made of polyacrylic with a window made of X-ray amorphous polystyrene foil. The working electrode was a PIGE that was situated in such a way that the circular surface of the electrodes was parallel to the foil and in the axis of the diffractometer. The distance between the foil and the electrode surface must be kept small, because otherwise no X-ray reflections are observed owing to absorption by the electrolyte solution. With the help of this cell, the electrochemical reduction of red PbO (litharge) and that of Pb(OH)Cl (laurionite) has been studied by scanning only a very small angle range simultaneously to the

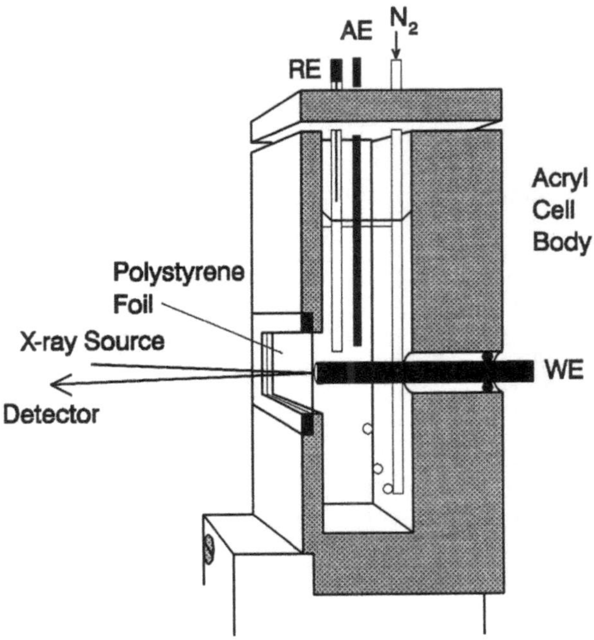

Fig. 4.30 Electrochemical cell for in situ X-ray diffraction. *AE* auxiliary electrode, *RE* reference electrode, *WE* working electrode [B36]. Copyright ©1995 with permission from Elsevier

electrochemical conversion. Figure 4.31 shows the voltammogram of PbO reduction and the simultaneously recorded reflections of PbO and Pb. The latter prove that, on the time scale of the experiment, there is a simultaneous decrease of PbO reflections and increase of Pb reflections. That means that the reaction does not pass through a range where the metallic lead exists in an X-ray amorphous state. This, however, was evident when Pb(OH)Cl was reduced to lead. Figure 4.32 shows the linear scan voltammogram of Pb(OH)Cl reduction and the corresponding in situ X-ray diffraction signals. The figure proves that the reduction of Pb(OH)Cl proceeds via an amorphous phase that slowly recrystallizes. The simultaneous growth of lead crystals in the course of reduction of tetragonal PbO has later also been studied by in situ AFM (see Sect. 4.2), where indeed it was possible to see that this transformation proceeds without disintegration of the crystal.

The in situ X-ray diffraction of immobilized microparticles has certainly a large potential, provided that high intensity X-ray sources are applied.

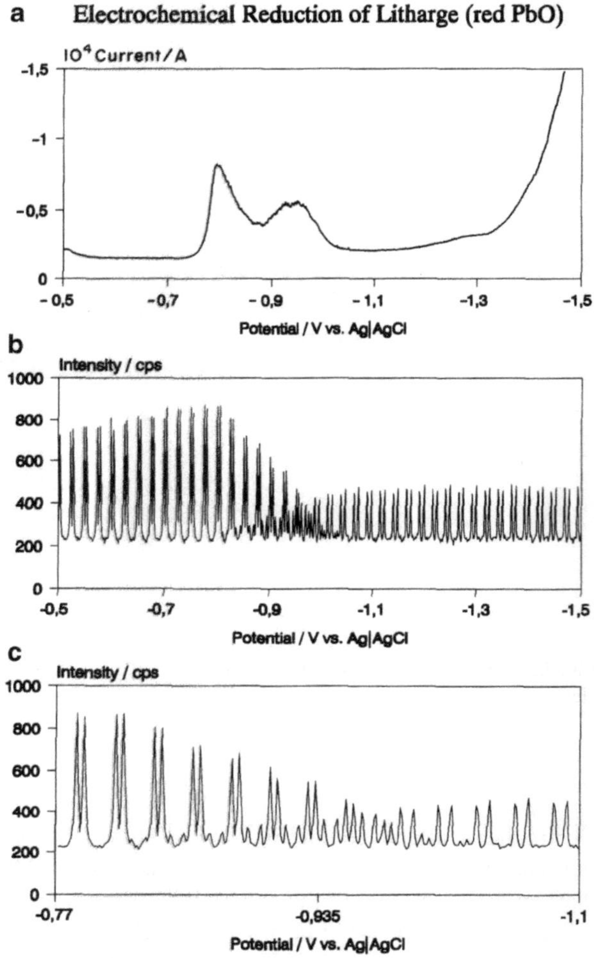

Fig. 4.31 Electrochemical reduction of red tetragonal PbO (litharge) in situ coupled with X-ray diffraction. PbO was mechanically attached to the surface of a PIGE, electrolyte 1 mol L^{-1} KCl: (**a**) linear sweep voltammogram (scan rate 0.1 mV s^{-1}); (**b**) diffraction signal pairs between –0.5 and –0.8 V are due to PbO ($d = 2.51$ Å), signal pairs between –1.1 and –1.5 V are due to Pb ($d = 2.48$ Å); (**c**) enlarged part of (**b**) showing the simultaneous decrease in the PbO reflection and the increase in the Pb reflection [B36]. Copyright ©1995 with permission from Elsevier

4.6 In Situ Calorimetry

Calorimetric measurements were the basic experiments to establish the fundamentals of chemical thermodynamics. In the nineteenth and twentieth centuries, electrochemical thermodynamics was developed and the relations between calorimetric and electrochemical data were discovered. Very often, electrochemical measurements give much more facile and reliable access to thermodynamic data. In theory, these are two equally suited methods for obtaining the same information. Hence, one might ask why one should attempt an in situ measurement of thermal effects of electrochemical reactions. Historically such measurements were of importance to give experimental proof for the interrelation between electrical and caloric data. But

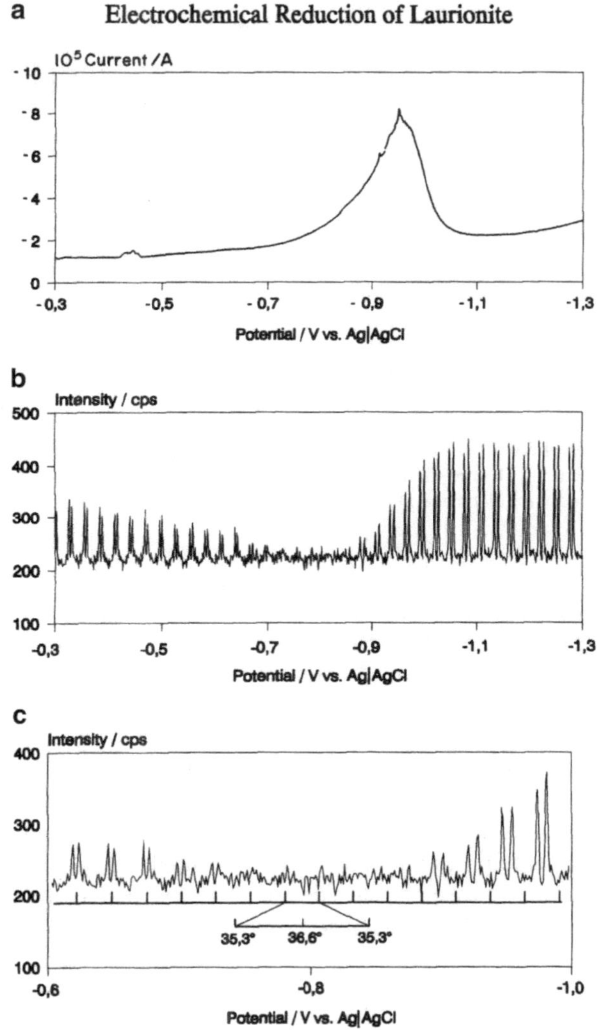

Fig. 4.32 Electrochemical reduction of orthorhombic Pb(OH)Cl (laurionite) in situ coupled with X-ray diffraction. PbO was mechanically attached to the surface of a PIGE, electrolyte 1 mol L^{-1} KCl: (**a**) linear sweep voltammogram (scan rate 0.1 mV s^{-1}); (**b**) diffraction signal pairs between –0.3 and –0.65 V are due to Pb (OH)Cl ($d = 2.52$ Å), signal pairs between –0.9 and –1.3 V are due to Pb ($d = 2.48$ Å); **c** enlarged part of (**b**) showing a potential (time) range with an amorphous phase [B36]. Copyright ©1995 with permission from Elsevier

even nowadays, there are still reasons for performing in situ calorimetry–electrochemistry: in the case of reversible electrochemical reactions, the calorimetric data give unambiguous proof or disproof for the correctness of the electrochemical data. The other good reason for such a combination is the study of irreversible systems, because it makes the irreversible heat effects accessible.

To study the heat effects of electrochemical reactions of immobilized microparticles, special electrodes have had to be developed so that the minute temperature effects, usually in the mK range, can be recorded with sufficient reliability. Figure 4.33 depicts thermistor electrodes that allowed measurements with amounts of 5×10^{-9} up to 5×10^{-8} mol of solid metal hexacyanoferrates.

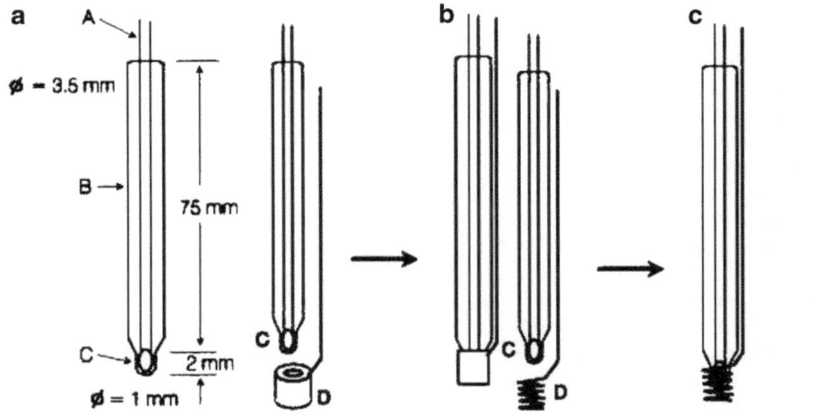

Fig. 4.33 (**a**) The thermistor electrode: (A) electric wires, (B) glass tube housing, (C) thermistor element; (**b**) the thermistor electrode with graphite cylinder D; (**c**) the thermistor electrode with gold wire electrode D [B219]. Copyright ©2002 with permission from Elsevier

The solid microparticles have been immobilized either on a very small graphite cylinder (made from a paraffin-impregnated graphite rod) or on a gold wire coil. The graphite cylinder and the gold coil were fitted to the thermistor that was housed in a glass tube. To calibrate the thermistor electrode, a Kanthal wire (22.3 Ω m^{-1} and isolated with a 0.004-mm-thick polyamide film) was wound around the graphite cylinder.

The heat release or absorption occurs at the surface of the electrode, because that is where the electrochemically active compound is immobilized. To quantify this heat, it is necessary to understand the relation between the thermal balance and the temperature, as the latter is the measured quantity. For calculating the amount of heat that corresponds to a certain change in temperature, the thermal behavior of the thermistor electrode has to be modeled. Figure 4.34 shows the heat production and

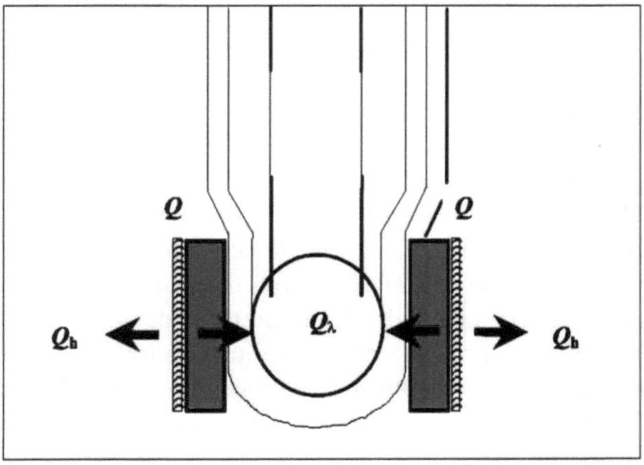

Fig. 4.34 The heat Q produced at the electrode surface is transferred partly to the thermistor element (Q_λ) and partly to the solution (Q_h) [B219]. Copyright ©2002 with permission from Elsevier

dissipation at the electrode. One part of the heat will flow to the thermistor by heat conduction, while some other part will be lost to the electrolyte solution, partly by convection. The temperature profile at the electrode can be estimated using the following assumptions: in principle, heat transport from the electrode surface to the solution could occur via free convection. However, an analysis of the hydrodynamic conditions showed that at least 100 times more heat is exchanged by conductance than by convection to the solution. To model the heat transfer at the thermistor electrode, it is necessary to calculate the ratio of heat transfer by conduction in the solid parts to the amount of heat transported by conduction into the solution. Because of the very high heat conductivity of the graphite and the low heat conductivity of water, the electrode will be the so-called lumped system, i.e., the temperature in the electrode will be uniform. Within the thermistor element, a logarithmic temperature drop will occur, as is known to occur for radial heat conduction in cylinders. Figure 4.35a depicts the most probable temperature profile at the thermistor electrode, and Fig. 4.35b, c depict two different models that have been used in the calculations.

Figure 4.35b depicts the assumed temperature profile in model I. In this model, the heat balance is

$$Q_{\mathrm{acc,\,electrode}} = P - Q_{\mathrm{loss}} \tag{4.3}$$

where $Q_{\mathrm{acc,\,electrode}}$ is the heat accumulated in the electrode, P is the produced heat, and Q_{loss} is the heat lost via the electric contact and convection. These two values can be calculated as follows:

$$Q_{\mathrm{acc,\,electrode}} = mC_{\mathrm{p}}(\mathrm{d}T^*/\mathrm{d}t) \tag{4.4}$$

and

$$Q_{\mathrm{loss}} = KT^* \tag{4.5}$$

where T^* is the temperature change with respect to the solution temperature ($T^* = T_{\mathrm{T}} - T_{\mathrm{S}}$, T_{T} is the temperature of the thermistor and T_{S} that of the solution), K is a global heat transfer coefficient due to conduction and convection, and mC_{p} is the heat capacity of the thermistor electrode, i.e., the sum of all heat capacities of the different parts of the thermistor electrode. From Eqs. (4.3)–(4.5), a first-order differential equation is obtained:

$$mC_{\mathrm{p}}(\mathrm{d}T^*/\mathrm{d}t) = P - KT^* \tag{4.6}$$

The solution of this equation for $T^* = 0$ at $t = t_0$ is

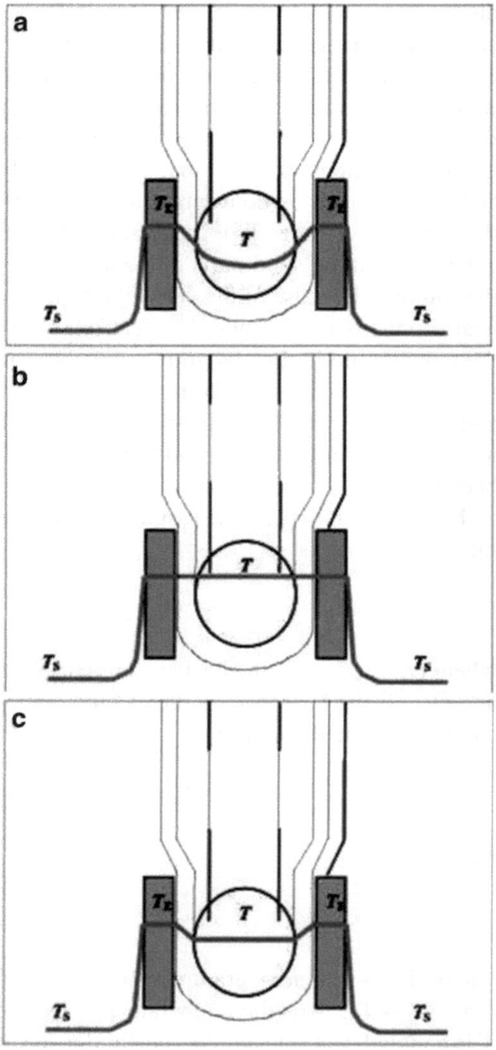

Fig. 4.35 (**a**) Temperature profile at the thermistor electrode: T, temperature in the thermistor element; T_E, temperature within the electrode; T_S, temperature within the solution. (**b**) Assumed temperature profile in Model I. T^* is the homogeneous temperature in the electrode and the thermistor element and the glass shaft separating both. TS is the temperature in the solution. (**c**) Assumed temperature profile in Model II. TE is the homogeneous temperature in the electrode, T is the homogeneous temperature in the thermistor element, and TS is the temperature in the solution [B219]. Copyright ©2002 with permission from Elsevier

$$T^* = \frac{P}{K}\left(1 - e^{-\frac{K}{mC_p}(t-t_0)}\right) \tag{4.7}$$

where $\frac{K}{mC_p}$ is the reciprocal time constant of the temperature response. The parameters of this equation can be determined as follows: when the heat pulse is sufficient to establish a steady-state temperature T_{st}, this value equals P/K. Since P is known in a calibration experiment, K can be easily calculated from T_{st}. With T_{st} follows from Eq. (4.7)

$$\ln\left(1 - \frac{T^*}{T_{st}}\right) = -\frac{K}{mC_p}(t - t_0) \tag{4.8}$$

This equation allows the determination of K/mC_p from a linear regression. Both constants K and mC_p have to be determined for each experiment (each electrode preparation) as minute differences in the electrode assembly will influence their values.

In model II, it was assumed that the temperature within the thermistor electrode is not homogeneous, but, due to the lower heat conductivity of glass (≈1.2 W/m K), the temperature within the thermistor element will be lower than within the electrode. Figure 4.35c depicts this model. The heat balance was assumed as

$$Q_{acc,\,electrode} = P - Q_T - Q_{loss,\,1} \tag{4.9}$$

where $Q_{acc,electrode}$ is the heat accumulated in the electrode, P is the heat produced, Q_T is the heat exchanged with the thermistor, and Q_{loss} is the heat lost via the electric contact and convection. The heat balance for the thermistor is

$$Q_{acc,\,thermistor} = Q_T - Q_{loss,\,2} \tag{4.10}$$

The differential equations for the two heat balances are

$$mC_{P,E}\frac{dT_E^*}{dt} = P - K_1\left(T_E^* - T^*\right) - K_2T_E^* \tag{4.11}$$

$$mC_{P,T}\frac{dT^*}{dt} = K_1\left(T_E^* - T^*\right) - K_3T_E^* \tag{4.12}$$

with $T_E^* = T_E - T_S$ (T_E is the temperature of the electrode and T_S the temperature of the solution bulk). In these equations, K_1 and K_2 are the heat transfer coefficients at the thermistor–electrode interface and for the heat loss at the electrode via wires and convection, respectively. K_3 is the coefficient for the heat transfer from the thermistor to the glass, electric wires, and loss by convection. $mC_{P,E}$ and $mC_{P,T}$ are the heat capacities of the electrode and the thermistor, respectively. All these parameters are unknown. From Eqs. (4.11) and (4.12), the following second-order differential equation with three unknowns is obtained:

$$\frac{d^2T^*}{dt^2} + a_1\frac{dT^*}{dt} + a_2T^* = a_3P \tag{4.13}$$

The new parameters a_1, a_2, and a_3 are combinations of the heat capacities and heat transfer coefficients. The solution of Eq. (4.13) is as follows:

$$T^* = T_{st} + A_1e^{\alpha_1(t-t_0)} + A_2e^{\alpha_2(t-t_0)} \tag{4.14}$$

where T_{st} is the steady-state temperature, α_1 and α_2 ($\alpha_1 \neq \alpha_2$) are constants depending on a_1 and a_2, and A_1 and A_2 are constants that have to fulfill the initial conditions. The following equations show the interrelation of these values:

$$T_{st} = \frac{a_3 P}{\alpha_1 \alpha_2} \tag{4.15}$$

$$\alpha_1 = -\frac{a_1}{2} + \sqrt{\frac{a_1^2}{4} - a_2} \tag{4.16}$$

$$\alpha_2 = -\frac{a_1}{2} - \sqrt{\frac{a_1^2}{4} - a_2} \tag{4.17}$$

A comparison of Eqs. (4.14) and (4.7) shows that Eq. (4.14) contains an additional exponential term. Since the thermistor electrode is a system with a finite mass and heat capacity, the initial conditions have been assumed to be $T^* = 0$ and $\frac{dT^*}{dt} = \gamma$. Plotting the experimental data as $\frac{dT}{dt} = f(t)$ for different electric powers showed that there is a power independent value of $\frac{dT}{dt}$ at the start of the heating pulse. This can be explained by the fairly high heat conductivity of the graphite electrode. With these experiments, an initial value $\frac{dT}{dt} = \gamma$ was determined that was used for solving the differential equation. Figure 4.36 shows the experimental data and the simulation curve. The latter exhibited a very good fit of the experimentally observed behavior of the thermistor electrode. For the determination of the parameters, it is important to use the temperature data before the electrode attains a steady state. Assuming n layers surrounding the thermistor, the calculation affords a system of differential equations that can be presented as a matrix:

$$d\underline{T}^*/dt = \underline{\underline{A}}\,\underline{T}^* + \underline{P} \tag{4.18}$$

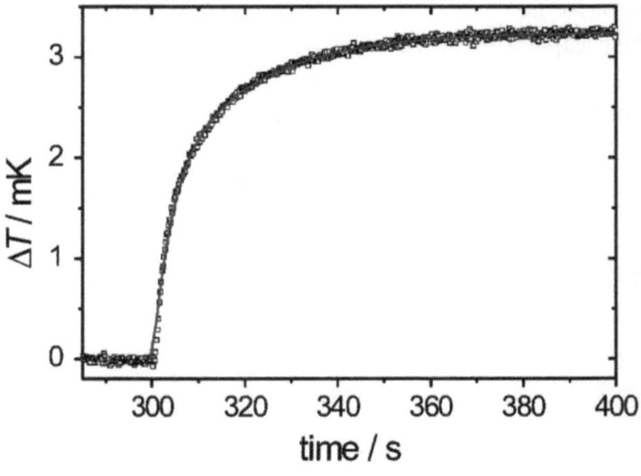

Fig. 4.36 Comparison of the experimental temperature-time curve (*dots*) with the calculated temperature-time curve based on model II. The heat pulse had a power of 38.08 μW, a duration of 300 s, and the parameters were estimated as follows: $\alpha_1/s^{-1} = -1/3.809$ and $\alpha_2/s^{-1} = -1/19.33$ [B219]. Copyright ©2002 with permission from Elsevier

where $\underline{T}^*$ is the vector of temperatures, $\underline{\underline{A}}$ is the $(n \times n)$ matrix of the thermal properties of the layers, and $\underline{P}$ is the vector of heat production.

The integral solution is

$$\frac{T^* - T_S}{T_{st} - T_S} = 1 - \sum_{i=1}^{n} a_i \exp\left(-\frac{t}{\tau_i}\right) \tag{4.19}$$

where τ is a constant.

Due to the finite amount of electroactive compound immobilized on the thermistor electrode, no steady-state response is attained and the calculation of heat produced or consumed is more complex. For that it is necessary to use differential equations. Since the power applied in calibration experiments was denoted as P, Q will be used to denote the heat change of an electrochemical reaction. Equations (4.6) and (4.13) have to be solved for the two models. The parameters have been determined in calibration experiments. To calculate the amount of heat associated with the electrochemical reaction, two different methods can be used, i.e., the differential method and the integral method. The differential method based on model I requires the determination of the parameters K and mC_p, as well as the first time derivative of temperature, since Eq. (4.6) will be used in the form

$$Q(t) = KT + mC_p \mathrm{d}T/\mathrm{d}t \tag{4.20}$$

In the case of model II, the parameters a_1, a_2, and a_3, as well as the first and second time derivatives of the temperature, are necessary, since Eq. (4.13) will be used in the form

$$Q(t) = \frac{1}{a_3}\frac{\mathrm{d}^2T}{\mathrm{d}t^2} + \frac{a_1}{a_3}\frac{\mathrm{d}T}{\mathrm{d}t} + \frac{a_2}{a_3}T \tag{4.21}$$

Application of the differential method is associated with a number of errors, i.e., errors in the determination of the parameters and errors due to the derivations. The advantage of the differential method is that it yields the heat transients. The integral method allows the calculation of the time derivatives of temperature to be avoided. Equations (4.20) and (4.21) can, in principle, be solved with the Runge–Kutta method or by Fourier transformation. However, to obtain a clear insight into the role of the different parameters, it is convenient to use the relation $Q(t) = \frac{dq}{dt}$ in Eqs. (4.20) and (4.21), which allows to write for model I

$$\mathrm{d}q = KT\mathrm{d}t + mC_p\mathrm{d}T \tag{4.22}$$

and for model II

$$\mathrm{d}q = \frac{1}{a_3}\mathrm{d}T' + \frac{a_1}{a_3}\mathrm{d}T + \frac{a_2}{a_3}T\mathrm{d}t \tag{4.23}$$

with $T' = \frac{dT}{dt}$. Integration of Eqs. (4.22) and (4.23) within certain limits allows the calculation of the heat change within these limits:

$$q = K\int_{t_1}^{t_2} T\mathrm{d}t + mC_\mathrm{p}(T_2 - T_1)(\mathrm{model\,I}) \tag{4.24}$$

$$q = \frac{1}{a_3}\left(T'_2 - T'_1\right) + \frac{a_1}{a_3}(T_2 - T_1) + \frac{a_2}{a_3}\int_{t_1}^{t_2} T\mathrm{d}t \;\; (\mathrm{model\,II}) \tag{4.25}$$

From a comparison of these two equations, it follows that *K* in model I and a_2/a_3 in model II are identical. The heat produced in an electrochemical reaction is the sum of the electrochemical Peltier heat and the irreversibly exchanged heat. The latter is at least the Joule heat and that associated with a possible overvoltage:

$$Q(t) = Q_\mathrm{Peltier}(t) + Q_\mathrm{Joule} + Q_\mathrm{overvoltage} \tag{4.26}$$

$$Q(t) = \frac{\Pi}{zF}I + RI^2 + \eta I \tag{4.27}$$

where Π is the molar Peltier heat, *R* the solution resistance (and also the resistance of the solid compound when the current *I* flows through it), and η the overvoltage of the electrochemical reaction. There is a simple way to determine the irreversible heat in these experiments because the Peltier heats of the oxidation and reduction must have equal absolute values but opposite signs, whereas the irreversible heat will have the same sign and absolute value. A measurement of the heat associated with the oxidation and subtraction of the heat associated with reduction thus gives the irreversible heat. Of course, small scan rates and small amounts of substance, i.e., small currents, will reduce the irreversible heat effects. Experimentally it was shown that the parameters of model I and model II do not depend on the amount of heat released. To quantify the heat of an electrochemical reaction, there are two possibilities: the integral method depends only on parameter *K* and is applicable to model I and II. For application of model II, the value a_2/a_3 was determined as *K* using model I. The differential method was not applied to model II because of the severe errors involved in calculating the second time derivative of temperature. Using the differential method and model I, the smaller time constant of model II, i.e., $-1/\alpha_1$ was used because mC_p/K, which is the time constant of model I, is a too bad estimate when relatively fast temperature changes occur. mC_p/K is a good estimate only for slower temperature changes. (The time constants of the gold coil electrode are about one order of magnitude smaller than those of the graphite electrode. Hence the gold electrode lends itself for studies with faster potential scan rates.) Table 4.1 gives electrochemical Peltier coefficients in J/C for the oxidation and reduction of copper(II) hexacyanoferrate(II) immobilized on the graphite surface of the thermistor electrode, using a 0.1 M KNO_3 electrolyte solution, for different scan rates and calculated with the integral as well as the differential method. The relation between the molar Peltier heat and the Peltier

Table 4.1 Electrochemical Peltier coefficients π for copper(II) hexacyanoferrate(II) as determined at different scan rates

	Integral method		Differential method	
Scan rate (mV s^{-1})	π_{Ox} (J C^{-1})	π_{Red} (J C^{-1})	π_{Ox} (J C^{-1})	π_{Red} (J C^{-1})
1	0.1953	0.1910	0.1911	0.1972
2	0.1913	0.1937	0.1934	0.1921
5	0.1904	0.1983	0.1957	0.1945
10	0.1943	0.1927	0.1983	0.1927
20	0.1713	0.1776	0.1943	0.1936
50	0.1478	0.1437	0.1989	0.1924

Electrolyte 0.1 M KNO_3 [B219]

coefficient π is $\Pi = \pi z F$. The entropy change follows as $\Delta S_{\text{reaction}} = \Pi / T$. The entropy change at an electrode has two sources, the reaction entropy and the transfer entropy of electrons and ions. The transfer entropy was assumed to be zero as a first approximation. From the table, it follows that the integral method yields reliable data up to 10 mV s^{-1}, whereas at higher scan rates, deviations are serious and the differential method has to be used. It seems that at scan rates above 10 mV s^{-1}, an overlapping between the released and consumed heat occurs. The data given in Table 4.1 clearly show that the irreversible heat must be negligible as the Peltier coefficients for oxidation and reduction are almost identical and show no systematic differences. Figure 4.37 depicts a cyclic voltammogram of a graphite thermistor electrode with immobilized copper(II) hexacyanoferrate(II) and the corresponding thermogram. The two traces, taken together, provide a clear picture of the electrode process. In the cyclic voltammogram, the absolute peak currents of oxidation and reduction are equal, indicating reversibility.

The maximal change in temperature was 3 mK, corresponding to a current of about 0.2 mA for the following reaction: $\{KCu[Fe(CN)_6]\}_s + K^+_{(aq)} + e^- \leftrightarrows \{K_2Cu[Fe(CN)_6]\}_s$. The subscript s denotes the solid phase and the subscript aq indicates that this ion is dissolved in the aqueous phase. From Fig. 4.37, it

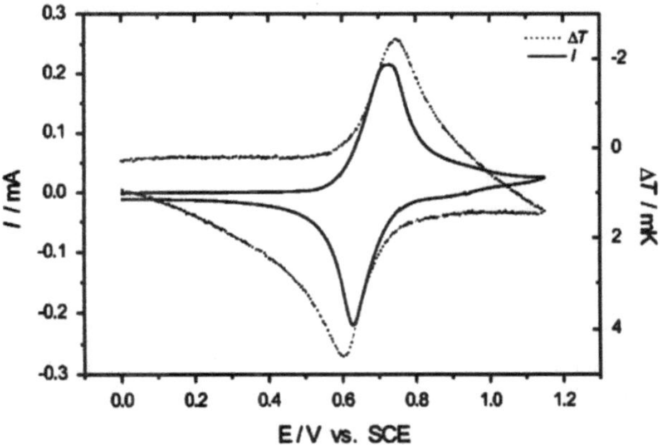

Fig. 4.37 Cyclic voltammogram (*solid lines*) and thermogram (*dotted lines*) recorded simultaneously at a graphite thermistor electrode with immobilized copper (II) hexacyanoferrate(II). The scan rate was 10 mV s^{-1} [B219]. Copyright ©2002 with permission from Elsevier

follows that the oxidation is endothermic, while the reduction is exothermic. The thermogram allows a rough estimate of the time constant by measuring the time delay between the current and temperature peaks. This time delay is about 3 s in the example shown.

The described thermistor electrode has been applied for the determination of the standard entropies of insertion electrochemical reactions of solid metal hexacyanoferrates. For that purpose, the standard entropies have been additionally determined by temperature variations of a thermostated cell. Table 4.2 gives the published data. A comparison of the data of different metal hexacyanoferrates and for sodium and potassium ions as inserting ions shows that the desolvation of the inserting ions

Table 4.2 Standard reaction entropies of copper(II) and iron(III) hexacyanoferrate as determined with the help of the thermistor electrode and with a temperature variation of a thermostated cell (*hcf* = hexacyanoferrate) [B220]

Reaction	$\Delta S°$ (J mol^{-1} K^{-1}) as determined with the help of the thermistor electrode	$\Delta S°$ (J mol^{-1} K^{-1}) as determined by temperature variation with the help of the thermostated cell
Fe(III)hcf(III): red. of Fe_{ls}(III), K^+ insertion	−41.0	−37.6
Fe(III)hcf(II): oxid. of Fe_{ls}(II), K^+ insertion	39.8	–
Fe(III)hcf(II): red. of Fe_{hs}(III), K^+ insertion	−42.1	–
Fe(II)hcf(II): oxid. of Fe_{hs}(II), K^+ insertion	40.8	–
Cu(II)hcf(II): oxid. of Fe_{ls}(II), K^+ insertion	40.0	38.6
Cu(II)hcf(II): oxid. of Fe_{ls}(II), Na^+ insertion	72.0	–
Ni(II)hcf(II): oxid. of Fe_{ls}(II), K^+ insertion	–	32.8
Ni(II)hcf(II): oxid. of Fe_{ls}(II), Na^+ insertion	–	72.4
Co(II)hcf(II): oxid. of Fe_{ls}(II), K^+ insertion	–	45.3
Co(II)hcf(II): oxid. of Fe_{ls}(II), Na^+ insertion	–	66.6

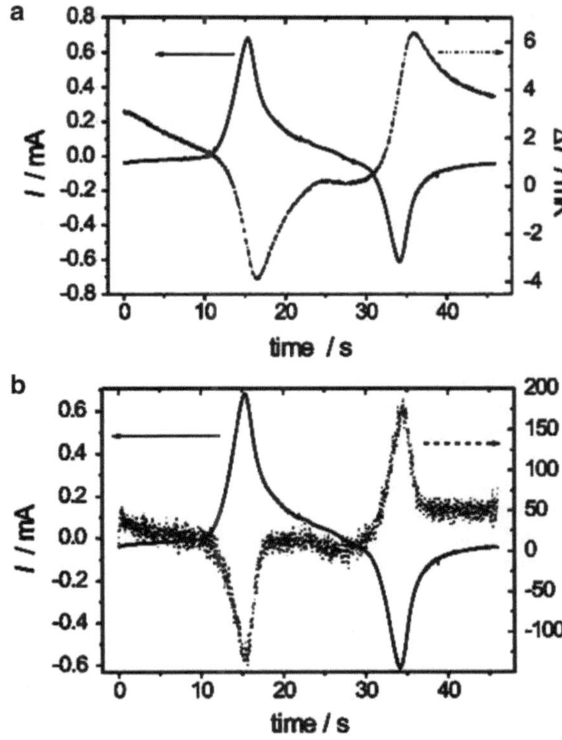

Fig. 4.38 (**a**) Cyclic voltammogram (*solid line*) and thermogram (*dotted line*) plotted as function of time for the second cycle of oxidation reduction of copper(II) hexacyanoferrate (II). ΔT is the temperature difference between the thermistor and the electrolyte solution. (**b**) Cyclic voltammogram (*solid line*) and heat versus time curves for the second cycle of oxidation–reduction of copper (II) hexacyanoferrate(II). Electrolyte 0.1 M KNO_3, scan rate 50 mV s^{-1} [B220]. Copyright ©2002 with permission from Elsevier

seems to give the major contribution to the entropies of the overall electrochemical process.

This chapter demonstrates that in situ calorimetric studies of electrochemical reactions of solid microparticles are possible with extremely small amounts of compounds. This is an advantage, because at least in some cases, similar measurements with large quantities of compounds are impossible, since the reaction rate would be much too small to achieve a complete conversion in a reasonable time span. This advantage is obvious from Fig. 4.38, where voltammograms and simultaneous thermograms are shown that were recorded with a scan rate of 50 mV s^{-1}, with which a complete conversion of the entire immobilized compound has been achieved.

References

[B29] Bond AM, Marken F (1994) Mechanistic aspects of the electron and ion transport processes across the electrode|solid solvent (electrolyte) interface of microcrystalline decamethylferrocene attached mechanically to a graphite electrode. J Electroanal Chem 372:125–135

[B33] Dostal A, Schröder U, Scholz F (1995) Electrochemistry of chromium (II) hexacyanochromate(III)) and electrochemically induced isomerization of solid iron (II) hexacyanochromate(III)) mechanically imobilized on the surface of a graphite electrode. Inorg Chem 34:1711–1717

[B35] Dostal A, Meyer B, Scholz F, Schröder U, Bond AM, Marken F, Shaw SJ (1995) Electrochemical study of microcrystalline solid Prussian blue particles mechanically attached to graphite and gold electrodes: electrochemically induced lattice reconstruction. J Phys Chem 99:2096–2103

[B36] Meyer B, Ziemer B, Scholz F (1995) In situ X-ray diffraction study of the electrochemical reduction of tetragonal lead oxide and orthorhombic Pb(OH)Cl mechanically immobilized on a graphite electrode. J Electroanal Chem 392:79–83

[B53] Shaw SJ, Marken F, Bond AM (1996) Simultaneous electrochemical and quartz crystal microbalance studies of nonconducting microcrystalline particles of *trans*-$Cr(CO)_2(dpe)_2$ and *trans*-$[Cr(CO)_2(dpe)_2]^+$ (dpe = $Ph_2PCH_2CH_2PDh_2$) attached to gold electrodes. J Electroanal Chem 404:227–235

[B54] Shaw SJ, Marken F, Bond AM (1996) Detection of new features associated with the oxidation of microcrystalline tetrathiafulvalene attached to gold electrodes by the simultaneous application of electrochemical and quartz crystal microbalance techniques. Electroanalysis 8:732–741

[B56] Bond AM, Fletcher S, Marken F, Shaw SJ, Symons PG (1996) Electrochemical and X-ray diffraction study of the redox cycling of nanocrystals of 7,7,8,8-tetracyanoquinodimethane. Observation of a solid phase transformation controlled by nucleation and growth. J Chem Soc Faraday Trans 92(20):3925–2933

[B58] Schröder U, Meyer B, Scholz F (1996) A cell for in-situ incident-light microscopy for the study of electrochromism of solid state electrochemical reactions. Fresenius J Anal Chem 356:295–298

[B65] Bond AM, Colton R, Mahon PJ, Tan WT (1997) Tetrabutylammonium cation expulsion versus perchlorate electrolyte anion uptake in the electrochemical oxidation of microcrystals of $[(C_4H_9)_4N][Cr(CO)_5I]$ mechanically attached to a gold electrode: a voltammetric and quartz crystal microbalance study. J Solid State Electrochem 1:53–61

[B67] Schröder U, Scholz F (1997) Microscopic in situ diffuse reflectance spectroelectrochemistry of solid state electrochemical reactions of particles immobilized on electrodes. J Solid State Electrochem 1:62–67

[B70] Bond AM, Fiedler DA (1997) *In situ* electrochemical and electron spin resonance studies of microcrystals mechanically attached to an electrode surface. J Electrochem Soc 144:1566–1574

[B72] Bond AM, Marken F, Hill E, Compton RG, Hügel H (1997) The electrochemical reduction of indigo dissolved in organic solvents and as a solid mechanically attached to a basal plane pyrolytic graphite electrode immersed in aqueous electrolyte solution. J Chem Soc Perkin Trans 2:1735–1742

[B74] Marken F, Webster RD, Bull SD, Davies SG (1997) Redox processes in microproplets studied by voltammetry, microscopy, and ESR spectroscopy: oxidation of *N,N,N′,N′*-tetrahexylphenylene diamine deposited on solid electrode surfaces and immersed in aqueous electrolyte solution. J Electroanal Chem 437:209–218

[B96] Kulesza PJ, Malik MA, Berrettoni M, Giorgetti M, Zamponi S, Schmidt R, Marassi R (1998) Electrochemical charging, countercation accomodation, and spectrochemical identity of microcrystalline solid cobalt hexacyanoferrate. J Phys Chem B 102:1870–1876

[B101] Bond AM, Fletcher S, Symons PG (1998) The relationship between the electrochemistry and the crystallography of microcrystals – the case of TCNQ (7,7,8,8-tetracyanoquinodimethane). Analyst 123:1891–1904

[B102] Bond M, Deacon GB, Howitt J, MacFarlane DR, Spiccia L, Wolfbauer G (1998) Voltammetric determination of the reversible redox potential for the oxidation of the

highly active polypyridyl ruthenium photovoltaic sensitizer *cis*–$Ru^{(II)}(dcbpy)_2(NCS)_2$. J Electrochem Soc 146:648–656
[B119] Suárez MF, Marken F, Compton RG, Bond AM, Miao WJ, Raston CL (1999) Evidence for nucleation-growth, redistribution, and dissolution mechanisms during the course of redox cycling experiments on the C-60/NBu_4C_{60} solid-state redox system – voltammetric, SEM, and in-situ AFM studies. J Phys Chem 103:5637–5644
[B124] Suárez MF, Bond AM, Compton RG (1999) Significance of redistribution reactions detected by in situ atomic force microscopy during early stages of fast scan rate redox cycling experiments at a solid 7,7,8,8-tetracyanoquinodimethane-glassy carbon electrode-aqueous (electrolyte) interface. J Solid State Electrochem 4:24–33
[B128] Eklund JC, Bond AM (1999) Photocatalytic reactions at microcrystalline *fac*-$Mn(CO)_3(\eta^2$-$Ph_2PCH_2PPh_2)$Cl-electrode-aqueous (electrolyte) interfaces. J Am Chem Soc 121:8306–8312
[B136] Schröder U, Scholz F (2000) The solid state electrochemistry of metal octacyanomolybdates, octacyanotungstates, and hexacyanoferrates explained on the basis of dissolution and reprecipitation reactions, their lattice structures and crystallinities. Inorg Chem 39:1006–1015
[B154] Bond AM, Miao W, Raston CL (2000) Identification of processes that occur after reduction and dissolution of C_{60} adhered to gold, glassy carbon, and platinum electrodes placed in acetonitrile (electrolyte) solution. J Phys Chem B 104:2320–2329
[B184] Hasse U, Scholz F (2001) In situ atomic force microscopy of the reduction of lead oxide nanocrystals immobilized on an electrode surface. Electrochem Commun 3:429–434
[B219] Bárcena Soto M, Kubsch G, Scholz F (2002) Cyclic voltammetry of immobilized microparticles with *in situ* calorimetry. Part I: The thermistor electrode. J Electroanal Chem 528:18–26
[B220] Bárcena Soto M, Scholz F (2002) Cyclic voltammetry of immobilized microparticles with *in situ* calorimetry. Part II: Application of a thermistor electrode for in situ calorimetric studies of the electrochemistry of solid metal hexacyanoferrates. J Electroanal Chem 528:27–32
[B276] Hasse U, Nießen J, Scholz F (2003) Atomic force microscopy of the electrochemical reductive dissolution of sub-micrometer sized crystals of goethite immobilized on a gold electrode. J Electroanal Chem 556:13–22
[B278] Hasse U, Wagner K, Scholz F (2004) Nucleation at three-phase junction lines: *in situ* atomic force microscopy of the electrochemical reduction of sub micrometer size silver and mercury(I) halide crystals immobilized on gold electrodes. J Solid State Electrochem 8:842–853
[B279] Hasse U, Scholz F (2004) In situ AFM evidence of the involvement of an oversaturated solution in the course of oxidation of silver nanocrystals to silver iodide crystals. Electrochem Commun 6:409–412
[B280] Komorsky-Lovrić Š, Mirčeski V, Kabbe C, Scholz F (2004) An *in situ* microscopic spectroelectrochemical study of a three-phase electrode where an ion transfer at the water|nitrobenzene interface is coupled to an electron transfer at the interface ITO|nitrobenzene. J Electroanal Chem 566:371–377
[B319] Doménech-Carbó A, Doménech-Carbó MT (2006) Chronoamperometric study of proton transfer/electron transfer in solid state electrochemistry of organic dyes. J Solid State Electrochem 10:949–958
[B320] Doménech-Carbó A, Doménech-Carbó MT (2008) In situ AFM study of proton-assisted electrochemical oxidation/reduction of microparticles of organic dyes. Electrochem Commun 10:1238–1241
[B321] Doménech-Carbó A, Doménech-Carbó MT, Calisti M, Maiolo V (2010) Identification of naphtoquinonic and anthraquinonic dyes via sequential potential steps applied to the voltammetry of microparticles methodology. J Solid State Electrochem 14:465–477

[B322] Doménech-Carbó A, Doménech-Carbó MT, Calisti M, Maiolo V (2010) Sequential identification of organic dyes using the voltammetry of microparticles approach. Talanta 81:404–414

[B323] Doménech-Carbó A, Martini M, de Carvalho LM, Doménech-Carbó MT, Silva M (2013) Screening of pharmacologic adulterant classes in herbal formulations using voltammetry of microparticles. J Pharm Biomed Anal 74:194–204

[B324] Doménech-Carbó A, Martini M, de Carvalho LM, Viana C, Doménech-Carbó MT, Silva M (2013) Standard additions-dilution method for absolute quantification in voltammetry of microparticles. Application for determining psychoactive 1,4-benzodiazepine and antidepressants drugs as adulterants in phytotherapeutic formulations. J Pharm Biomed Anal 80:159–163

[B325] O'Mullane AP, Neufeld AK, Bond AM (2012) Scanning electrochemical microscopy study of the solid-solid interconversion of TCNQ to phase I and phase II CuTCNQ. Electrochem Commun 22:21–24

[B326] Doménech-Carbó A, Montoya N, Alarcón J (2011) Electrochemical characterization of praseodymium centers in $Pr_xZr_{1-x}O_2$ zirconias using electrocatalysis and photoelectrocatalysis. J Solid State Electrochem 16:963–975

[B327] Doménech-Carbó A, Montoya N, Alarcón J (2011) Electrochemical characterization of praseodymia doped zircon. Catalytic effect on the electrochemical reduction of molecular oxygen in polar organic solvents. Electrochim Acta 56:7104–7111

[B328] Karnicka K, Eckhard K, Guschin DA, Stoica L, Kulesza PJ, Schuhmann W (2007) Visualisation of the local bio–electrocatalytic activity in biofuel cell cathodes by means of redox competition scanning electrochemical microscopy (RC-SECM). Electrochem Commun 9:1998–2002

[B329] Szot K, Nogala W, Niedziolka-Jonsson J, Jonsson-Niedziolka M, Marken F, Rogalski J, Kirchner CN, Wittstock G, Opallo M (2010) Hydrophilic carbon nanoparticle-laccase thin film electrode for mediatorless dioxygen reduction: SECM activity mapping and application in zinc-dioxygen battery. Electrochim Acta 54:4620–4625

[B330] Guadagnini L, Malijusch A, Chen X, Neugebauer S, Tonelli D, Schuhmann W (2009) Visualization of electrocatalytic activity of microstructured metal hexacyanoferrates by means of redox competition mode of scanning electrochemical microscopy (RC-SECM). Electrochim Acta 54:3753–3758

Chapter 5
Immobilized Particles

The first prerequisite for studying the electrochemistry of particles immobilized on an electrode surface is their insolubility in the used electrolyte solution. Of course, no particle is absolutely insoluble, and so it is a matter of a practical viewpoint what solubility can be tolerated. There is no general answer to this question; however, it makes sense to discuss here only those compounds that do not dissolve to a detectable extent during their electrochemical reactions or only prompted by their electrochemical reactions. What are the general possibilities of electrochemical behavior of an immobilized particle? Figure 5.1 depicts some frequently observed cases. Case 1 illustrates the reduction of an insoluble metal salt to the metal, e.g., AgCl to Ag, the released anions diffusing into the bulk of the electrolyte solution. For the reduction of silver halides and also of lead(II) oxide, it has been shown that the solid educt is directly transformed to the solid product (see the AFM results discussed in Sect. 4.2). Case 2 is the oxidation of a metal salt with release of metal ions into the electrolyte solution. An example would be the oxidation of copper (I) sulfide to copper(II) sulfide. Case 3 depicts the oxidation of a metal and formation of an insoluble metal compound, e.g., salt, oxide, or complex. This reaction can proceed via an oversaturated solution of the compound MX as in the case of silver oxidation to AgBr and AgI (see AFM results given in Sect. 4.2). In case 4 the "simple" anodic oxidative dissolution of a metal particle is considered. Case 5 concerns the insertion electrochemistry of a particle where ions are exchanged between the solid particle and the electrolyte solution when electrons are exchanged between the particle and the electrode. Plenty of examples are discussed in Sects. 5.3, 5.5, and 5.6. Case 6 depicts the case where the electrochemical reaction is confined to the surface layer only. Examples are given in Sect. 5.4. In case 7 it is assumed that the electrochemical reactions occur only via a preceding dissolution of the solid particle (see the electrochemistry of metal dithiocarbamate complexes discussed in Sect. 5.6), and case 8 symbolizes the complete electrochemical dissolution of a particle resulting from its oxidation or reduction. An example of the latter type is the reductive dissolution of iron(III) oxides (see Sect. 5.7).

F. Scholz et al., *Electrochemistry of Immobilized Particles and Droplets*,
DOI 10.1007/978-3-319-10843-8_5

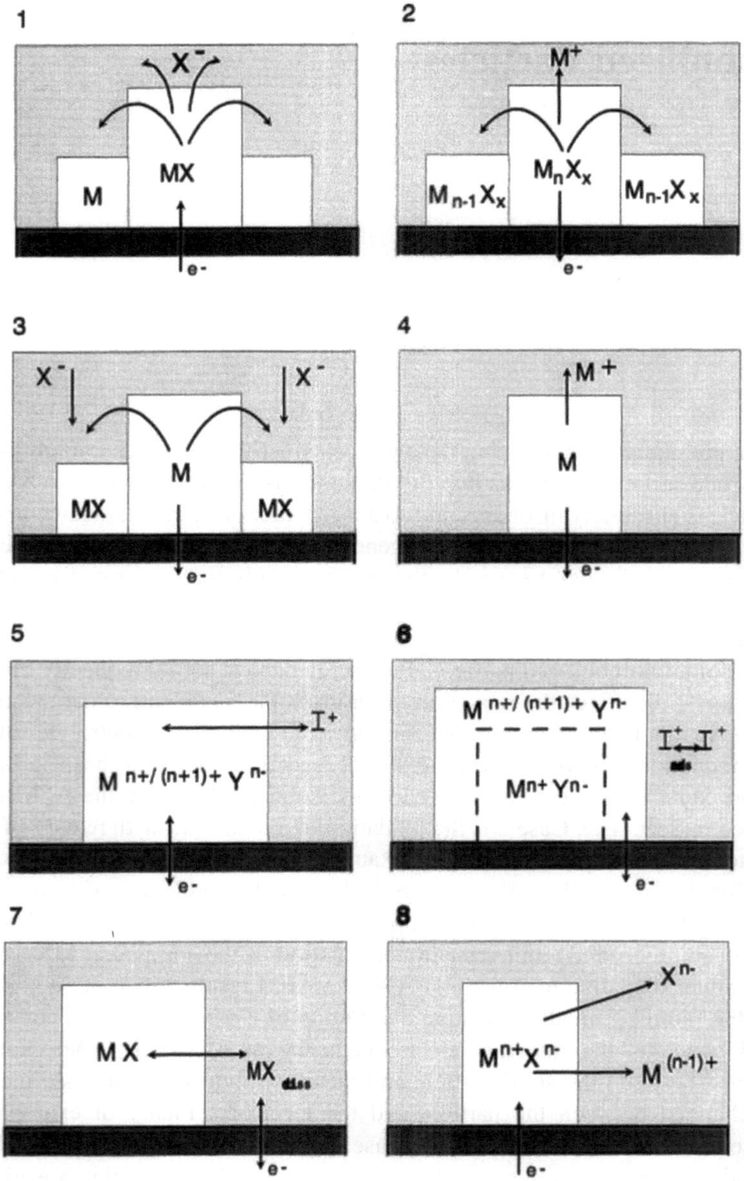

Fig. 5.1 Scheme of possible electrochemical reactions of solid particles immobilized on a solid electrode and in contact with an electrolyte solution

A special case of immobilization of particles is when a solution layer is sandwiched between the electrode (e.g., graphite) and the particles. For this case, it was observed that the electron transfer between the electrode material (glassy carbon) and the immobilized microparticles (Cu_2S, Cu_2Se) can be catalyzed by certain ions and compounds (e.g., thiocyanate ions, thiourea) that are already well known as catalysts in electron transfer reactions of dissolved ions (e.g., Ni^{2+}) [B60]. Such studies may help understand also *interparticle* electron transfer processes.

Some general remarks may suffice at this point to explain the specific features of studies of immobilized particles:

1. Due to the small amounts of immobilized compounds, the electrochemical measurements can be performed on the same time scale as measurements with dissolved species. This feature was elegantly demonstrated by Fiedler et al. for the screening of potential battery materials [B73, B86] where measurements could be performed in the mV s^{-1} range instead of using μV s^{-1} for the usual composite electrodes.
2. In general, all solid compounds that can be synthesized can be directly measured. The measurements do require the precipitation of these compounds directly on an electrode surface. Thus, for example, the latter procedure limited very much the number of studied metal hexacyanometalates, whereas it was demonstrated that with the immobilization technique all synthetically accessible compounds of that type can be investigated by electrochemical techniques (e.g., [B33, B41, B137, B247]).
3. The mechanical immobilization of particles avoids using liquid binders that may alter the electrochemical response, e.g., by film formation in case of organic binders and by chemical dissolutions in case of aqueous electrolyte binders.
4. The immobilization technique allows measurements with very small amounts of compounds, at least down to 10^{-12} mole, and even with single microparticles [B10, B78].
5. Generally it is difficult to deliberately control the amount of substance immobilized on the electrode, and this seems to be a drawback. However, it is rather simple to deduce the amounts of electrochemically active compound from the electrochemical data, e.g., by integration of voltammograms or by direct coulometry. This is most simple in case of reversible systems. When the characteristic voltammetric potentials, e.g., peak potentials, depend on the amount of electrochemically active compound, it is necessary to take this into account by determining these amounts as outlined. Examples are given in Sect. 5.3 [B57].

5.1 Metals and Alloys

5.1.1 *Abrasive Voltammetry of Metals and Alloys*

The idea of immobilizing microparticles on an electrode surface for the purpose of an electrochemical analysis came to the mind of one of the book's authors (F.S.) from an old gold probe technique where the metal sample was scratched on a special stone and the transferred metal traces were then probed for their stability toward nitric acid. Since voltammetric techniques can detect extremely small currents, caused by oxidations or reductions of invisibly small amounts of substances, it was clear that a mechanical transfer of a solid to an electrode surface must be capable of depositing sufficient amounts of a metal, an alloy, or any other solid material on the electrode surface as to detect them electrochemically. The first publication based on that idea reported the analysis of lead–antimony alloys [B1]. Figure 5.2 shows differential pulse voltammograms of the anodic oxidation of Pb–Sb alloy of various compositions. The transfer was accomplished by gently (!) rubbing the lower circular surface of a paraffin-impregnated graphite electrode (PIGE) rod on the surface of the alloys. Because of this *abrasive* transfer and deposition of the alloy, in early publications the term *abrasive stripping voltammetry* was coined to be consistent with terms like *adsorptive stripping voltammetry*, where *adsorption* is the process of attaching a compound to the electrode. Later, the more general term *voltammetry of immobilized microparticles* was preferably used. Figure 5.3 depicts the calibration graphs derived from the voltammograms shown in the Fig. 5.2. Since the absolute amounts of alloy traces transferred by rubbing cannot be controlled and vary to some extent from experiment to experiment, the evaluation of the voltammograms was based on the percentage of each peak current in relation to the sum of peak currents.

This first example proved that the idea of an abrasive transfer of traces of a material from a compact sample to an electrode worked quite well and a number of similar publications followed [B3, B7, B8, B18, B177, B193, B261]. The great advantage of performing electrochemical measurements with trace amounts of an alloy is the high resolution that can be achieved at rather high scan rates, due to the complete anodic oxidation and dissolution of the constituents of the alloy without overlapping of the signals. The latter is observed whenever compact pieces of alloys are subjected to an anodic oxidation. Of course, the position of the signals and their shape depend in a complex manner on the thermodynamics of alloys, the thermodynamics of anodic oxidation reactions including complex formation of the metal ions in solution, and on the kinetics of all involved reaction steps. Figure 5.4 shows differential pulse voltammograms of the anodic oxidation of tin–mercury alloys recorded following a mechanical transfer of traces of the alloys to the surface of a PIGE [B18]. Also shown are the calibration plots. This figure shows the results for two different electrolytes, and it is obvious that the electrolyte has a large influence on the shape, position, and relative intensities of the signals.

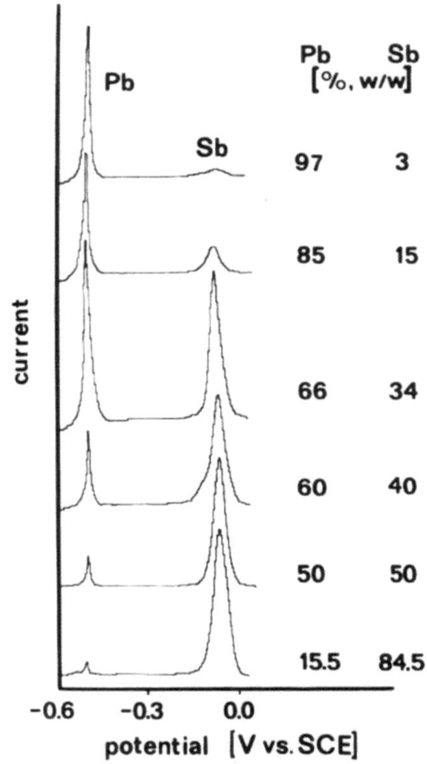

Fig. 5.2 Differential pulse voltammograms of the anodic oxidation of traces of Pb–Sb alloys that have been transferred onto a PIGE by mechanical rubbing of the electrode on pieces of the alloys. The electrolyte solution was a 0.1 M oxalic acid [B1]. Copyright ©1989 with permission from Springer

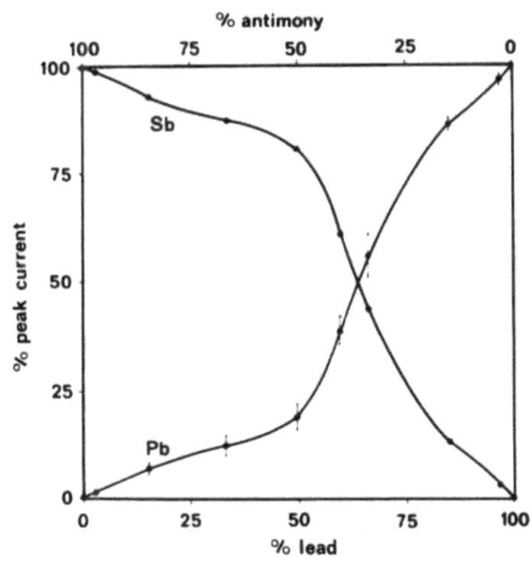

Fig. 5.3 Percentage of peak currents for lead and antimony oxidation as a function of the alloy composition. Data derived from the voltammograms shown in Fig. 5.2 [B1]. Copyright ©1989 with permission from Springer

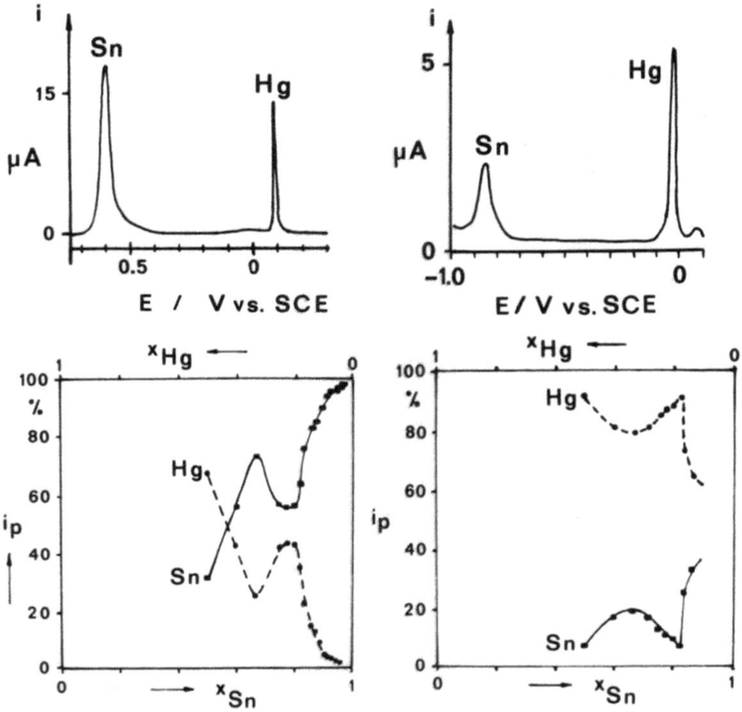

Fig. 5.4 *Top*: differential pulse voltammograms of the anodic oxidation of tin–mercury alloys with 47 % tin. *Bottom*: calibration plots of the percentage peak current of tin and mercury oxidation. *Left side*: electrolyte 0.05 mol L^{-1} KCl, 0.05 mol L^{-1} oxalic acid. *Right side*: electrolyte 1 mol L^{-1} NH_3, 1 mol L^{-1} NH_4Cl [B18]. Copyright ©1992 with permission from Wiley-VCH

Cepriá et al. [B177] have demonstrated how the mechanical transfer of alloy traces to the surface of an electrode can be used for a quick identification of metal alloys (cf. Fig. 5.5). These authors have used a pyrolytic graphite rod electrode which they impregnated with paraffin as described in Sect. 3.1.

Quantitative analyses have so far been reported only for binary alloys. If there are more than two constituents in an alloy, major complications may primarily arise from intermetallic compounds and phase formation. However, especially, this aspect seems to be very interesting to be studied in the future with the technique of immobilizing traces of alloys on electrodes.

Corrosion properties of metals and alloys can be assessed for bulk material and for powders. However, the applicability of the technique to bulk material studies strongly depends on how much the sampling affects the corrosion properties. If the abrasive transfer of alloy traces to an electrode results in alloy patches on the electrode that possess the same or very similar corrosion properties as the compact piece of metal, then there are no objections for an application of this technique. This is obviously the case for many amalgams, the electrochemical corrosion of which has been studied and appeared to be as that known from studies of compact alloy

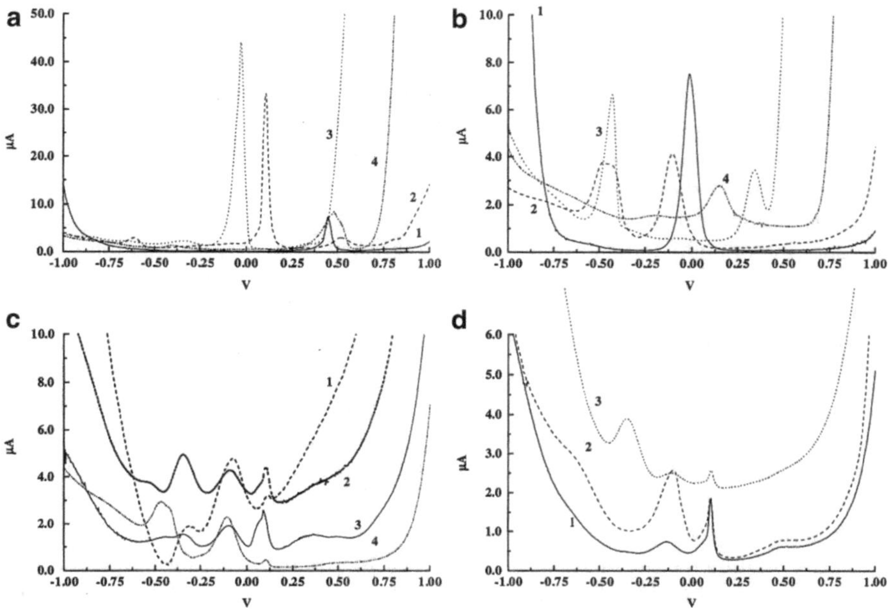

Fig. 5.5 Differential pulse voltammograms of anodic dissolution of different metals and alloys. (**a**) Silver; (**b**) copper; (**c**) 1, Ag(91)–Cu(9) alloy; 2, Ag(70)–Cu(30) alloy; 3, Ag(40)–Cu(60) alloy; 4, Ag(10)–Cu(90) alloy; (**d**) jewelry items: (1) earring A, (2) earring B, (3) pendant. Electrolytes: (**a** and **b**) 1, 0.4 M oxalic acid; 2, 0.1 M ammonia buffer; 3, 0.4 M thiocyanate solution; 4, 0.2 M EDTA. (**c** and **d**) 0.1 M ammonia buffer [B177]. Copyright ©2001 with permission from Wiley-VCH

pieces [B18]. The only, however, very advantageous difference was that the signal resolution was considerably improved in studies of mechanically transferred alloy traces. The detection of the rather corrosion-sensitive so-called γ_2-phase in dental amalgams was extremely easy to perform since the presence of this phase produced a well-developed anodic peak. This phase produced just a faint shoulder when compact pieces of that alloy were used as electrodes in anodic oxidation studies. Electrochemical corrosion studies of metal or alloy powders are well known to be difficult to perform for the simple reason of electrode construction. Žežula and Galova have shown that the corrosion potential of iron powder can be determined by immobilizing the powder on a PIGE and performing the usual potentiodynamic measurements [B108]. As to be expected, the authors observed a small shift in the corrosion potential for repetitive polarizations, most probably due to a reduction of the primarily present oxide layer on the iron particles. The reactivity of iron powders has been assessed by pulsed chronoamperometry [B307].

5.1.2 Age Determination

Dating is an important insight in archaeometry, conservation, and restoration, constituting in general a significant analytical problem. Relative dating, i.e., establishing the time sequence of appearance of different objects, can be obtained from stylistic analysis combined with different techniques of physicochemical analysis. In several instances, such as in paleomagnetism or dendrochronology, the duration of an historical event can be established, although its situation in time remains unknown. Absolute dating, i.e., determining the age of an object relative to conventional time scale, can be obtained from radiocarbon, uranium–lead radioactive series, and luminescence analysis,[1,2] but such methods have limitations dealing with the materials suitable for dating and the accessible times. A complementary set of dating methods is based on the chemical analysis of the extent of alteration processes on archaeological artifacts. This is the case of the obsidian method developed by Friedman and Smith to estimate the age of ceramics[3] and the Meissner effect-based method proposed by Reich et al.[4] for dating archaeological lead.

The voltammetry of microparticles can also be used for dating purposes. The first application of this kind, proposed by Scholz and Brainina et al. [B34] for dating ceramic materials, consisted of electrochemical monitoring the electrocatalytic response of such materials as electrode modifiers relative to oxygen-based electrochemical processes. The basic idea is that natural radioactivity and cosmic radiation causes accumulation of defects in nonconducting materials, so that, assuming that the studied objects are exposed to a continuous and uniform radiation, the number of point defects should increase accordingly and, assuming that the electrocatalytic ability of such materials depends on the concentration of point defects suitable to act as catalytic sites, the catalytic response could be related to the age of the object. The appearance of such catalytic effects and their disappearance upon heating the sample could be used for authentication purposes of, for instance, ceramic materials.

More recently, Doménech-Carbó et al. [B331, B332] have applied the same philosophy for dating archaeological lead, also exploiting the capabilities of the VMP. This method is based on the analysis of voltammetric signatures specifically attributed to different PbO_n and PbO species formed during prolonged corrosion. It is known that under ordinary conditions of manufacturing and using of metal artifacts, the metal reacts with its environment yielding a primary patina of corrosion products, which can be accompanied by a secondary patina, formed during the last period of utilization and/or the initial stages of long-term alteration giving rise

[1] Aitken MJ (1990) Science-based dating in archaeology. Longman, New York, chapter 8

[2] Renfrew C, Bahn P (1991) Archaeology: theory, methods and practice. Thames & Hudson, London, chapter 4

[3] Friedman I, Smith RL (1960) Am Antiq 25:476–493

[4] Reich S, Leitus G, Shalev S (2003) New J Phys 5:99.1–99.9

to pedological processes (recrystallization, segregation, desegregation, cementation, monolithizations). In most cases, a tertiary or contamination patina, formed as a result of processes of diffusion–segregation–deposition involving the interaction of the materials of the secondary patina with the environment, appears.

In the case of lead artifacts, the primary patina is typically formed by a continuous layer of litharge (PbO), whereas the secondary patina consists of "porous" litharge and/or $PbSO_4$ (anglesite) and $Pb(SbO_4)_2$, $PbCO_3$ (cerussite), $PbCl_2$ (cotunnite), and $2PbCO_3 \cdot Pb(OH)_2$ (lead white or *albayalde*), among others [B333]. The electrochemical reduction of microparticulate deposits of lead oxides has been studied by Zakharchuk et al. [B148], the reduction of litharge to lead metal occurring as a topotactic solid-to-solid conversion process [B36, B184]. The electrochemical response of PbO passivation layers differs from one to another depending on the size of the pores. When relatively small pores are formed, "thermodynamic" passivation occurs, the formed patina acting as a semipermeable membrane that allows only slow permeation by H_2O, H^+, OH^-, and Pb^{2+} ions. When relatively large pores are formed, "kinetic" passivation permits the motion of all ions through the pores so that when the flow of lead ions is sufficiently high, the reaction proceeds at the pore orifice near the electrolyte, thus resulting in electrode potentials more negative than those for the reduction of the semipermeable patina [B334, B335, B336]. This is reflected in voltammograms recorded for microsamples of lead artifacts abrasively transferred to graphite electrodes in contact with aqueous electrolytes, as shown in Fig. 5.6. In this figure, square-wave voltammograms for graphite electrodes modified for contemporary lead and Visigothic sample (sixth century AD) from the Punta de l'Illa, Spain, immersed into aqueous acetate buffer are depicted. Peak I can be attributed to the reduction of the primary patina of PbO, while signals II and III are assignable to the reduction of different porous PbO layers.

$$PbO_{(solid)} + 2H^+_{(aq)} + 2e^- \rightarrow Pb_{(solid)} + H_2O \qquad (I)$$

In the above voltammogram, overlapping peaks IVa and IVb can be partly attributed to the electrochemical reduction of PbO_2 electrochemically generated at potentials above +1.0 V by oxidation of the PbO patina but also to the presence of α- and β-PbO_2 and substoichiometric lead oxides forming the pristine patina [B337]. The electrochemical reduction of PbO_2 can be formulated as $(0.6 < x < 1)$ [B338, B339]:

$$PbO_{2(solid)} + 2xH^+_{(aq)} + 2xe^- \rightarrow PbO_{2-x(solid)} + xH_2O \qquad (II)$$

Just the presence of well-developed signals IV–V and II–III can be taken as a criterion for authentication purposes [B340] because such processes become ill defined in forgeries where lead has been eventually subjected to "accelerated" corrosion procedures. Another criterion for detecting lead of archaeological origin is the appearance, in voltammograms where the potential is scanned in the positive

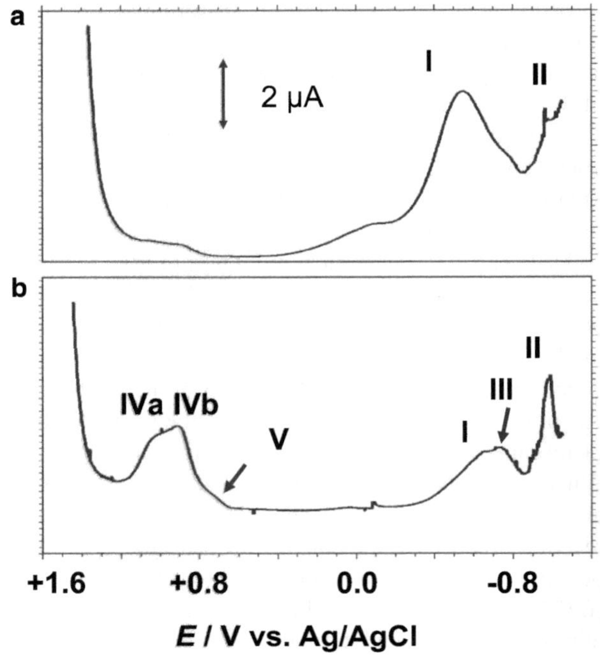

Fig. 5.6 Square-wave voltammograms for graphite electrodes modified by means of "one-touch" methodology for (**a**) contemporary lead; (**b**) Visigothic sample (sixth century AD) from the Punta de l'Illa, Spain, immersed into 0.50 M HAc/NaAc, pH 4.85. Potential scan initiated at +1.45 V in the negative direction. Potential step increment 4 mV, square-wave amplitude 25 mV, frequency 5 Hz [B331]. Copyright ©2011 with permission from American Chemical Society

direction from sufficiently cathodic potentials, of stripping peaks for the oxidative dissolution of metals and semimetals accompanying lead in minerals: antimony, arsenic, bismuth, and copper [B340]. Interestingly, using a "one-touch" sampling methodology [B333, B340, B341] permits an essentially noninvasive analysis of archaeological pieces.

Dating can be obtained from electrochemical data assuming that, under essentially uniform corrosion conditions, the formation of the porous patina is subsequent to the formation of the continuous patina and growths uniformly with time. Since the relative amount of porous and continuous patina can be estimated from the relative height of the corresponding voltammetric signals, the time variation of, for instance, the peak current ratio between processes II and I, $i_p(\text{II})/i_p(\text{I})$, can be correlated with theoretical modeling of long-term corrosion processes assuming that a potential law applies [B331, B332, B341]. Then, the ratio between the peak currents satisfies the relationship

$$\frac{i_p(\text{II})}{i_p(\text{I})} = K[(1+\alpha)t]^{\frac{1}{1+\alpha}} \tag{5.1}$$

where K represents a rate constant and α the exponent of the potential rate law. Similar considerations can be applied using the quotient between the PbO_n-characteristic voltammetric signals [B331] and also corrosion and electrochemical impedance spectroscopy data [B332]. Figure 5.7 shows the calibration graph

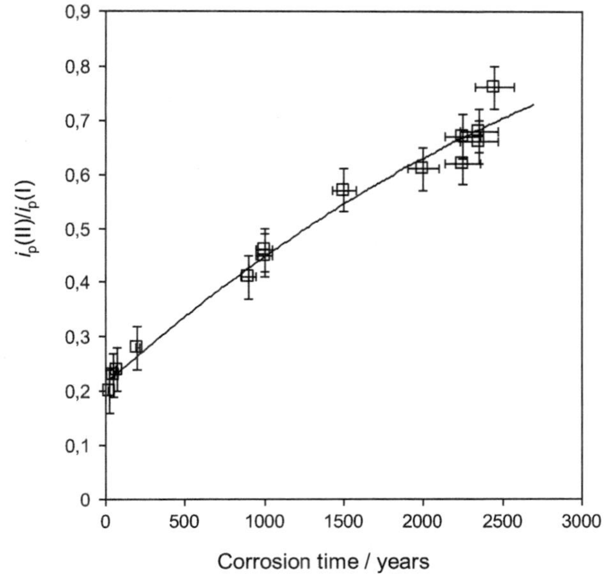

Fig. 5.7 Variation of the $i_p(II)/i_p(I)$ ratio for square-wave voltammograms of sample-modified paraffin-impregnated graphite electrodes immersed in 0.50 M aqueous acetate buffer, pH 4.85 using conditions such as in Fig. 5.6. Experimental data for a set of archaeological samples and theoretical line using Eq. (5.1) with $\alpha = 0.07$ [B331]. Copyright ©2011 with permission from American Chemical Society

obtained from a set of samples from different archaeological sites in the Comunitat Valenciana (Spain) dated between the Iberian times (fifth–second centuries BC) and the modern age (twentieth century), all found under burial conditions in calcareous soils and not submitted to cleaning/restorative treatment nor exposition. Experimental data fit satisfactorily with theoretical curves from Eq. (5.1), as can be seen in Fig. 5.7. Interestingly, linear fit between $i_p(II)/i_p(I)$ and $t^{\frac{1}{1+\alpha}}$ is obtained for $\alpha = 0.07$, just the value obtained by Reich et al. (see footnote 4) on studying lead dating by means of measurements on the Meissner fraction in the superconducting state of lead.

A second set of data for age determination can be obtained, based on Scholz, Brainina, Zakharchuk et al.'s [B34] catalytic scene, upon measuring the catalytic currents for selected electrochemical processes [B331]. In the case of lead samples, the essential idea is that more or less patinated lead attached to graphite electrodes should produce different catalytic effects depending on the composition of the sample. Hydrogen evolution reaction (HER), oxygen evolution reaction (OER), and oxygen reduction reaction (ORR) in aqueous media, among other processes, were found suitable, at least qualitatively, for dating purposes [B331]. Figure 5.8 compares the voltammetric responses for contemporary lead- and archaeological sample-attached graphite electrodes immersed into phosphate buffer at pH 7.00. One can see that the archaeological sample enhances both the current for the OER and the ORR relative to electrodes modified with contemporary lead samples. These features would be consistent with the recognized catalytic ability of PbO_2 for different oxidation processes [B342, B343], in particular, the OER [B344].

The contrary effect was observed for the HER, where modern lead samples exhibit a catalytic effect increased relative to that of archaeological lead. This

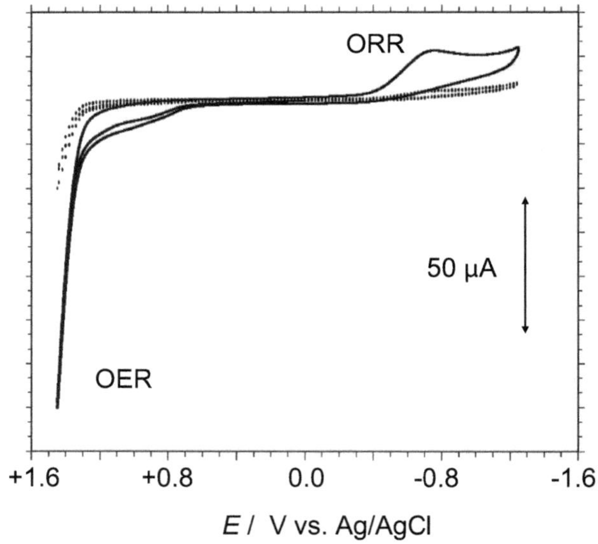

Fig. 5.8 Cyclic voltammograms for graphite electrodes modified by means of "one-touch" methodology for contemporary lead (*dotted line*) and archaeological sample from the Tossal de Sant Miquel, Spain (dated fourth–second century BC, continuous line), immersed into 0.50 M potassium phosphate buffer, pH 7.00. Potential scan rate 100 mV s^{-1} [B331]. Copyright ©2011 with permission from American Chemical Society

feature can be explained as a result of the catalytic ability of lead metal for the HER [B345], because the patina should block more or less intensively the catalytic effect due to the "clean" metal surface. This can be seen in Fig. 5.9 where cyclic voltammograms for graphite electrodes modified with contemporary lead and archaeological lead samples in contact with 5.0 mM salicylic acid solution in 0.50 M potassium phosphate buffer are shown [B331]. One can see in this figure that the cathodic current at ca. −1.2 V is larger for contemporary lead than for archaeological lead, while the contrary effect was observed for the anodic peak corresponding to salicylic acid oxidation. Here, $K_4Fe(CN)_6$ was added to the solution as an internal standard for comparing peak currents because no significant catalytic effects were observed for the $Fe(CN)_6^{3-}/Fe(CN)_6^{4-}$ couple.

It is pertinent to note that, in the above catalytic experiments, few nanograms of lead sample were transferred to the graphite surface by means of "one-touch" methodology, so that it appears as (more or less patinated) metallic grains adhered to the graphite layers. In these circumstances, the observed catalytic currents result not only as a combination of the catalytic effects due to the "free" metal and the patinated regions exposed to the electrolyte solution but also from the catalytic sites generated by scratching the graphite surface, where different oxygen-based functionalities can be formed [B346, B347]. This means that the measured catalytic currents are sensitive to the characteristics of the abrasive sampling process used for electrode modification so that maintaining the more uniform sampling conditions as possibly is needed.

Figure 5.10 shows the variation of the catalytic currents for HER and OER reactions, i_{HER}/i_{OER}, measured in cyclic voltammetric experiments such as in Figs. 5.8 and 5.9.

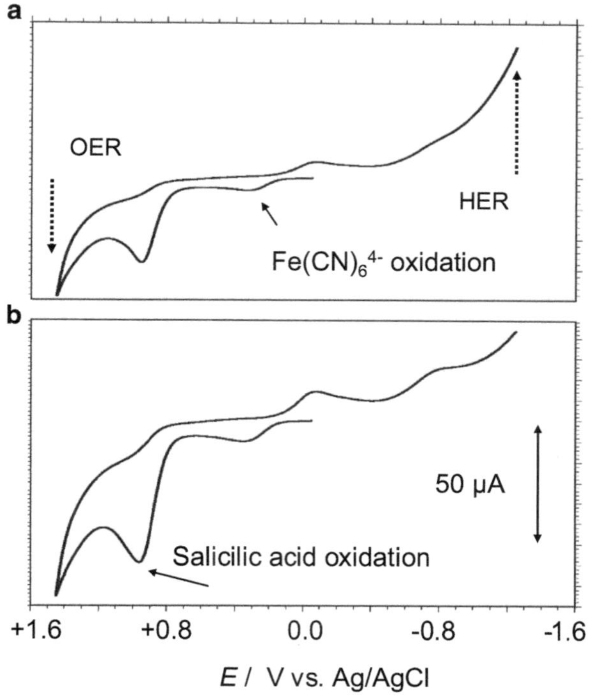

Fig. 5.9 Cyclic voltammograms for graphite electrodes modified by means of "one-touch" methodology for (**a**) contemporary lead and (**b**) archaeological sample from the Tossal de Sant Miquel, Spain (dated fourth–second century BC), immersed into 5.0 mM salicylic acid solution in 0.50 M potassium phosphate buffer, pH 7.00. Potential scan rate 100 mV s^{-1} [B331]. Copyright ©2011 with permission from American Chemical Society

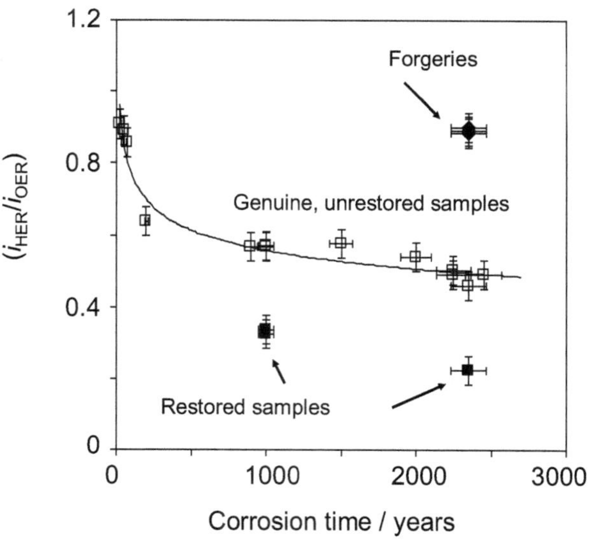

Fig. 5.10 Variation of the i_{HER}/i_{OER} ratio for square-wave voltammograms of sample-modified paraffin-impregnated graphite electrodes immersed in 0.50 M aqueous acetate buffer, pH 4.85 using conditions such as in Fig. 5.40. Experimental data for a set of unrestored and restored archaeological samples and forgeries and theoretical curve from Eq. (5.2) [B331]. Copyright ©2011 with permission from American Chemical Society

Under fixed electrochemical conditions, such ratio should be representative of the relative amounts of PbO_2 and the total lead (the parent deposit plus those electrochemically generated by reduction of lead oxides) accumulated in the electrochemical run. Assuming that a potential rate law is applicable for both the formation of PbO and PbO_2, one can write

$$\frac{i_{\mathrm{HER}}}{i_{\mathrm{OER}}} = A\frac{1 + K_1 t^{\frac{1}{1+\alpha}}}{K_2 t^{\frac{1}{1+\beta}}} \tag{5.2}$$

A similar procedure has been reported for dating copper and bronze artifacts based on the record of the voltammetric signatures of cuprite (Cu_2O) and tenorite (CuO) [B451]. This equation predicts a linear dependence between $(i_{\mathrm{HER}}/i_{\mathrm{OER}})t^{1/(1+\beta)}$ and $t^{1/(1+\alpha)}$ where one can expect that $\alpha \approx \beta$. Experimental data for unrestored samples in Fig. 5.10 show a satisfactory agreement with expectances, again taking $\alpha = 0.07$. Remarkably, data points for restored samples, i.e., lead samples submitted to cleaning treatments, as well as forgeries in museum reservoirs, provide data points clearly separated from the "archaeological" curve.

It should be noted that the calibration curves in Figs. 5.7 and 5.10 can be used for dating purposes provided that a set of simplifying assumptions are valid. In particular, it is assumed that (1) the primary patina was similar for all studied artifacts; (2) each sample suffered a uniform corrosion process, i.e., that the corrosion conditions of that piece do not vary during time; and (3) similar corrosion conditions operated for all compared archaeological objects [B331, B332]. These are, obviously, oversimplifications, because there is possibility of nonuniform corrosion along the "history" of the object and, even under uniform corrosion conditions, multilayer corrosion may occur. As a result, relatively large data dispersion appears when archaeological data are used.

Another important aspect to remark is that the above calibration curves are not universal, as far as they depend on the assumed common corrosion conditions for all samples used for calibration. Thus, one can expect that the aging conditions for samples in Figs. 5.7 and 5.10, all found under burial conditions in calcareous soils in the Mediterranean area, differ considerably from those occurring, for instance, for lead artifacts in submarine archaeological sites [B348]. In spite of these limitations, the above methods can provide interesting results for authentication and dating which extend available methods for age determination with applications in different fields [B349], including, for instance, dating porcine blood-based binding media used in Taiwanese architectural polychromies [B350]. Age determination of copper and bronze artifacts based on the characteristic voltammetric signatures of cuprite and tenorite has also been reported [B451].

5.1.3 Solution Analysis by Electrochemical Analysis of a Solid Precipitate

Gulppi et al. [B445] have proposed a new strategy to analyze very small solution volumes with rather low concentrations of analytes. The following scheme depicts the sequence of manipulations for the case of chloride determination in a small μL-size droplet:

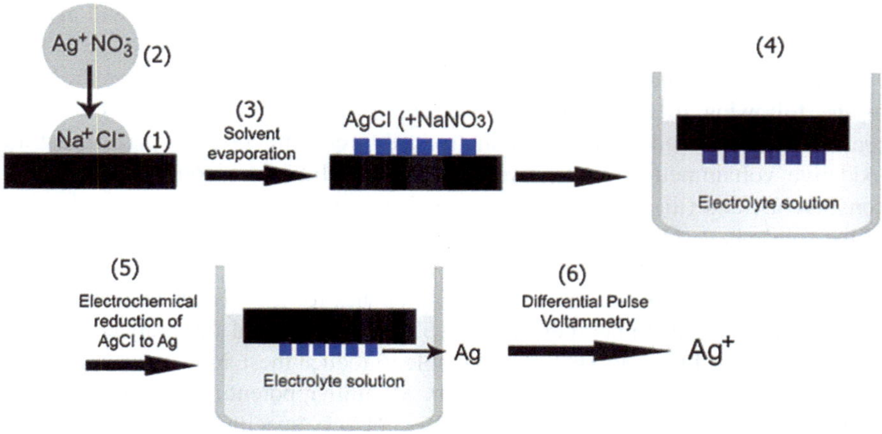

The analyte droplet is attached to an upside-down-turned electrode surface (1). A silver nitrate containing droplet is added (2), AgCl is precipitated, and the solvent is evaporated (3). The electrode is introduced to an electrolyte solution (4) where the precipitated AgCl remains attached to the electrode. The attached AgCl is reduced to Ag (5) and the finally a differential pulse voltammetric scan (6) is recorded for quantification. This approach allows the analysis of μL (may be even nL-)-sized solution volumes, and it overcomes the well-known concentration restrictions of anodic (or cathodic) stripping voltammetric techniques which are based on the precipitation of sparingly soluble salts, as the short time interval between introduction of the electrode to the solution and the electrochemical conversion will hardly prompt any significant dissolution.

5.2 Minerals and Pigments

The identification of minerals and pigments is conventionally performed with the help of microscopic techniques, including infrared and Raman spectroscopy, X-ray diffraction, and several microprobe techniques. Here we will show that the voltammetry of immobilized microparticles can be used as a powerful alternative technique, both for the identification of phases and for their quantitative analysis. The only prerequisite is electrochemical activity, which, however, is shown by the majority of minerals and pigments.

5.2.1 Phase Identification

Generally, solid phases can be identified when a characteristic fingerprint is obtainable by application of a certain analytical technique. It is not necessary to have a complete understanding of the signals of the fingerprint, although this will always be desirable in a truly scientific study.

Voltammetric fingerprints of minerals and pigments can be obtained on the basis of the following measurements (i) performing a reductive voltammetric scan, provided that the solid phase contains reducible constituents; (ii) performing an oxidative voltammetric scan, provided that the solid phase contains oxidizable constituents; and (iii) performing an oxidative scan after a preliminary reduction of the solid phase. Of course, other approaches are possible as well; however, the published applications all fall into these three categories. Figure 5.11 shows voltammograms exemplifying the three approaches for the case of boulangerite, a lead antimony sulfide mineral $Pb_5Sb_4S_{11}$ [B2]. Since lead and antimony are present in the mineral as Pb^{2+} and Sb^{3+}, it is possible to reduce these ions to the respective metals. Obviously, their reduction occurs at similar potentials and no separate signals are obtained. The reduction signal is not very specific and cannot be used for an unambiguous identification. The same holds true for the oxidative scan where the sulfide ions are oxidized to yet unidentified products. Much more specific signals can be obtained with approach (iii), i.e., an oxidative scan following a preliminary reduction. In the case of boulangerite, upon reduction at sufficiently negative electrode potential, the two metals, lead and antimony, remain on the electrode, whereas the sulfide ions escape into the electrolyte solution. In the oxidative scan, the two metals give two well-separated signals that are specific for each metal in the used electrolyte solution. Approach (iii) very much resembles that of conventional stripping voltammetry where metals are deposited on an electrode from a solution and oxidatively stripped off during an anodic scan. Figure 5.12 gives further examples for reductive and Fig. 5.13 for oxidative voltammograms of minerals. Both figures illustrate that, although the signals are different, they are not very suitable for unambiguous phase identification. Figure 5.14 shows examples of the highly specific voltammograms that are obtained after a preliminary reduction of the metal ions to metals [approach (iii)]. The respective metal signals can be easily identified. A closer look on the voltammograms reveals that the signals of anodic metal oxidation are frequently structured and not as symmetric as in anodic stripping voltammetry using mercury electrodes. The structured signals are due to the fact that the metals are present as solids on the electrode, usually a PIGE. The fact that one and the same metal shows differently structured signals depending on the starting mineral shows that the resulting metal deposits also differ. To avoid such structured signals, one can add a small concentration of a Hg(II) salt, e.g., 10^{-4} mol L^{-1} $HgCl_2$, to the electrolyte solution. During the pre-electrolysis of the minerals, a simultaneous plating of mercury occurs, and the metals dissolve in the mercury droplets that are deposited on the surface of the PIGE. In this case, the anodic dissolution of the metals occurs from the amalgam state and very clean anodic peaks are measured. Figure 5.15 shows anodic differential pulse voltammograms recorded after a preliminary electrolysis of different thallium–tin sulfides using a mercury(II)-containing electrolyte [B32]. The relative heights of the thallium and tin peaks are clearly depending on the stoichiometry of the minerals, and they can serve for an unambiguous

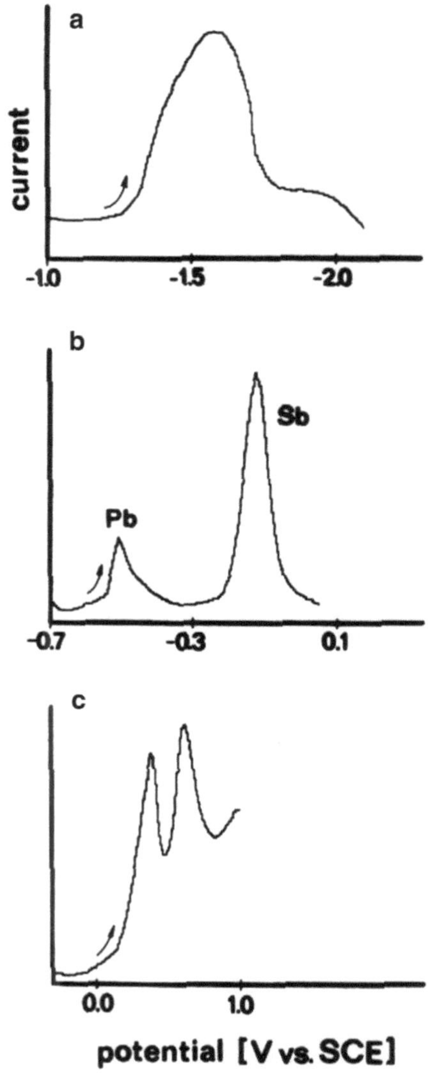

Fig. 5.11 Differential pulse voltammograms of boulangerite ($Pb_5Sb_4S_{11}$) obtained after mechanical transfer of traces of the mineral onto the surface of a paraffin-impregnated graphite electrode. (**a**) Reductive scan; electrolyte 0.1 mol L^{-1} sodium oxalate, pH 5; (**b**) inverse voltammogram after reduction at −1.5 V; electrolyte 0.1 mol L^{-1} sodium oxalate, pH 2; (**c**) oxidative scan; electrolyte 0.1 mol L^{-1} oxalic acid [B2]. Copyright ©1989 with permission from Springer

identification. In Sect. 5.2.3, it will be shown how a quantitative determination of the Tl:Sn ratio can be performed with the help of chronocoulometry. The simplicity of measurements, the extremely small amounts of sample (down to 10^{-12} mole are sufficient), the high mineral specificity, and also the elemental specificity make this

$Tl_2As_2SnS_6$

Origin: synthetic

Method: differential pulse voltammetry, reduction
Electrolyte: 1 M HCl
Method parameters: E_{pulse} = 0.050 V t_{pulse} = 0.01 s
E_{step} = 0.0037 V t_{int} = 0.50 s

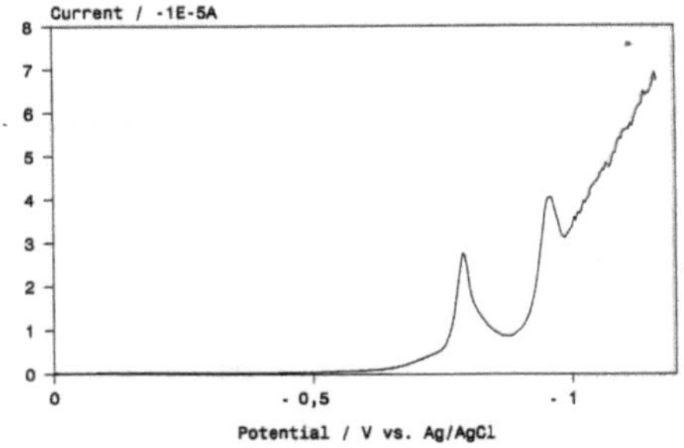

Results of peak search:

	E/ V vs. Ag/AgCl	height / A E - 04	area /AV E - 06	W½ / V
1	-0.787	-2.172	9.447	0.029
2	-0.952	-1.659	5.266	0.029

$Tl_2As_2Sn_2S_7$

Origin: synthetic

Method: differential pulse voltammetry, reduction
Electrolyte: 1 M HCl
Method parameters: E_{pulse} = 0.050 V t_{pulse} = 0.01 s
E_{step} = 0.0037 V t_{int} = 0.50 s

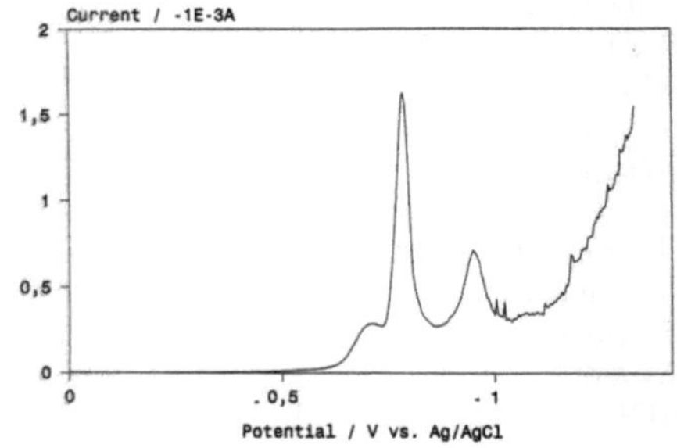

Results of peak search:

	E/ V vs. Ag/AgCl	height / A E - 03	area /AV E - 05	W½ / V
1	-0.696			
2	-0.784	-1.346	5.021	0.029
3	-0.948	-0.424	2.555	0.051

Tl_2SnS_3

Origin: synthetic

Method: linear sweep voltammetry, reduction
Electrolyte: 1 M HCl
Method parameters: E_{step} = 0.002 V scan rate = 0.010 V/s

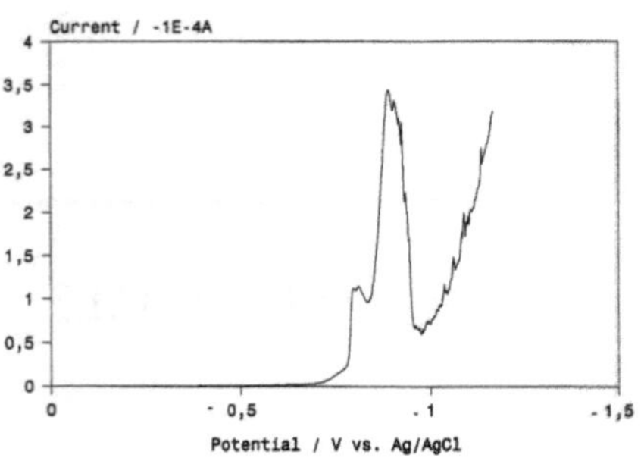

Results of peak search:

	E/ V vs. Ag/AgCl	height / A E - 04	area / C E - 03	W½ / V
1	-0.793	—	—	—
2	-0.891	-2.573	2.550	0.073

$Tl_2Sn_2S_3$

Origin: synthetic

Method: linear sweep voltammetry, reduction
Electrolyte: 1 M HCl
Method parameters: E_{step} = 0.002 V scan rate = 0.010 V/s

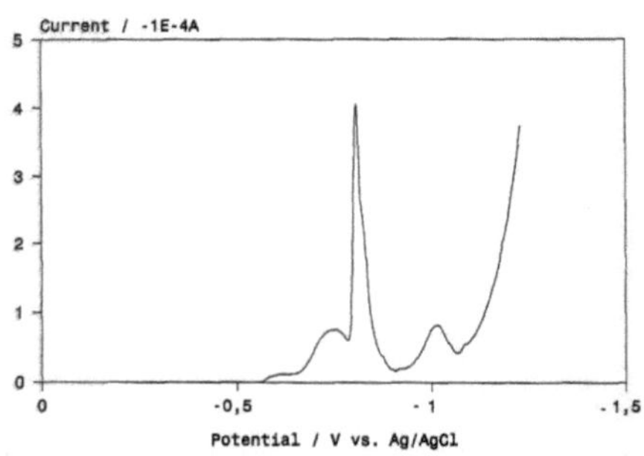

Results of peak search:

	E/ V vs. Ag/AgCl	height / A E - 04	area / C E - 03	W½ / V
1	-0.730	-0.314	0.211	0.071
2	-0.806	-3.540	1.049	0.027
3	-1.011	-0.481	0.2923	0.059

Fig. 5.12 Reductive voltammograms of several mineral phases [B46]

Chalcopyrite ($CuFeS_2$)

Origin: Daaden, Altenkirchen

Method: differential pulse voltammetry, oxidation
Electrolyte: 1 M KCl
Method parameters: E_{pulse} = 0.050 V t_{pulse} = 0.01 s
E_{step} = 0.004 V t_{int} = 0.50 s

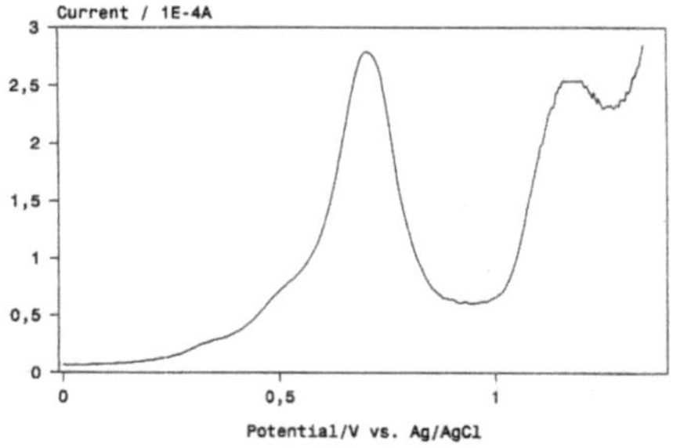

Results of peak search:

	E/ V vs. Ag/AgCl	height / A E - 04	area /AV E - 05	$W_{½}$ / V
1	0.703	2.315	4.258	0.146
2	1.187	—	—	—

Chalcostibite ($CuSbS_2$)

Origin: Abfaltersbach (Austria)

Method: differential pulse voltammetry, oxidation
Electrolyte: 1M KCl
Method parameters: E_{pulse} = 0.050 V t_{pulse} = 0.07 s
E_{step} = 0.005 V t_{int} = 0.50 s

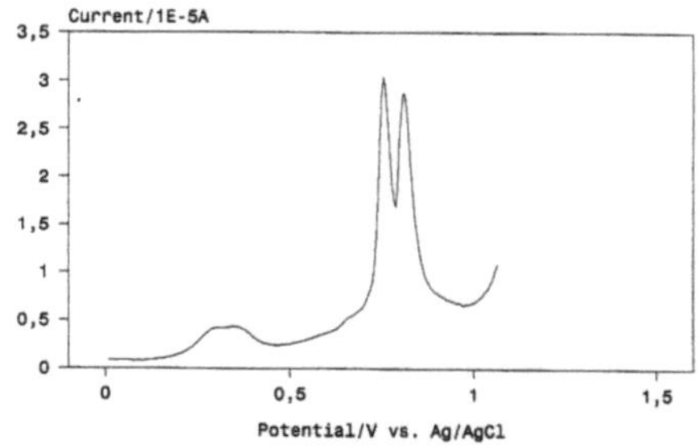

Results of peak search:

	E/ V vs. Ag/AgCl	height / A E - 05
1	0.303	0.255
2	0.752	2.575
3	0.811	2.344

Chalcostibite ($CuSbS_2$)

Origin: synthetic

Method: differential pulse voltammetry, oxidation
Electrolyte: 1 M KCl
Method parameters: E_{pulse} = 0.050 V t_{pulse} = 0.07 s
E_{step} = 0.005 V t_{int} = 0.50 s

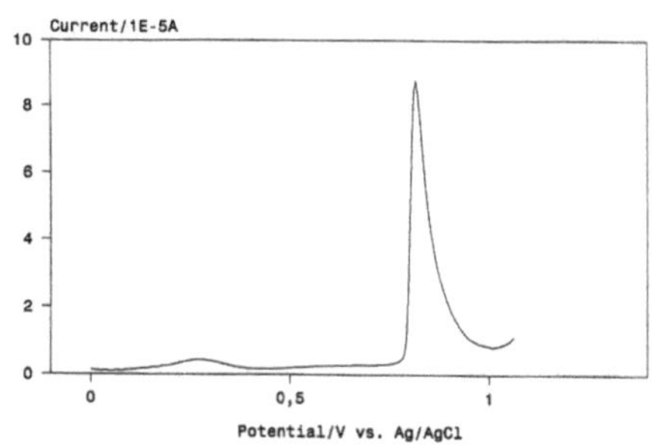

Results of peak search:

	E/ V vs. Ag/AgCl	height / A E - 05	area / AV E - 06	$W_{½}$ / V
1	0.269	0.297	0.4633	0.142
2	0.815	8.314	5.1200	0.049

Cuprite (Cu_2O)

Origin: Gumerchewsk, Ural (Russia)

Method: differential pulse voltammetry, stripping mode
Electrolyte: 1 M KCl
Method parameters: E_{pulse} = 0.050 V t_{pulse} = 0.01 s
E_{step} = 0.004 V t_{int} = 0.50 s
E_{dep} = -1.400 V t_{dep} = 60 s

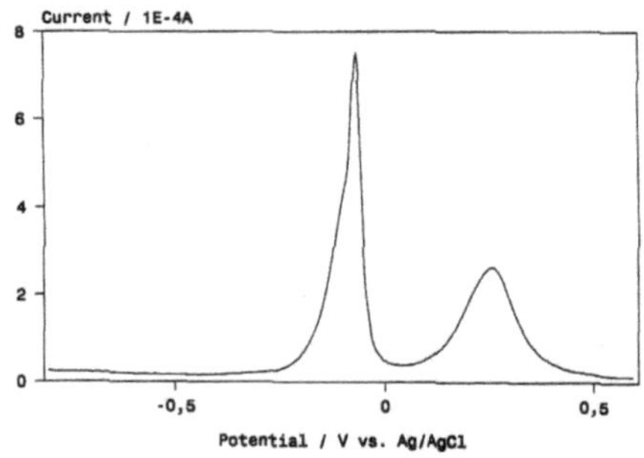

Results of peak search:

	E/ V vs. Ag/AgCl	height / A E - 04	area /AV E - 05	$W_{½}$ / V
1	-0.068	7.1836	4.8064	0.055
2	0.255	2.3612	3.5910	0.132

Fig. 5.13 Oxidative voltammograms of several mineral phases [B46]

Stephanite (Ag_5SbS_4)

Origin: Freiberg, Erzgebirge (Germany)

Method: differential pulse voltammetry, stripping mode
Electrolyte: 1M HCl
Method parameters: E_{pulse} = 0.010 V, t_{pulse} = 0.07 s; E_{step} = 0.005 V, t_{int} = 0.50 s; E_{dep} = -1.200 V, t_{dep} = 60 s

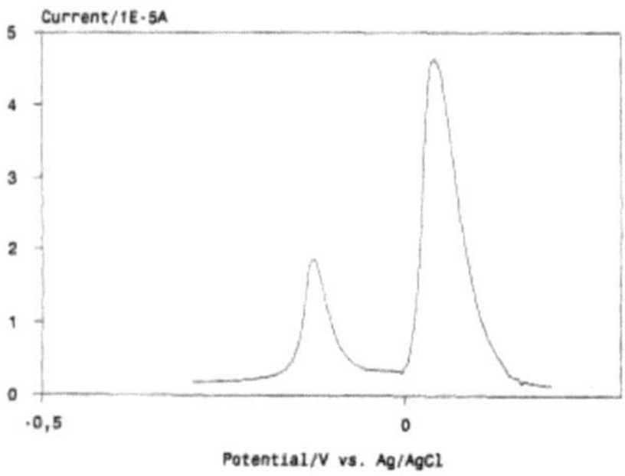

Results of peak search:

	E/ V vs. Ag/AgCl	height / A E - 05	area / C E - 06	$W_{1/2}$ / V
1	-0.127	1.597	0.712	0.034
2	0.039	4.331	2.650	0.053

Stibnite (Sb_2S_3)

Origin: Shikoko (Japan)

Method: differential pulse voltammetry, stripping mode
Electrolyte: 1 M HCl
Method parameters: E_{pulse} = 0.050 V, t_{pulse} = 0.01 s; E_{step} = 0.004 V, t_{int} = 0.50 s; E_{dep} = -1.200 V, t_{dep} = 60 s

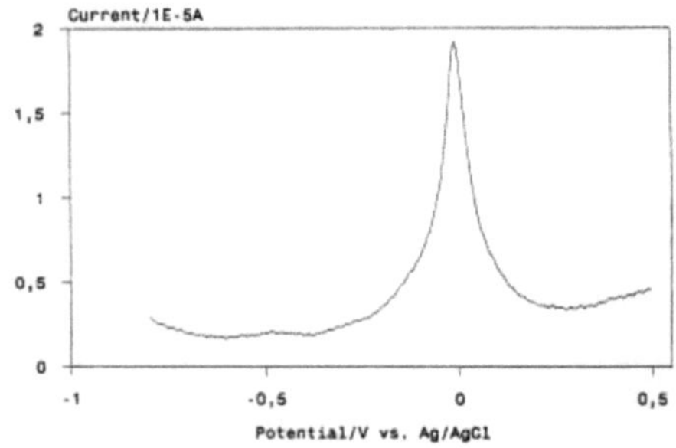

Results of peak search:

	E/ V vs. Ag/AgCl	height / A E - 05	area /AV E - 06	$W_{1/2}$ / V
1	-0.009	1.6125	1.9413	0.077

$Tl_2As_2SnS_6$

Origin: synthetic

Method: differential pulse voltammetry, stripping mode
Electrolyte: 1 M KCl
Method parameters: E_{pulse} = 0.050 V, t_{pulse} = 0.01 s; E_{step} = 0.004 V, t_{int} = 0.50 s; E_{dep} = -1.700 V, t_{dep} = 60 s

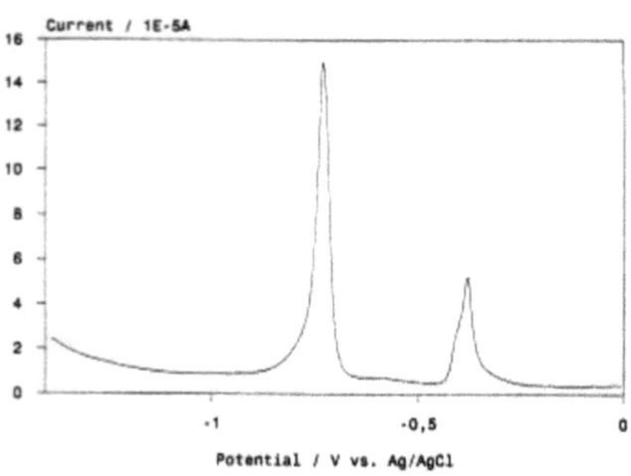

Results of peak search:

	E/ V vs. Ag/AgCl	height / A E - 04	area /AV E - 06	$W_{1/2}$ / V
1	-0.730	1.419	6.483	0.033
2	-0.378	0.476	2.352	0.040

Pyrargyrite (Ag_3SbS_3)

Origin: St. Andreasberg, Harz (Germany)

Method: differential pulse voltammetry, stripping mode
Electrolyte: 1M KCl
Method parameters: E_{pulse} = 0.010 V, t_{pulse} = 0.07 s; E_{step} = 0.005 V, t_{int} = 0.50 s; E_{dep} = -1.200 V, t_{dep} = 60 s

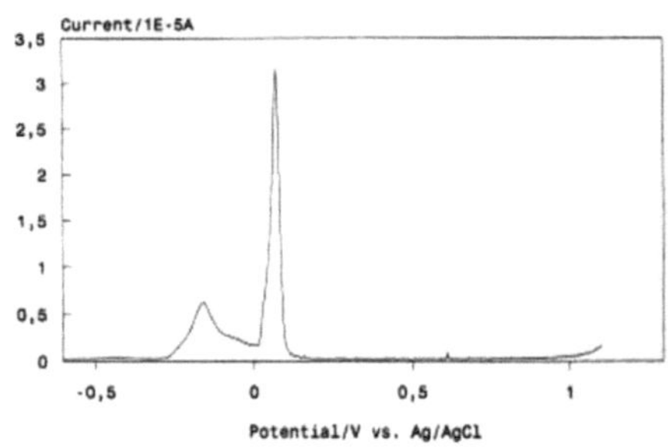

Results of peak search:

	E/ V vs. Ag/AgCl	height / A E - 05	area / C E - 06	$W_{1/2}$ / V
1	-0.161	0.5305	0.5415	0.078
2	0.074	3.0474	1.0295	0.029

Fig. 5.14 Oxidative voltammograms recorded following to a preliminary reduction of several mineral phases [B46]

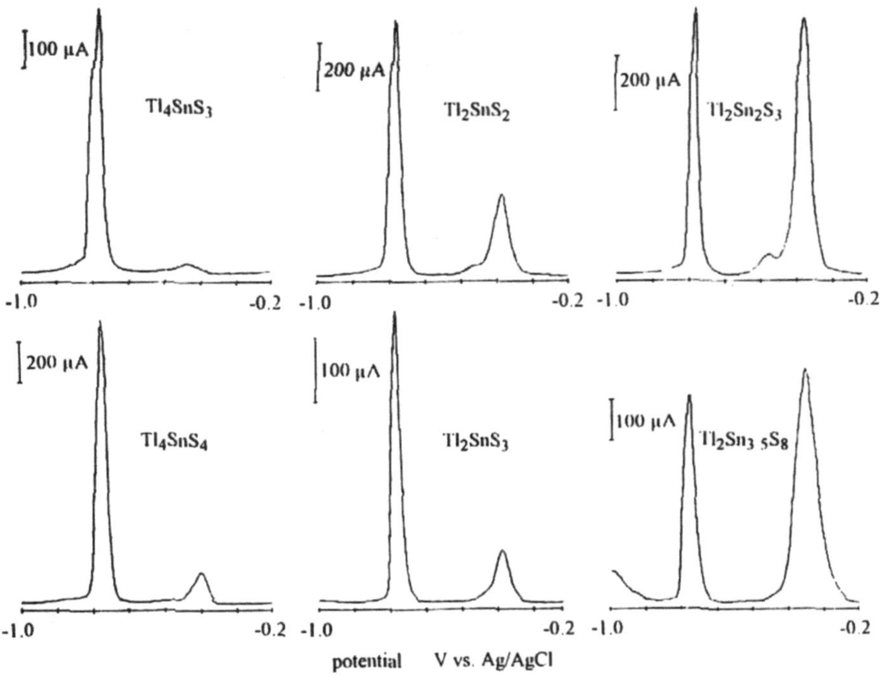

Fig. 5.15 Anodic differential pulse voltammograms of thallium–tin sulfides of different stoichiometry after a preliminary reduction of the solid phases at –1.0 V versus Ag/AgCl for 60 s. The electrolyte solution contained 1 mol L^{-1} HCl and 10^{-4} mol L^{-1} $HgCl_2$. The scan rate was 10 mV s^{-1} [B32]. Copyright ©1995 with permission from Elsevier

analytical approach highly attractive for mineral and pigment identification, both in the laboratory and for field analysis [B17]. Figure 5.16 shows voltammograms of several inorganic pigments [B25]. Pigment identification is an important analytical target for archaeometry, conservation, and restoration applied to archaeological artifacts and works of art. Identification in real samples is made difficult by the low concentration of pigment in the sample and the presence of other components such as bindings, varnishes, and alteration products. The paint samples usually consist of a succession of strata from the preparative layer to an external layer frequently resulting from dust accumulation and residuals of chemical and even biological aggression to the painting. This means that important interference and matrix effects can occur [B351]. Another important aspect is that pigments used in works of art, apart from minerals or mixtures of minerals, frequently consists of natural products of animal and vegetal origin, usually constituting multicomponent systems. In spite of these difficulties, identification of pigments and/or alteration products in real samples is possible, even when mixtures of pigments appear. This is

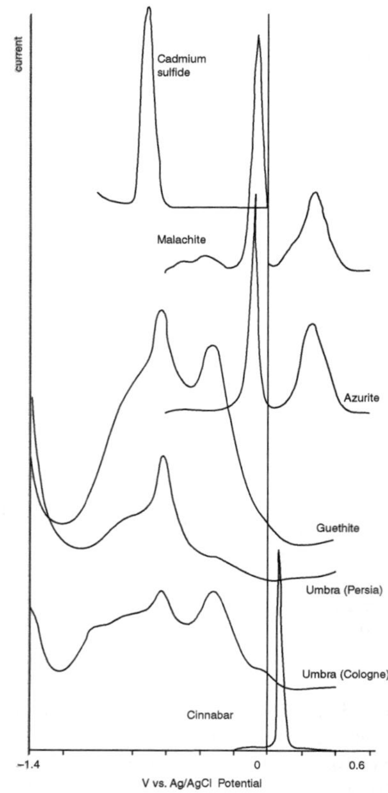

Fig. 5.16 Differential pulse voltammograms of pigments (pulse amplitude 10 mV, pulse duration 0.07 s, scan rate 0.526 V min^{-1}, pre-electrolysis for CdS at –1.2 V versus Ag/AgCl, for all other pigments at –1.5 V versus Ag/AgCl). The electrolyte solution was a 0.1 mol L^{-1} oxalic acid for CdS and 1 mol L^{-1} KCl for all other compounds [B25]. Copyright ©1993 with permission from publisher

the case shown in Fig. 5.17, where square-wave voltammograms for three samples extracted from the ceiling vault of the *Sant Joan del Mercat* church in Valencia (Spain) in contact with aqueous phosphate buffer are shown [B352]. The samples consisted originally of a mixture of azurite and cobalt blue, displaying overlapping signals at ca. −0.10 V. These signals are accompanied by the tenorite (CuO) reduction peak at ca. −0.50 V, which would be the main product of the oxidation of the above during the fire suffered by the vault in 1936. Depending on the position in the vault, the proportion of tenorite relative to azurite, which can be quantified from peak current measurements, was different [B352].

Doménech-Carbó et al. [B143] have studied the Pb(II) and Pb(IV) content in medieval glazes. They used three different techniques for immobilizing the

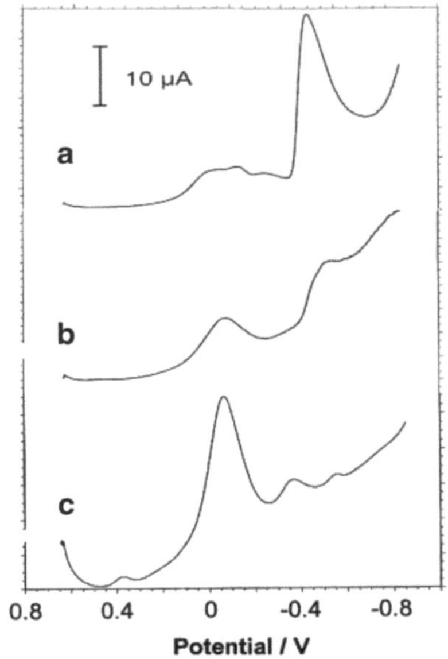

Fig. 5.17 Square-wave voltammograms for three pictorial samples from the ceiling vault of the *Sant Joan del Mercat* church in Valencia (Spain), painted by Antonio Acisclo Palomino ca. 1700, attached to paraffin-impregnated graphite electrodes immersed into 0.50 M phosphate buffer, pH 7.4. Potential scan initiated at +0.45 or +0.65 V in the negative direction. Potential step increment 4 mV, frequency 2 Hz [B352]. Copyright ©2008 with permission from American Chemical Society

microparticles obtained after milling small amounts of samples in a mortar: (i) they prepared modified paste electrodes by mixing graphite powder (35 wt%), paraffin oil (35 wt%), and the sample powder (30 wt%). (ii) Aliquots of a dispersion of the sample in a 0.5 wt% solution of Paraloid B 72 in acetone were applied to the surface of a freshly polished glassy carbon electrode to prepare the modified film. (iii) 0.1 mg of the sample powder was mixed with 1 mg graphite powder and the mixture was pressed onto an electrode that was finally covered by a Paraloid B72 coating. Figure 5.18 shows the response of PbO, PbO_2, and Pb_3O_4 in the film electrodes [preparation (ii)].

Figure 5.19 depicts the responses of medieval glazes. The same authors applied the film electrodes for a wide range of pigments from polychrome sculptures and paintings. Figure 5.20 illustrates very impressively how specific the voltammetric fingerprints are [B146]. Extensive tables of peak potentials given in this publication support this statement. The film electrodes have also been used studying the canvas painting in a Basilica in Valencia, Spain [B175], and for the identification of the decomposition products, i.e., alteration products of copper-containing pigments [B176]. For this purpose, the authors have developed a very interesting approach: the pigments verdigris [a basic copper acetate, $Cu(C_2H_3O_2)_2 2Cu(OH)_2$], azurite [a basic copper carbonate, $2CuCO_3Cu(OH)_2$], malachite [another basic copper carbonate, $CuCO_3Cu(OH)_2$], and copper trihydroxychloride ($Cu_2(OH)_3Cl$) have rather similar reductive voltammograms, making their identification difficult, especially in their mutual presence. However, when Tafel plots are constructed from linear scan voltammograms and when the slopes of the Tafel plots are plotted versus the peak potential of reductive dissolution, a two-dimensional diagram is obtained with very clear separation of the pure phases and their binary mixtures.

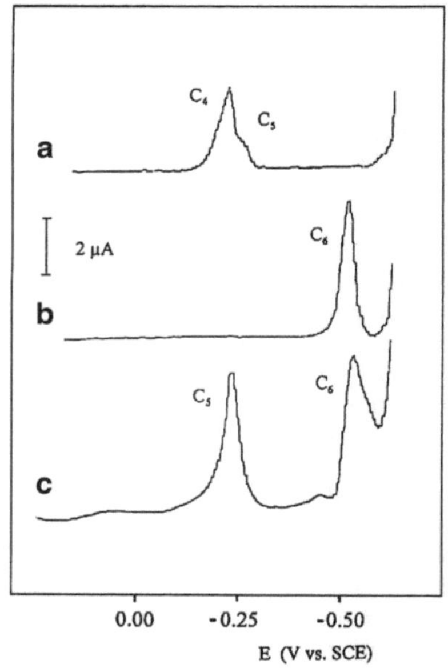

Fig. 5.18 Cathodic differential pulse voltammograms for modified film electrodes containing (**a**) PbO, (**b**) PbO_2, and (**c**) Pb_3O_4. Scan rate 10 mV s^{-1}, pulse height 10 mV [B143]. Copyright ©2000 with permission from Wiley-VCH

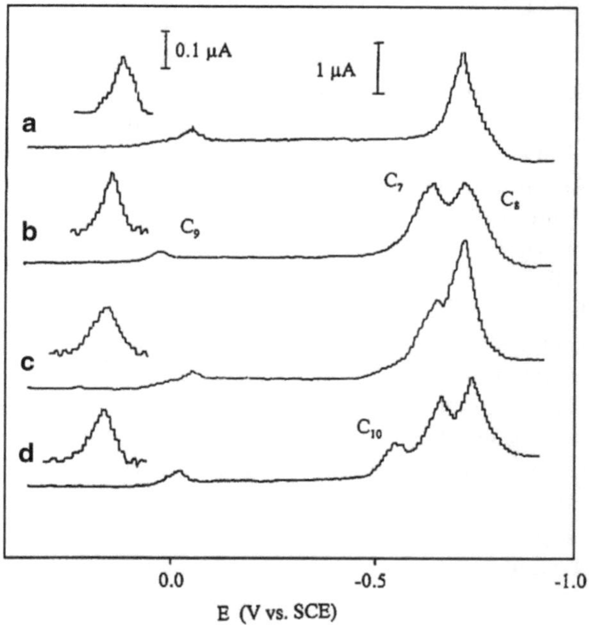

Fig. 5.19 Cathodic differential pulse voltammograms for modified film electrodes containing samples of test specimen and archaeological samples. Scan rate 10 mV s^{-1}, pulse height 10 mV, acetate buffer [B143]. Copyright ©2000 with permission from Wiley-VCH

Figure 5.21 shows such a plot including sample spots from green-blue pigments of the Basilica in Valencia. The composition that can be derived from this diagram for the pigment samples was supported by SEM/EDX and FT-IR measurements.

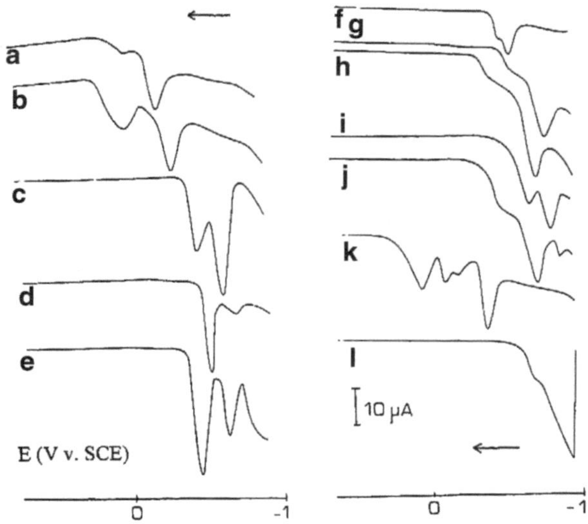

Fig. 5.20 Anodic differential pulse voltammograms following a preliminary reduction at –1.0 V in 1 M NaCl solution: (**a**) malachite, (**b**) azurite, (**c**) minium, (**d**) chrome yellow, (**e**) litharge, (**f**) lead white, (**g**) lead–tin yellow, (**h**) chrome orange, (**i**) cadmium red, (**j**) cadmium yellow, (**k**) vermilion, (**l**) zinc white [B146]. Copyright ©2000 with permission from Elsevier

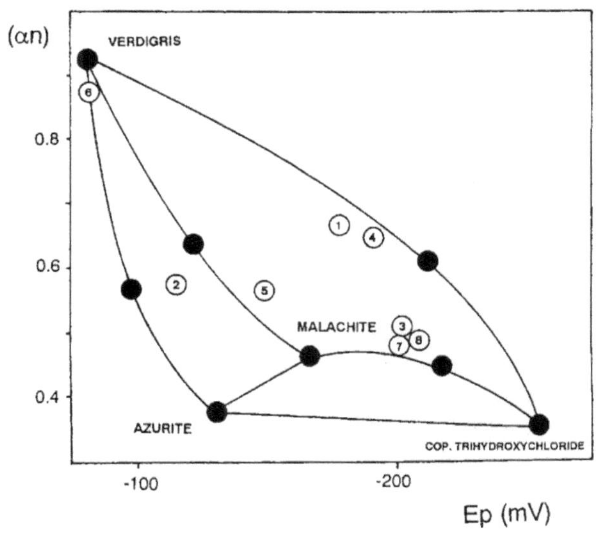

Fig. 5.21 Representation of the apparent charge transfer coefficients calculated from the slope of Tafel plots as a function of the peak potential from linear scan voltammograms recorded at 60 mV s^{-1}. *Black points*: samples of the pure pigments and their binary 1:1 mixtures. *Circled numbers*: samples from a basilica in Valencia, Spain [B176]. Copyright ©2001 with permission from Springer

In another study, Doménech-Carbó et al. [B189] have identified manganese (IV) centers in archaeological glass. The authors studied the relation between the voltammetric responses and the conditions of glass deterioration. The same authors identified iron oxide pigments (green earth, iron oxide red, Mars black, ochre yellow, Sienna raw, umber raw, Van Dyke brown) in microsamples extracted from polychromed sculptures, canvas paintings, wall paintings, altarpieces, and panel paintings from Spain, Ethiopia, and Italy from the twelfth to the twentieth centuries [B195]. The authors employed a mechanical immobilization of sample

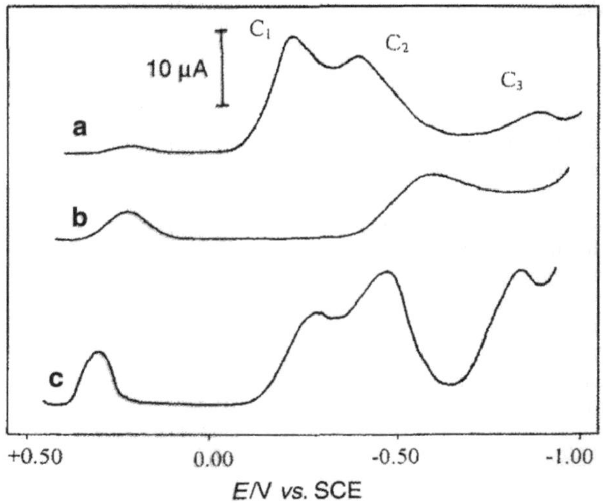

Fig. 5.22 Cathodic differential pulse voltammograms of (**a**) Mars black, (**b**) Van Dyke brown, and (**c**) umber raw modified electrodes immersed in 0.1 mol L^{-1} HCl [B195]. Copyright ©2001 with permission from Royal Society of Chemistry

particles on the surface of self-made graphite–polyester electrodes and also modified polyester–graphite electrodes. Figure 5.22 shows the unique shape of cathodic differential pulse voltammograms of three iron oxide pigments. Another example of the application of the voltammetry of immobilized microparticles is the detection of Co, Cu, Sb, Mn, Sn, and Fe in archaeological samples of Spanish ceramic glazes from the sixteenth to eighteenth centuries [B217, B304]. The authors used the mechanical immobilization of microparticles on the surface of a polyester–graphite electrode. Doménech-Carbó et al. have presented an ingenious application of the voltammetry of immobilized particles—the determination of the boron content of minerals [B318]. Other applications concern the phase identification of iron oxides [B312], copper and iron oxides [B314], and cobalt cordierites [B317]. Recent advances involve the layer-by-layer analysis of stratified systems, based on repetitive voltammetry [B353], and the use of "one-touch" sampling strategies to perform an essentially noninvasive analysis of works of art [B354] and mapping corrosion products in archaeological artifacts [B355].

5.2.2 Quantitative Determination of a Phase in a Phase Mixture

In the case that more than one electrochemically active phase is present in a powder mixture, it is possible to utilize the voltammetry of immobilized microparticles on the basis of a calibration procedure [B444]. This has been shown for two phase mixtures [B11]. When both phases give an electrochemical signal that is directly proportional to the absolute amount of immobilized substance, however having different slopes (sensitivities), a plot of the percentage peak currents versus the

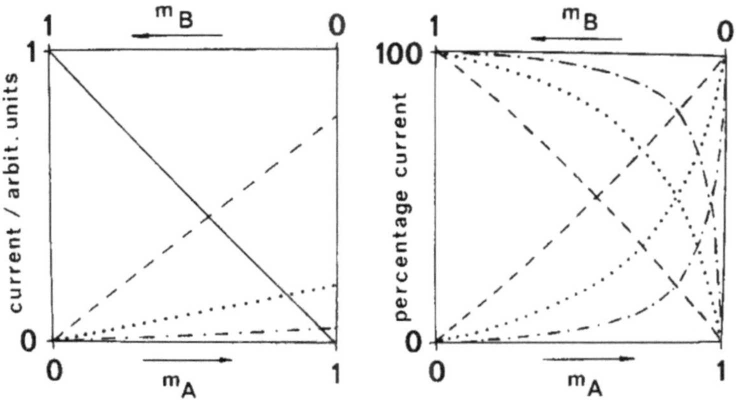

Fig. 5.23 *Left*: Theoretical calibration plots for the absolute peak currents when the amount of mixture is constant for each measurement. *Right*: Corresponding theoretical calibration plots for the percentage peak currents [B11]. Copyright ©1991 with permission from Springer

molar ratio of the compounds in the mixture is nonlinear (cf. Fig. 5.23). Because it is impossible to immobilize a known amount of the phase mixture on the electrode surface, the percentage peak currents are evaluated for a mixture of phase A and B in the following way:

$$i_{\%,A} = \frac{i_A}{i_A + i_B} 100\,\% \tag{5.3}$$

and

$$i_{\%,B} = \frac{i_B}{i_A + i_B} 100\,\% \tag{5.4}$$

where i_A and i_B are the peak currents of the phases A and B, respectively. Figure 5.24 shows calibration plots for mixtures of FeOOH/MnO_2, HgS/HgO, and PbO/HgO. The standard deviations for the determined powder compositions were in the range of 10–20 %. These relatively high values were attributed to inhomogeneities of the powder mixtures. It is clear that inhomogeneities play an increasing role the smaller the samples are.

If a solid phase contains a certain concentration of electroactive centers, and if this concentration is to be determined, it is possible to use another electroactive solid phase as inner standard. In solid phases of copper(II) hexacyanoferrate–hexacyanocobaltate, only the hexacyanoferrate unit is electroactive. To study the electroactivity of this moiety in solid solutions of different compositions, i.e., different ratios of hexacyanoferrate to hexacyanocobaltate, this solid phase was mixed in a 1:1 ratio with solid nickel(II) hexacyanoferrate. The latter exhibits a hexacyanoferrate signal well away from that of the solid solutions. In this way, it

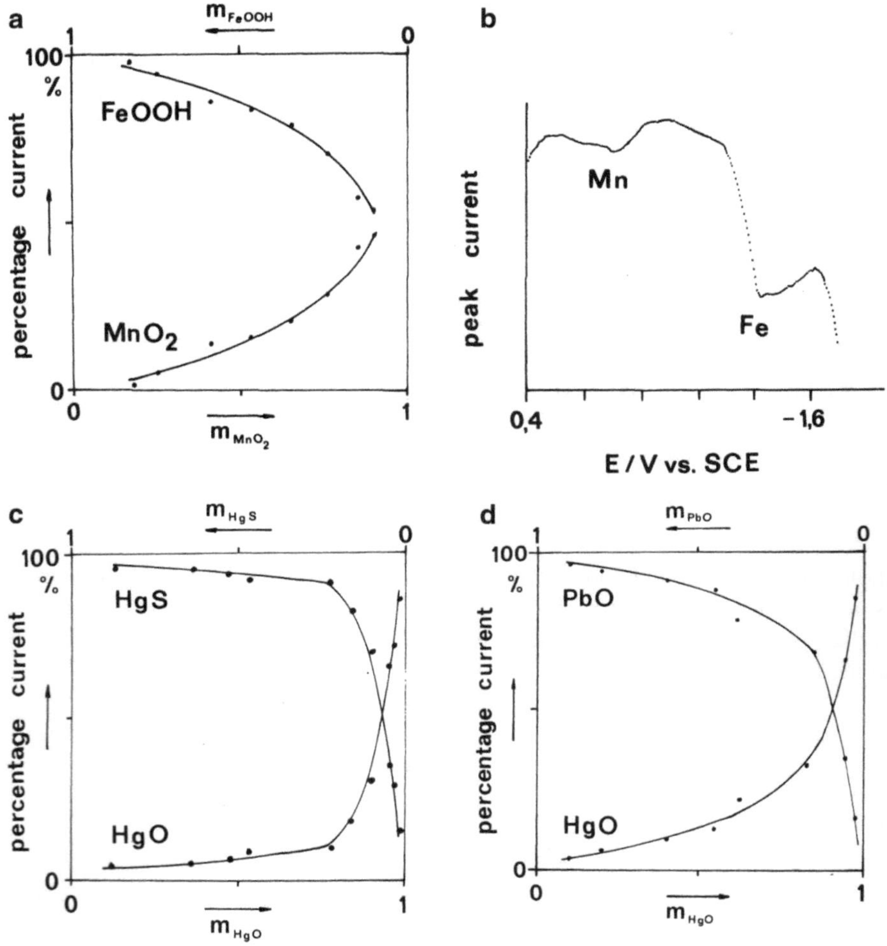

Fig. 5.24 Calibration plots of percentage peak currents of the following powder mixtures: (**a, b**) MnO_2/α-FeOOH, (**c**) HgS(*black*)/HgO(*yellow*), and (**d**) PbO/HgO(*yellow*) [B11]. Copyright ©1991 with permission from Springer

was possible to evaluate the peak current of the hexacyanoferrate of the solid solution with respect to the constant signal of the nickel(II) hexacyanoferrate [B247]. The same approach of an inner standard should be applicable to the case where the concentration of an electroactive solid phase has to be determined in the matrix of an electroinactive powder mixture. Indeed this has been done by van Oorschot et al. [B198] who determined iron(oxy)(hydr)oxides in soils and sediments using pyrolusite (β-MnO_2) as inner standard. The authors report this method as being semiquantitative. In a later publication, Grygar et al. reported the determination of ferric oxides in clay with the help of electroactive carbon paste electrodes [B222]. The paste was a mixture of graphite powder, the sample, and an acetate buffer solution. A thin layer of the paste was spread on the surface of a carbon electrode and covered by a foil. The iron oxide was first reductively dissolved, and then the iron(II) that must remain in the paste was oxidized to iron (III) ions. The oxidation charge was used to determine the iron content in the samples.

Absolute quantification of a phase in a phase mixture is available to VMP measurements. It should be noted that, as previously indicated, there is no possibility, under ordinary VMP experimental conditions, to control exactly the amount of sample transferred to the electrode surface. This means that, in general, non-repeatable values of the peak currents or peak areas will be obtained in replicate experiments so that only the mass ratio and the ratio of currents and/or peak areas can be used for calculations. Let us consider a material containing an unknown mass fraction f_X of an electroactive compound, X. Now, let us assume that a mass m_M of the material is mixed with a weighted amount, m_R, of a solid reference compound, R, so that the mass ratio between the reference compound and the mass of the problem material, $f_R(=m_R/m_M)$, is known. Assuming that, upon immersion into a suitable electrolyte, X and R compounds display well-separated, independent voltammetric signals, voltammograms of the R-spiked mixture must provide peaks corresponding to their respective electrochemical processes. Then, peak currents (or peak areas), $i_P(X), i_p(R)$, recorded for such voltammetric processes can in general be taken as proportional to the amount of each one of the compounds deposited on the electrode, as described by Lovrić et al. for SWV of reactants confined to electrode surfaces [B22]. Assuming that the sample is homogenized so that the composition of the mass transferred to a graphite electrode is representative of the composition of the R-enriched mixture, the peak current ratio between the signals for X and R should satisfy the relationship:

$$\frac{i_p(X)}{i_p(R)} = \frac{g_X f_X}{g_R f_R} = G \frac{f_X}{f_R} \tag{5.5}$$

In the above equation, g_R, g_X represent the electrochemical "coefficients of response," for the X- and R-localized electrochemical processes under the used chemical (electrolyte, pH, base electrode, etc.) and electrochemical (potential scan rate, square-wave frequency, etc.) conditions used for the voltammetric measurements. Such coefficients will depend on the molecular mass and redox properties of the analyte (number of electrons transferred, reversibility, diffusion coefficients for electrons and ions). On first examination, the mass (or, equivalently, mole) fraction of X in the problem material can be determined from peak current or peak area measurements providing that f_X is known by weighting and the g_X/g_R ratio is determined from independent calibration experiments in R plus X mixtures of known composition. Figure 5.25 compares the square-wave voltammograms recorded for herbal samples enriched with diazepam, a widely used 1,4-benzodiazepine anxiolytic, and different possible standards: indigo, Prussian blue, and ferrocene, in contact with aqueous acetate buffer [B324]. In all cases, the

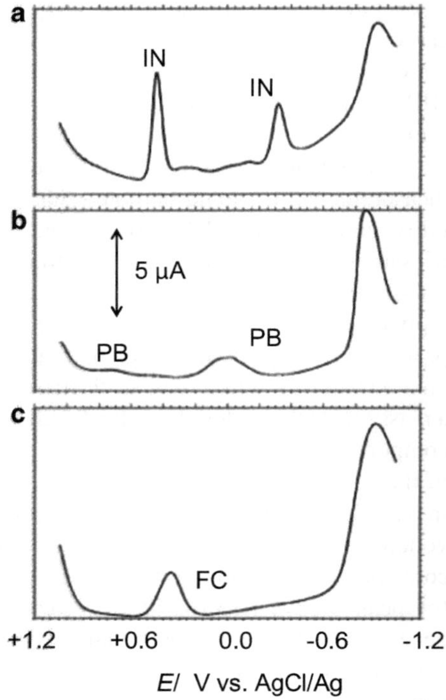

Fig. 5.25 Square-wave voltammograms for herbal samples spiked with diazepam (1 wt%) and sepiolite (29.60 wt%) and of (**a**) indigo (IN, 54.41 wt%), (**b**) Prussian blue (PB, 54.41 wt%), and (**c**) ferrocene (FC, 1.30 wt%) attached to paraffin-impregnated graphite electrodes in contact with 0.25 M aqueous sodium acetate buffer at pH 4.75. Potential scan initiated at +1.05 V in the negative direction, Potential step increment 4 mV, square-wave amplitude 25 mV, frequency 5 Hz [B324]. Copyright ©2013 with permission from Elsevier

prominent reduction peak for diazepam at ca. −1.0 V behaves independently of the signals of the added reference compound. In order to facilitate that all electroactive compounds appear as separate microparticles, it is convenient to introduce an additional dilution using an inert solid (sepiolite clay in the case in Fig. 5.25).

For practical purposes, however, it is convenient to perform standard additions experiments because the other components of the sample, although assumed to be electroinactive, conform a matrix which can influence the voltammetric responses of X and R. Now, let us consider that a series of aliquots of the R-enriched sample are mixed with weighted amounts of an X standard. Now, each specimen will contain a mass m_M of pristine sample plus a mass m_R of reference compound and a mass m_X^{add} of the added X standard. Then, the $i_P(X)/i_p(R)$ ratio will be [B318, B324, B357]

$$\frac{i_p(X)}{i_p(R)} = \frac{g_X\left(f_X m_M + m_X^{add}\right)}{g_R m_R} = G\left(\frac{f_X}{f_R}\right) + G\left(\frac{m_X^{add}}{m_R}\right) \tag{5.6}$$

Since f_X and f_R are constant, the above equation predicts a linear variation of $i_P(X)/i_p(R)$ on the m_X^{add}/m_R of slope G and ordinate at the origin $G(f_X/f_R)$. Since the f_R ratio is known, the mass percentage of the compound X in the original sample can be determined from standard additions experiment [B318, B324, B357]. This methodology can in principle be extended to the simultaneous determination of two of more electroactive analytes in the problem material. A case of particular interest occurs when two analytes display overlapping voltammetric peaks. Under favorable conditions, the H-point standard additions methodology, originally developed for spectrophotometric measurements,[5] can be applied to determine the concentrations of two compounds displaying overlapping but independent voltammetric signals in the solid sample [B358, B359].

The standard additions method can be modified for cases where limited amounts of the sample and the X standard were available. Then, a standard addition–dilution method can be proposed [B324]. Let us consider again a homogenized mixture with a mass m_S of sample containing an unknown fraction f_X of the analyte and a known fraction f_R mass of the reference compound. Now, let us prepare a second mixture where weighted amounts of reference compound, m^*_R, are added so that the m^*_R/m_S ratio is known, defining a ratio f^*_R. Then, the $i_P(X)/i_p(R)$ ratio will be

$$\frac{i_p(X)}{i_p(R)} = G\left(\frac{f_X}{f_R + f^*_R}\right) = G\left(\frac{f_X/f_R}{1 + f^*_R/f_R}\right) \tag{5.7}$$

This equation predicts that the plot of $i_P(X)/i_p(R)$ versus $(1 + f^*_R/f_R)^{-1}$ should be a straight line passing by the origin and slope $(SL) = G(f_X/f_R)$. External calibration would allow to determine the electrochemical coefficient of response G and then the value of f_X. As far as matrix effects cannot in general be discarded, for practical purposes, it is convenient to perform a second dilution experiment replacing the original sample by a sample spiked with a weighted amount of standard of the analyte, m^*_X, and, as in the precedent experiment, obtaining the voltammetric response of aliquots of that X-enriched sample upon addition of different amounts of the reference compound R. Now, the $i_p(X)/i_p(R)$ ratio becomes

$$\frac{i_p(X)}{i_p(R)} = G\left(\frac{f_X + f^*_X}{f_R + f^*_R}\right) = G\left(\frac{f_X/f_R + f^*_X/f_R}{1 + f^*_R/f_R}\right) \tag{5.8}$$

In this equation, f^*_X denotes the fixed m^*_X/m_S ratio. According to this last expression, a plot of $i_P(X)/i_p(R)$ versus $(1 + f^*_R/f_R)^{-1}$ should fit to a straight line at the origin of slope $(SL)* = G(f_X/f_R + f^*_X/f_R)$. The quotient between the slopes (SL)* and (*SL*) will be

$$\frac{(SL)*}{(SL)} = \left(\frac{f_X/f_R + f^*_X/f_R}{f_X/f_R}\right) = 1 + \frac{f^*_X}{f_X} \tag{5.9}$$

Accordingly, the problem f_X value can be calculated from the f^*_X one provided that the slopes in the addition–dilution experiments are obtained from voltammetric data. Figure 5.26 illustrates the application of this method for quantifying

[5] Bosch-Reig F, Campins-Falcó P (1988) Analyst 113:1011–1016

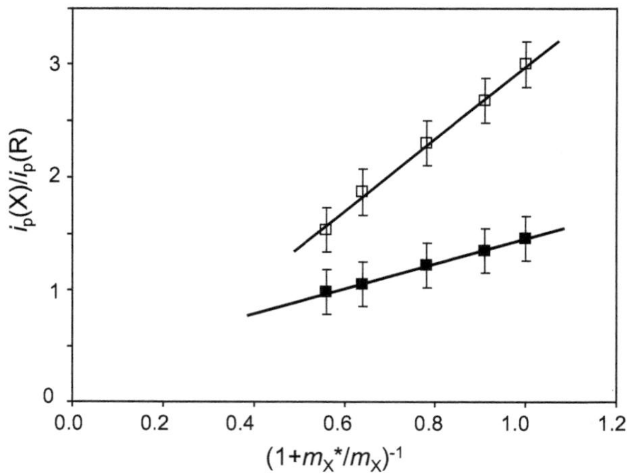

Fig. 5.26 Plots of $i_P(X)/i_p(R)$ versus $(1+f_R^*/f_R)^{-1}$ for an herbal sample spiked with alprazolam ($f_X = 8.66\,\%$) (*solid squares*) and after addition of alprazolam standard ($f_X^* = 9.45\,\%$) (*squares*) and different amounts of ferrocene standard, using square-wave voltammograms such as in Figure 5.25 [B324]. Copyright ©2013 with permission from Elsevier

alprazolam, a 1,4-benzodiazepine anxiolytic, introduced as an adulterant in herbal formulations [B324].

5.2.3 *Quantitative Analysis of the Composition of a Phase*

Here, we understand the term "quantitative analysis of a solid phase" as a determination of either the complete stoichiometry or at least the determination of one or some phase constituents. Examples for the determination of the complete stoichiometry are Tl_2S, TlS, Tl_4S_3, and Tl_2S_5 [B32]. Examples for the determination of the ratios of the metal contents of sample phases are (1) the Tl:Sn ratio in Tl_4SnS_4, $Tl_2Sn_{3.5}S_8$, Tl_2SnS_3, Tl_4SnS_3, $Tl_2Sn_2S_3$, and Tl_2SnS_2 [B32] and (2) the Bi:Cu:Pb ratio in $Bi_{1.8}Pb_{0.39}Sr_{1.99}Ca_{2.06}Cu_{3.15}O_{10.5}$ [B9].

For the quantitative analyses of solid phases, it is highly important that the measured charges are caused by strictly defined electrochemical reactions (no side reactions). Any loss of the compounds from the electrode surface has to be prevented during the electrochemical steps. Especially, when solid metal compounds immobilized on the electrode surface are reduced to the metals, it can occur that the metal particles do not adhere on the electrode surface. Further, it was found in the case of thallium sulfides [B32] that the thallium deposit obtained by reduction shows a very peculiar behavior when it is electrochemically oxidized: the oxidation was accompanied by a black cloud that was ejected from the electrode surface. Most probably the thallium deposit was partly lost due to electrostatic charging effects. Of course, such behavior must be circumvented, as well as the early hydrogen evolution on some metal deposits or compound phases. A very appropriate solution for all these problems is the co-deposition of metallic mercury. Two alternative approaches have been reported: (1) the solid sample phase can be thoroughly mixed with an excess of calomel (Hg_2Cl_2) [B9] or (2) the electrolyte

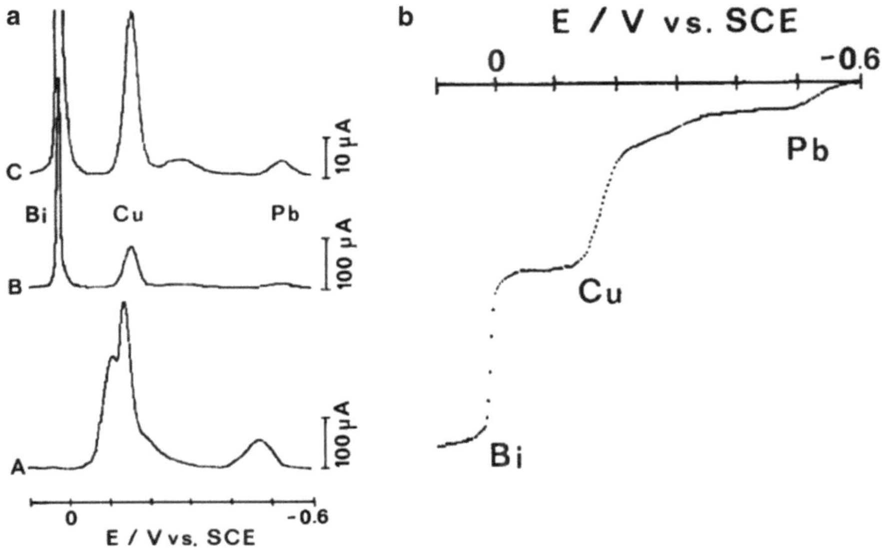

Fig. 5.27 (**a**) Differential pulse voltammograms of the compound $Bi_{1.8}Pb_{0.39}Sr_{1.99}Ca_{2.06}Cu_{3.15}O_{10.5}$ after a preliminary reduction at −1.5 V versus SCE for 30 s, A without calomel, B and C with calomel; (**b**) integrated curve B of Fig. 5.19a, in arbitrary units [B9]. Copyright ©1990 with permission from Springer

solution can be spiked with a small concentration of Hg^{2+} ions [B32]. In the case of the high-temperature superconductor $Bi_{1.8}Pb_{0.39}Sr_{1.99}Ca_{2.06}Cu_{3.15}O_{10.5}$, small amounts of the compound were thoroughly mixed with a 100- to 500-fold excess of solid calomel and a droplet of ethanol. After evaporation of the ethanol, the powder mixture was immobilized on a PIGE in the usual way, and anodic differential pulse voltammograms were recorded following a preliminary reduction at −1.5 V versus SCE for 30 s (Fig. 5.27a). Figure 5.27b depicts the integrated differential pulse voltammograms. The height of the steps was taken as proportional to the charges consumed for the oxidation of the deposited bismuth, copper, and lead. Taking the sum of the charges of the bismuth and lead signals as 100 %, the relative standard deviation of the bismuth signal was as low as 0.5 %, and taking the sum of the charges of the copper and lead signal as 100 %, the copper signal had a relative standard deviation of 1 %. These are very precise determinations, bearing in mind that they were obtained with μg amounts of the sample on the electrode surface.

In the case of the thallium sulfides Tl_4SnS_4, $Tl_2Sn_{3.5}S_8$, Tl_2SnS_3, Tl_4SnS_3, $Tl_2Sn_2S_3$, and Tl_2SnS_2, the stoichiometry of the composition can be derived from the determination of the charge necessary to reduce the phases to metallic thallium and the charge necessary for oxidation of the latter. This has been done with the help of triple step chronocoulometry [B32]. After immobilizing the sample onto the electrode surface, the first potential step was applied to charge the electrode at a potential shortly before the electrochemical reduction. After this, an appropriate reduction potential was applied for 60 s followed by an oxidation potential of −0.5 V versus Ag/AgCl. The appropriate reduction potential was chosen on the basis of reductive voltammograms. Figure 5.28 depicts some examples of chronocoulograms and Table 5.1 gives the theoretical and experimental charge ratios for four different compounds. Although the precision of the found charge ratios is certainly too low to be used for high-precision analyses, the ratios can be used to distinguish clearly the otherwise very similar thallium sulfides. It is practically impossible to distinguish them on the basis of simple reductive or oxidative voltammograms.

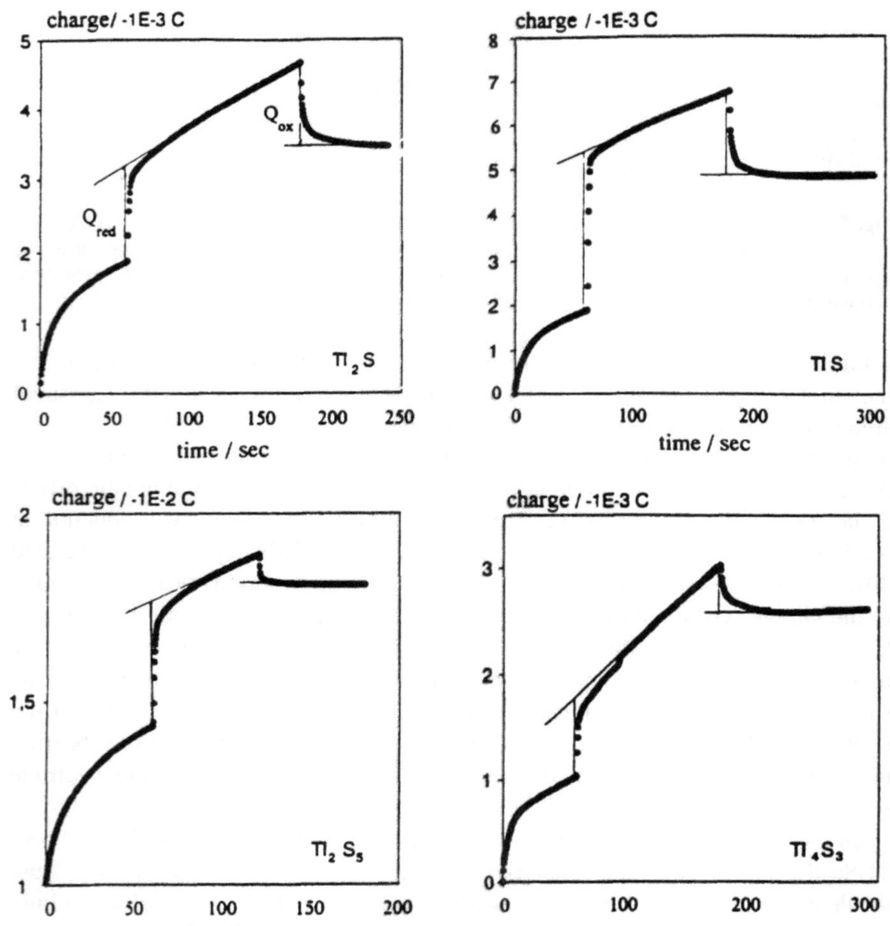

Fig. 5.28 Chronocoulograms of thallium sulfides: electrolyte 2 mol L^{-1} NaOH $+2 \times 10^{-4}$ mol L^{-1} $HgCl_2$. For details, see [B32]. Copyright ©1995 with permission from Wiley-VCH

Table 5.1 Reactions and ratios of charges necessary for the reduction of a thallium sulfide to the charge for the oxidation of thallium [B32]

Reduction of the metal sulfide	Oxidation of thallium metal	$q_{red}:q_{ox}$ (theory)	$q_{red}:q_{ox}$ (experiment)
$Tl_2S + 2e^- \rightarrow 2Tl + S^{2-}$	$2Tl \rightarrow 2Tl^+ + 2e^-$	2:2 = 1	1.18
$TlS + 2e^- \rightarrow Tl + S^{2-}$	$Tl \rightarrow Tl^+ + e^-$	2:1 = 2	2.0
$Tl_4S_3 + 6e^- \rightarrow 4Tl + 3S^{2-}$	$4Tl \rightarrow 4Tl^+ + 4e^-$	6:4 = 1.5	1.65
$Tl_2S_5 + 10e^- \rightarrow 2Tl + 5S^{2-}$	$2Tl \rightarrow 2Tl^+ + 2e^-$	10:2 = 5	4.68

5.2.4 Determination of Thermodynamic Data of Minerals

In a few cases, it has been shown that voltammetric measurements with immobilized microparticles of minerals can give access to the transformation enthalpies of polymorphic forms of minerals [B24, B62]. The basis of these measurements is the determination of the reversible potential of the following reactions of the two polymorphs of Ag_3AsS_3 xanthoconite and proustite, of Ag_3SbS_3 pyrostilpnite and pyrargyrite, and of $AgAsS_2$ trechmannite and smithite:

$$Ag_3AsS_3 + 3e^- \rightleftarrows 3Ag + {AsS_3}^{3-} \quad \text{(III)}$$

$$Ag_3SbS_3 + 3e^- \rightleftarrows 3Ag + {SbS_3}^{3-} \quad \text{(IV)}$$

$$AgAsS_2 + 3e^- \rightleftarrows Ag + {AsS_2}^{-} \quad \text{(V)}$$

Measuring of the formal potentials of these reactions needs to be performed very fast since the thioanions tend to decompose. It has been shown that square-wave voltammograms show all criteria of an electrochemically reversible process. From the differences in the square-wave peak potentials, i.e., formal potentials, the difference in the free energy was calculated. The reaction enthalpy of phase transition can be determined according to

$$\Delta_T G^\theta = \Delta_T H^\theta - T\Delta_T S^\theta \quad (5.10)$$

Assuming a temperature-independent entropy (1st Uhlich approximation), it follows with

$$\Delta_T S^\theta = \Delta_T H^\theta / T_T \quad (5.11)$$

(T_T is the transformation temperature) that the standard enthalpy is given by

$$\Delta_T H^\theta = \Delta_T G^\theta / [1 - (T/T_T)] \quad (5.12)$$

With the help of the thus determined transformation enthalpies, it is possible to calculate the standard formation enthalpy of one polymorphic form when that of the other is known. In the experiments to determine the formal potentials of the polymorphic forms, a small dependence of the square-wave peak potentials on the absolute amounts of electroactive mineral reacting on the electrode must be taken into account. This dependence is similar to that discussed in the case of solid solutions of CuS_xSe_{1-x} (Sect. 5.3.1). Figure 5.29 shows plots of square-wave peak potentials versus the charge underneath the peaks. The variation of charge is just the result of the sampling technique where unavoidably the amounts of mineral

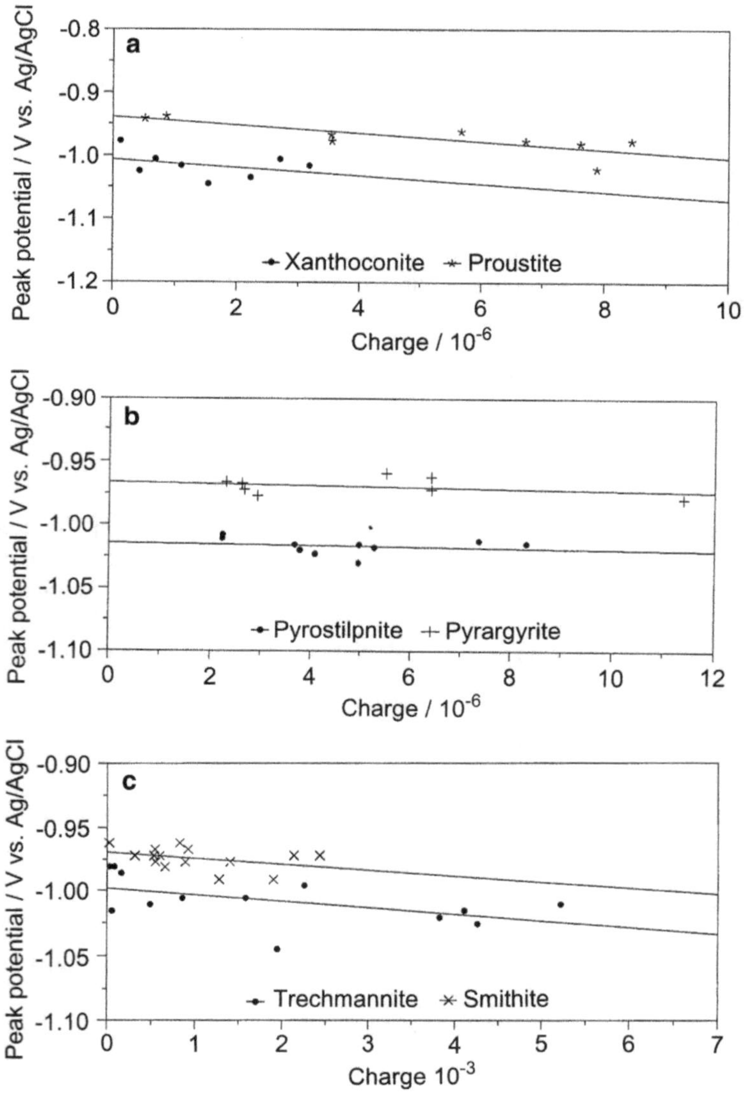

Fig. 5.29 Plot of the square-wave peak potentials of the three pairs of polymorphic minerals xanthoconite/proustite, pyrostilpnite/pyrargyrite, and trechmannite/smithite versus the charge consumed in the electrochemical reactions [B62]. Copyright ©1997 with permission from Springer

Table 5.2 Peak potential differences, transformation free energies, and enthalpies determined for the three pairs of polymorphic minerals

Mineral pair	ΔE_p (mV)	$\Delta_T G$ (kJ mol^{-1})	$\Delta_T H$ (kJ mol^{-1})
Xanthoconite → Proustite	-68 ± 12	19.7 ± 3.5	54.8 ± 9.6
Pyrostilpnite → Pyrargyrite	-50 ± 7	14.5 ± 20	40.3 ± 5.6
Trechmannite → Smithite	-30 ± 9	2.9 ± 0.9	15.65 ± 1.7

transferred to the electrode vary. Like in the case of CuS_xSe_{1-x} (Sect. 5.3.1), the dependencies of the peak potentials on the charge are linear and run parallel for the pairs of polymorphic forms. For the evaluation of the differences of the formal potentials, a linear regression analysis has been used. Because the dependencies are parallel to each other for each pair, the difference of formal potentials is independent on charge. Table 5.2 shows the transformation free energies and enthalpies determined for the three pairs of polymorphic minerals.

The outlined approach for determining thermodynamic data of minerals is, of course, only applicable when the formal potentials can be measured under reversible conditions.

5.3 Solid Solutions: Identification and Quantitative Analysis

Solid solutions are of tremendous importance in materials chemistry and physics. Electrochemical measurements can give access to vital information on solid solutions as they are closely linked to thermodynamics. Mixed phase thermodynamics predicts that the electrochemical properties of solid solutions differ in an unambiguous way from that of the pure phases. Therefore, it is obvious that from electrochemical measurements, two important questions can be answered in a very straightforward way:

1. Is a certain material a solid solution or is it just a mechanical mixture of two phases?
2. If it is a solid solution, what is its composition?

Figure 5.30 shows the electrochemical signals of two phases A and B and the signal of a solid solution containing A and B in a certain ratio. If a material is a mechanical mixture of the two phases A and B, then one will obtain two separate electrochemical signals. These signals may be any kind of characteristic potentials, provided that they are depending on the formal potentials (standard potentials) of the respective systems. Thus, it is possible to use peak potentials derived from linear scan voltammetry, from square-wave or pulse voltammetry, etc., or even the mid-peak potentials of cyclic voltammetry if the system exhibits sufficient reversibility. In any case, it will be necessary to have samples of the pure compounds A and B.

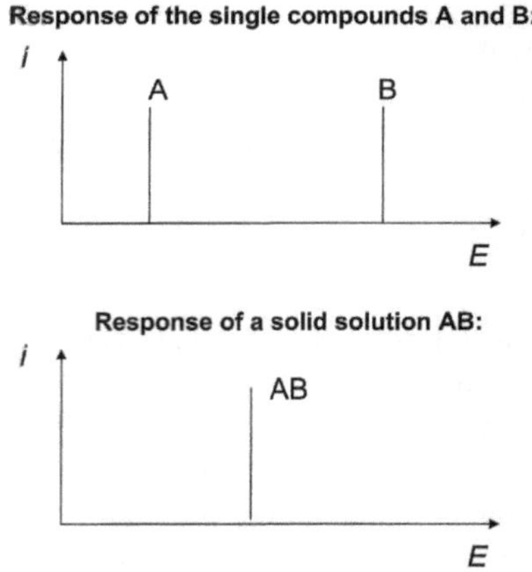

Fig. 5.30 Peak positions of single electroactive compounds A and B and of their solid solution A_xB_{1-x} [B204]. Copyright ©2002 with permission from publisher

When E_A and E_B are the characteristic potentials of these pure compounds, then it follows that the characteristic potential E_{AB} of the solid solution $AB_{(x)}$ with the molar ratio $x_B = \frac{n_B}{n_A+n_B}$ will shift with the molar ratio as follows:

$$E_{AB} = E_A + x_B(E_B - E_A) - \frac{RT}{zF}[(x_A \ln x_A + x_B \ln x_B) - x_A x_B \varepsilon] \tag{5.13}$$

The term $RT/zF(x_A \ln x_A + x_B \ln x_B)$ represents the deviation from linearity caused by the mixing entropy. ε is a parameter for the non-ideality.[6] The mixing entropy adds rather little to the nonlinearity of the dependence of E_{AB} on x_B. It just amounts to 17.7 mV for $x_B = x_A = 0.5$, $T = 25$ °C, and $z = 1$. When the reversibility of the electrochemical processes of the pure phases and the mixed phases differ, further deviations from Eq. (5.13) must be expected.

5.3.1 The Solid Solution System CuS–CuSe

Copper(II) sulfide and selenide form a continuous series of solid solutions. Immobilized on a graphite electrode, the pure compounds and the solid solutions can be easily reduced to metallic copper, the sulfide and selenide ions being liberated (Reactions IV and V) and, depending on the pH of the solution, protonated.

$$CuS_{(s)} + 2e^- \rightleftarrows Cu_{(s)} + S^{2-} \tag{VI}$$

$$CuSe_{(s)} + 2e^- \rightleftarrows Cu_{(s)} + Se^{2-} \tag{VII}$$

Figure 5.31 depicts the linear sweep voltammograms of the reduction of the pure phases CuS and CuSe, of the solid solution $CuSe_{0.4}S_{0.6}$, and of a mechanical

[6] Heusler KE (1996) Electrochim Acta 41:411–418

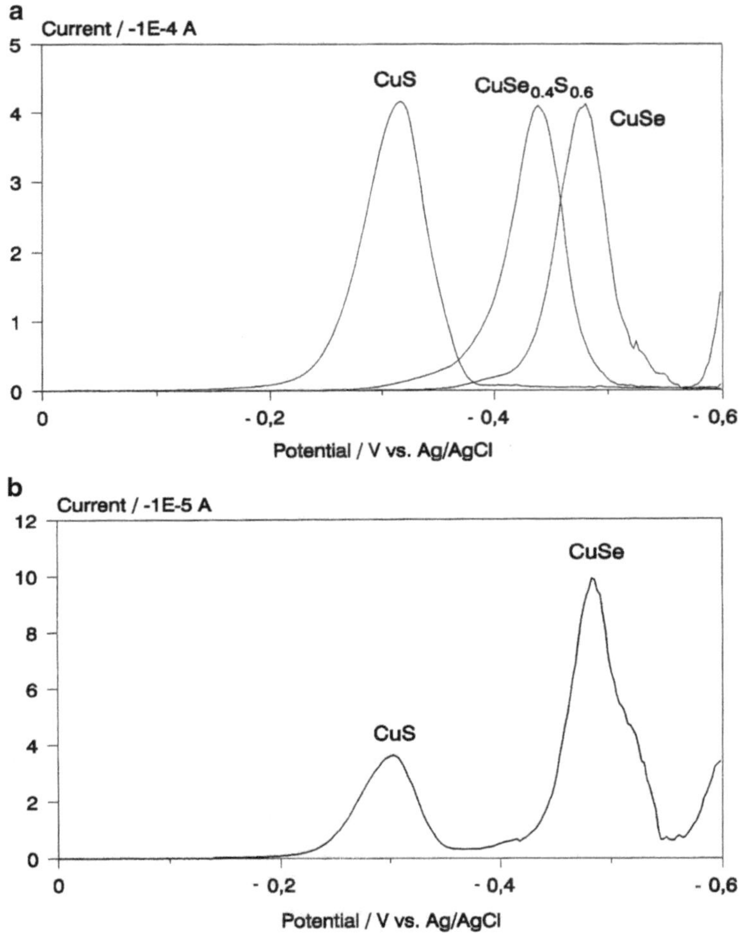

Fig. 5.31 Cathodic voltammograms of (**a**) CuS, CuSe, and $CuS_{0.6}Se_{0.4}$ and (**b**) a mechanical mixture of CuS and CuSe (1:2). Electrolyte 1 mol L^{-1} H_2SO_4, scan rate 0.011 V s^{-1} [B57]. Copyright ©1996 with permission from Springer

mixture of CuSe and CuS [B57]. The voltammogram of the solid solution $CuSe_{0.4}S_{0.6}$ exhibits a single peak in accordance with the expectations from mixed phase thermodynamics. For an exact determination of the composition of the solid solution from a measurement of the peak potential, an empirical calibration is necessary using a set of samples of known composition. Further it is necessary to establish whether the peak potential significantly depends on the amount of particles immobilized on the electrode surface. In the case of the copper selenides and sulfides, this dependence, which is specific for different immobilized species, has to be taken into account: When each sample is measured several times by trying to deliberately vary the amount and in each case the peak potential is

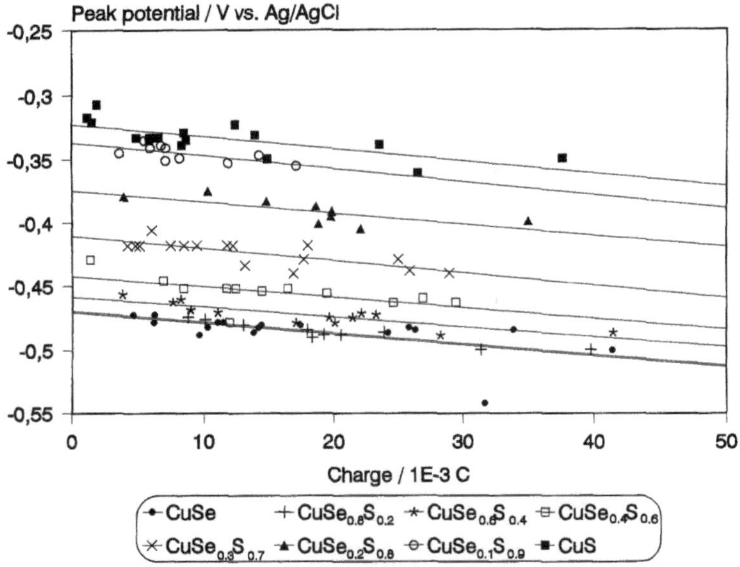

Fig. 5.32 Plot of the peak potential E_p of different solid solutions CuS_xSe_{1-x} versus charge Q, consumed in the electrochemical reactions when deliberately varying amounts of the compounds have been immobilized [B57]. Copyright ©1996 with permission from Springer

measured and the voltammograms is integrated, one can easily determine the dependence of the peak potentials on the charge, i.e., a value proportional to the amount of reduced substance. Figure 5.32 shows these plots.

Interestingly, they are all linear and parallel, and one can extrapolate the peak potentials for zero charge and use them as values standardized for zero amount (any other standardization is suited as well). Plotting these extrapolated peak potentials versus the molar ratio x_{Se} yields a smooth curve that can be used as a calibration plot for the analysis of samples of unknown composition (Fig. 5.33). Figure 5.34 shows voltammograms of $CuSbS_2$ and $CuBiS_2$ as well as voltammograms of a solid solution of these two compounds and of a mechanical mixture. Again it is obvious that a distinction between a solid solution and a phase mixture is easy to make. For this system, no quantification has been made.

5.3.2 *The Solid Solution System AgCl–AgBr*

Silver chloride and silver bromide form a continuous series of solid solutions. The reduction potential of the silver ions shifts as a function of the bromide content of the solid solutions that have been obtained by melting together AgCl and AgBr [B14]. The dependence (Fig. 5.35) is nonlinear because of a considerable non-ideality of the mixtures.

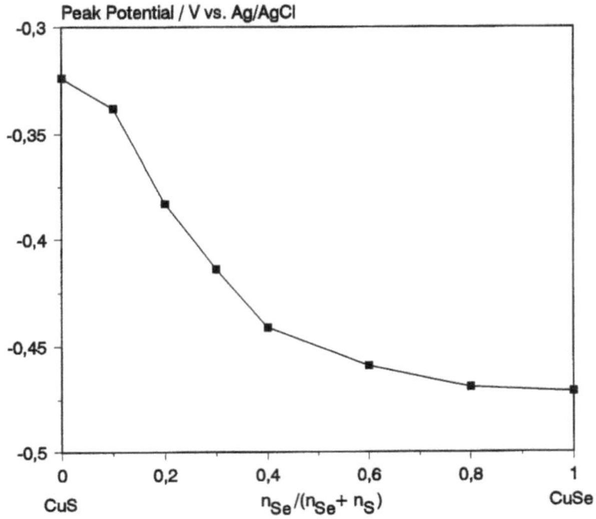

Fig 5.33 Plot of standardized peak potentials (extrapolated to values for $Q = 0$) of different solid solutions CuS_xSe_{1-x} versus the molar ratio $n_{Se}/(n_{Se} + n_S)$ [B57]. Copyright ©1996 with permission from Springer

5.3.3 Solid Solutions of Metal Hexacyanometalates

Prussian blue is the archetype of a large group of metal hexacyanometalates, and its electrochemistry has been extensively studied for the following reasons (1) It can be easily deposited by electrochemical procedures on an electrode surface. (2) It possesses two different kinds of electrochemically active iron ions, high-spin iron (III) that is nitrogen coordinated and low-spin iron(II) that is carbon coordinated. (3) The electrochemistry of both kinds of iron ions is reversible. (4) The reduction is coupled with a reversible uptake of cations from solution and the oxidation is accompanied by the expulsion of these cations. The ion transfer is possible due to the zeolitic structure of Prussian blue in which ions can be housed in cavities and move along channels. (5) Prussian blue exhibits electrocatalytic properties. (6) Prussian blue shows electrochromism. (vii) Prussian blue and its oxidized and reduced forms are highly insoluble in aqueous solutions. All these properties made it a favorable compound to be used for electrochemical studies. Since there exist dozens of other very similar metal hexacyanometalates, it was a temptation to study their electrochemistry as well. However, only a relatively small number of these compounds, e.g., copper, nickel, cobalt hexacyanoferrate, and some others, can be precipitated or synthesized on electrode surfaces. The majority of these compounds can be chemically synthesized, but a deposition on electrode surfaces was hampered until introduction of the technique of mechanical immobilization of microparticles.

Figure 5.36 depicts the structure that is representative for the vast majority of the hexacyanometalates. Three kinds of metal ions can be distinguished, and all three positions can be populated by mixtures of ions, provided that their radii fit to each other. Thus, it is possible to populate the nitrogen-coordinated sites by Cu^{2+} and Fe^{3+},

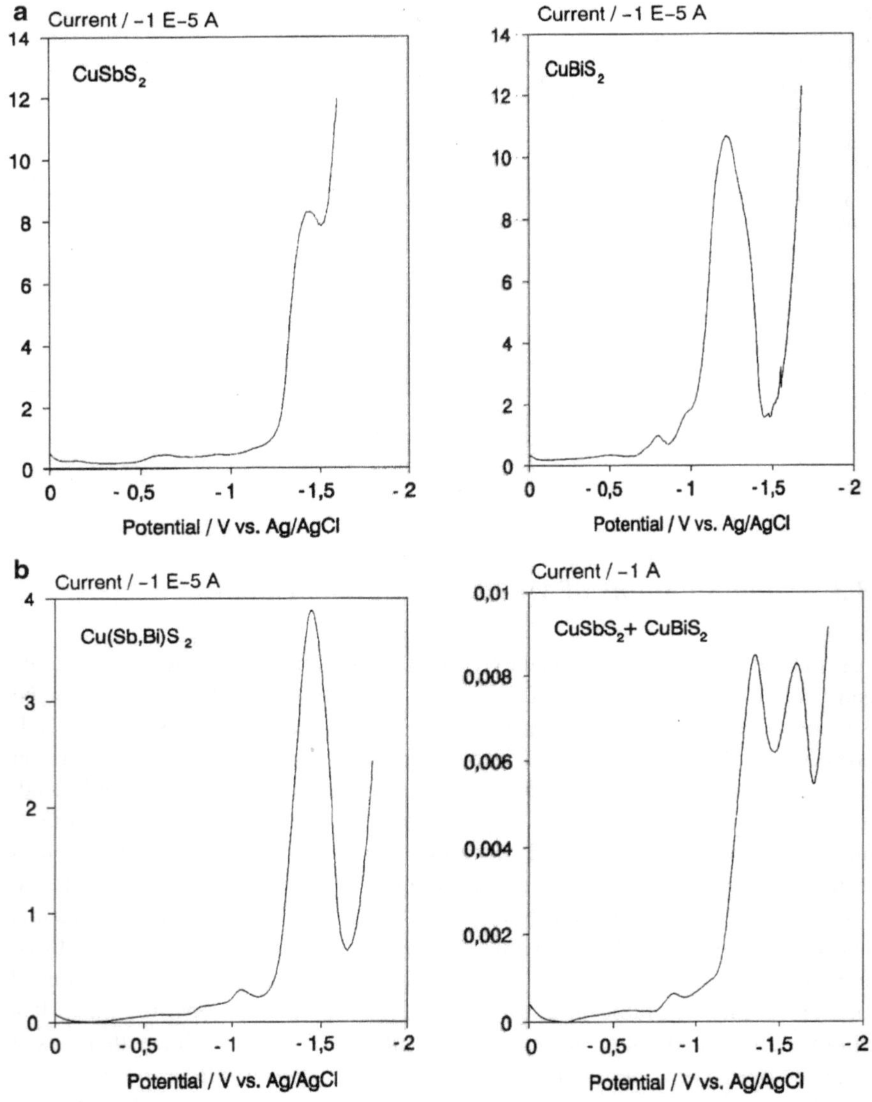

Fig. 5.34 Cathodic differential pulse voltammograms of (**a**) $CuBiS_2$ and $CuSbS_2$ and (**b**) $CuBi_{0.5}Sb_{0.5}S_2$ and a mechanical mixture of $CuBiS_2$ and $CuSbS_2$. Electrolyte 0.5 mol L^{-1} NaOH + 0.1 mol L^{-1} Na-K-tartrate, scan rate 6.25 mV s^{-1}, pulse height 25 mV, and pulse duration 0.05 s [B57]. Copyright ©1996 with permission from Springer

by Cu^{2+} and Ni^{2+}, etc., and the carbon-coordinated sites by $Fe^{2+/3+}$ and Co^{3+}, etc. The cations in the interstitial cavities can be populated by different alkali metals and also by transition metal ions. Because the cations on this position are mobile and can be exchanged with those in an adjacent solution, such solid solutions do not

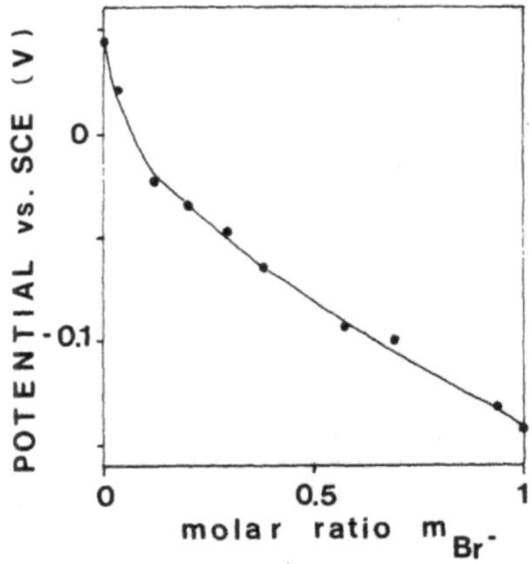

Fig. 5.35 Peak potential of differential pulse voltammograms obtained for reduction of solid solution of AgCl–AgBr as a function of the molar ratio of bromide. Electrolyte is 0.1 mol L^{-1} oxalic acid [B14]. Copyright ©1991 with permission from American Chemical Society

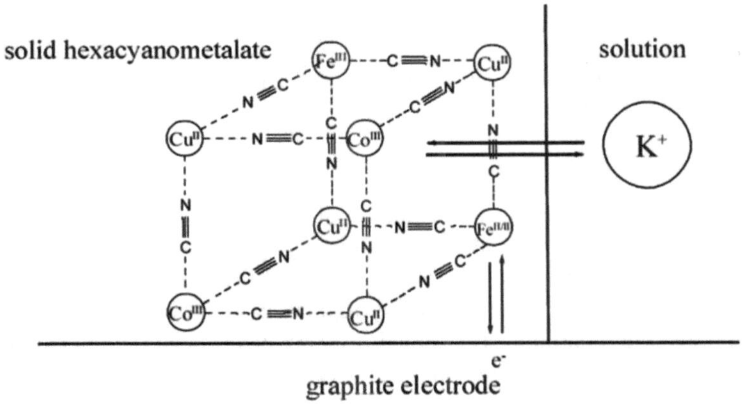

Fig. 5.36 Structure of a mixed copper–nickel hexacyanoferrate–hexacyanocobaltate and illustration of its insertion electrochemistry

possess much interest. However, they can be formed in the course of electrochemical reactions with mixed electrolyte solutions. Besides the substitutional solid solutions, metal hexacyanometalates can also form redox mixed solid solutions. Thus, a Prussian blue, where one half of the high-spin iron is in the +3 and one half of it is in the +2 oxidation state, is a solid solution provided that the redox centers are statistically distributed. This kind of solid solutions will be discussed later. Table 5.3 gives a schematic overview of the six basic types of solid solutions that can be formed in the case of metal hexacyanometalates. Pure solid solutions of all types are known with the exception of type IV. In many cases, higher-order solid solutions are formed, i.e., mixed types of II and III, etc. Type IV seems to exist only as a combination with type I and II [B111].

Solid solutions in which the nitrogen- and carbon-coordinated sites are mixed populated are very interesting systems as they give insight into the dependence of

Table 5.3 Basic types of solid solutions of metal hexacyanometalates

Type of solid solutions	Me′	Me″	Me‴
I	Different ions, e.g., K^+ and Na^+		
II		Different ions, e.g., Cu^{2+} and Fe^{3+}	
III			Different ions, e.g., Fe^{2+} and Co^{3+}
IV	One kind of ions in different oxidation states, e.g., Fe^{2+} and Fe^{3+}		
V		One kind of ions in different oxidation states, e.g., Fe^{2+} and Fe^{3+}	
VI			One kind of ions in different oxidation states, e.g., Fe^{2+} and Fe^{3+}

[Me′] [Me″ –NC- Me‴]; Me′, interstitial metal ions; Me″, nitrogen-coordinated metal ions; Me‴, carbon-coordinated metal ions

the electrochemical properties on the nature of metal ions. Voltammetric measurements on such solid solutions allow an extremely sensitive analysis of the composition of these solid solutions. Nickel(II) and iron(III) hexacyanoferrate(II) form a continuous series of solid solutions [B49]. In this series, the electroactive high-spin iron(III) ions are continuously diluted by the electroinactive nickel(II) ions, whereas the electroactive low-spin iron ions remain in place. Therefore, one observes that the peak currents of high-spin iron (III) decrease with increasing nickel content and the formal potential of the low-spin iron system shifts from the value of iron(III) hexacyanoferrate to the value of nickel hexacyanoferrate. This shift is almost linear, although, even in the ideal case, it must be slightly nonlinear because of the mixing entropy. The behavior of solid solutions of copper(II) and iron (III) hexacyanoferrate(II) is very similar [B137]. Figure 5.37 shows voltammograms of some representative examples and Fig. 5.38 depicts the dependence of the formal potential of the low-spin iron system on high-spin iron(III) content in the solid solutions. In the case of cadmium(II) and iron(III) hexacyanoferrate(II), it was observed that solid solutions are formed, too [B111]. This is rather surprising considering the large difference in ionic radii ($r_{Cd(II)} = 109$ pm, $r_{Fe(III)} = 78.5$ pm). The formation of solid solutions can be understood when both kinds of iron ions are in the +2 and +3 oxidation states, thus forming both substitutional and redox mixed solid solutions. For the case of metal hexacyanometalates with a mixed population of the carbon-coordinated sites, a careful study has been performed with solid solutions of hexacyanoferrates(III)/hexacyanocobaltates(III) [B247]. Figure 5.38 shows how sensitive the peak potentials

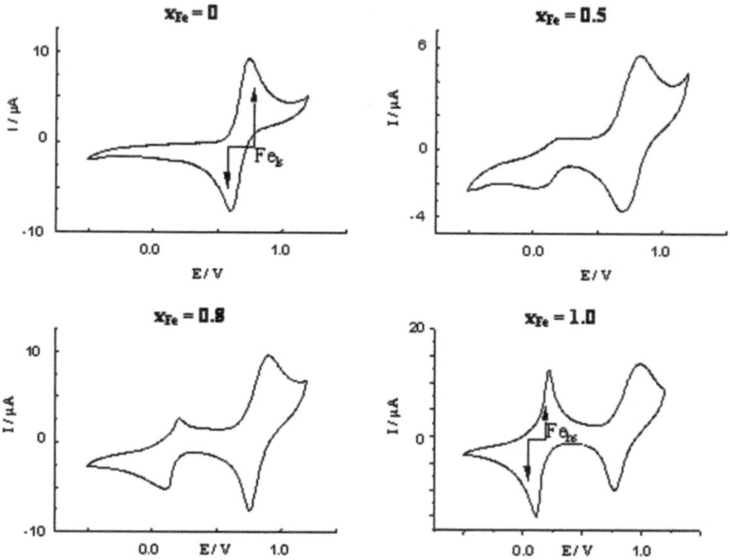

Fig. 5.37 Formal potential $E_f = (E_{p,\ anodic} + E_{p,\ cathodic})/2$ of the low-spin iron system of the mixed iron–copper hexacyanoferrates as derived from the cyclic voltammograms (scan rate 0.05 V s^{-1}) versus molar ratio of high-spin iron x_{Fe} [B137]. Copyright ©2000 with permission from Elsevier

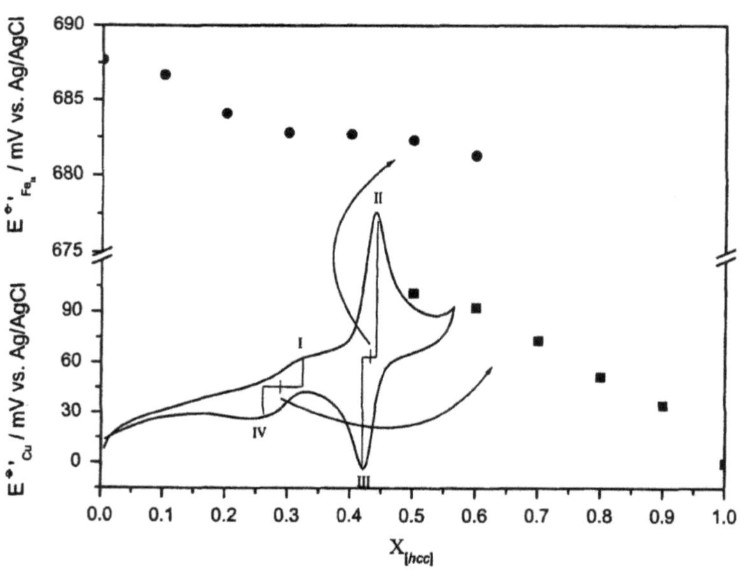

Fig. 5.38 Dependence of the formal potential of the Cu(I)/Cu(II) system (peaks I and IV) and of the hexacyanoferrate(II/III) system (peaks II and III) of $KCu^{II}[hcf]_{1-x}[hcc]_x$ on the molar ratio of cobalt (hcf, hexacyanoferrate; hcc, hexacyanocobaltate). The inset shows the cyclic voltammogram of $KCu^{II}[hcf]_{0.3}[hcc]_{0.7}$ in 0.1 mol L^{-1} KNO_3, with a scan rate of 50 mV s^{-1} [B247]. Copyright ©2002 with permission from American Chemical Society

of the $Cu^{1+/2+}$ and the $Fe^{2+/3+}$ systems of mixed copper hexacyanoferrate–hexacyanocobaltates are to the composition of the solid solutions. The iron system shows a relatively small shift because the diluting cobalt ions are rather far away from the iron ions. When the nitrogen-coordinated metal ions are continuously substituted by foreign metal ions, the effect is much stronger (see Fig. 5.37).

5.3.4 *Solid Solutions of Oxides*

The electrochemical dissolution of solid solutions of metal oxides provides access to information on their composition. Grygar et al. studied the following series of solid solutions: (a) C-Mn_2O_3 to α-Fe_2O_3, (b) $LiMn_2O_4$ to $LiFe_5O_8$, (c) $CaMnO_3$ through $Ca_3(Mn,Fe)_3O_{8+x}$ to $CaFe_2O_5$, and (d) MnO_x to FeOOH [B181], Li–Mn–O, and Li–Fe–Mn–O spinels [B191, B194, B311]. When an oxide is electrochemically dissolved by reduction of its metal ions to a lower valence state, this is a highly irreversible process, and the voltammograms are expected to strongly reflect the kinetics of the overall process (see Sect. 5.7). Therefore, it is not expected that the peak potentials of the dissolution voltammograms directly and solely depend on the thermodynamics. More interesting is the observation that the peak potentials show a clear dependence on the composition which makes them useful for analytical applications (see Fig. 5.39 for the system C-Mn_2O_3–α-Fe_2O_3). Figure 5.40 shows how the peak potential of reductive dissolution of a series of Li–Fe–Mn–O spinels shifts with the ratio Fe/(Fe + Mn).

5.4 Organometallics and Organic Compounds

Considering the importance of redox processes of organic and organometallic solids in catalysis, photovoltaics, analysis, etc., the voltammetric study of these materials is a highly interesting and rewarding task. In contrast to many inorganic solids, most organic and organometallic compounds are soluble in a suitable organic solvent, making them easily available for solution-phase electrochemical experiments. However, as it will be shown in the following chapter, the solid-state electrochemistry of these compounds can be very different than that of the dissolved species, which makes solid-state electrochemical studies indispensable. The ease of the electrode preparation and the short time scale of the voltammetric experiments (Chap. 3) favor the voltammetry of immobilized microparticles for these studies. In the following sections, an overview is given showing phenomena that are observed when microcrystalline organic and organometallic compounds are attached to an electrode surface and studied voltammetrically. This overview cannot give a complete picture of the work that has been done in the past decades, but by means of examples, it shall give an impression of how the voltammetry of

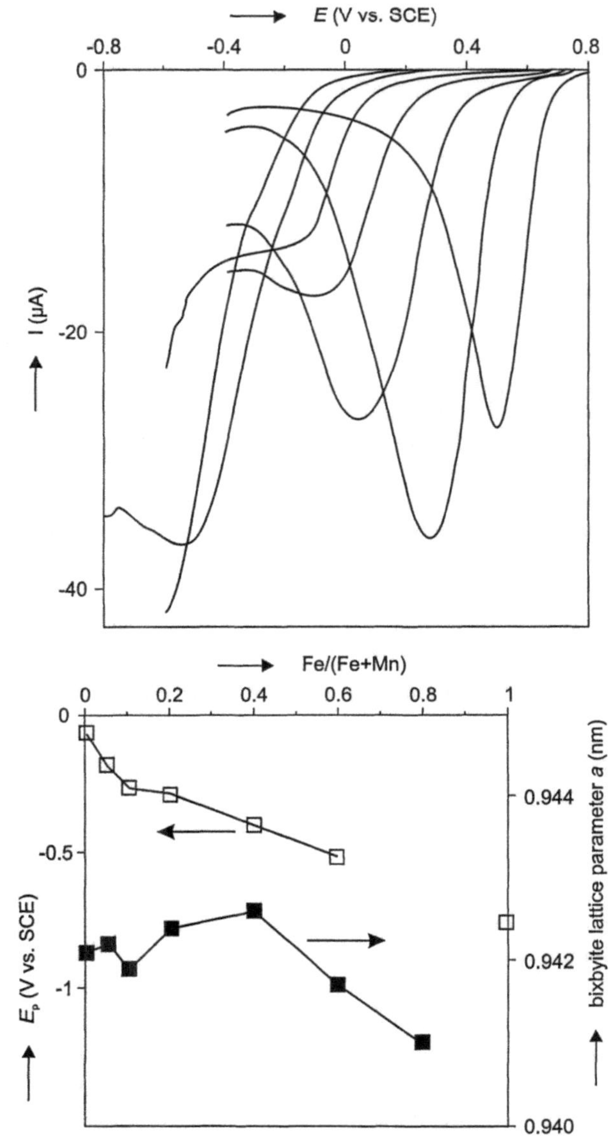

Fig. 5.39 Voltammograms (*top*) of reductive dissolution of C-Mn_2O_3–α-Fe_2O_3 solid solutions (electrolyte: acetate buffer at pH 4.4) and plot of peak potential versus molar ratio Fe/(Fe + Mn). Additionally are given the lattice parameters of the bixbyites [B181]. Copyright ©2001 with permission from publisher

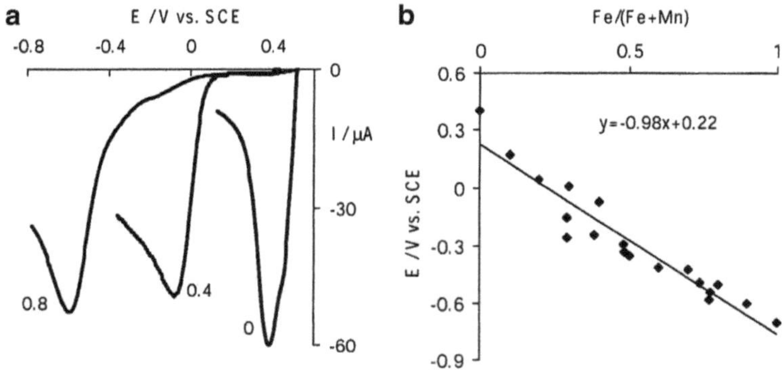

Fig. 5.40 (**a**) Linear-sweep voltammetric dissolution curves of several single-phase Li–Fe–Mn–O spinels (0.2 M acetate buffer: 1:1 acetate/acetic acid; scan rate 1 mV s^{-1}). The numbers denote the value Fe/(Fe + Mn). (**b**) Linear-sweep voltammetric peak potentials of single-phase Li–Fe–Mn–O spinels plotted versus Fe/(Fe + Mn) [B191]. Copyright ©2001 with permission from Springer

immobilized microparticles can be utilized to get access to physical, chemical, and mechanistic information on the studied compound.

5.4.1 Voltammetry of Organometallic Compounds

Based on their unique redox properties, organometallic complexes of the transition metals are of invaluable importance for catalytic and electron-mediating processes in nature and in technology. In particular, processes at solid organometallic compounds are not yet fully understood. In the following sections, we will present examples for the study of the different kinds of organometallic complexes, their redox properties, and reaction mechanisms by means of the voltammetry of immobilized microparticles.

Voltammetric studies of solid metal carbonyl compounds have been performed by Bond and coworkers [B28, B31, B53, B65, B90, B128]. After extensively investigating this class of compounds with respect to their solution electrochemical behavior (Fig. 5.41) [B77], the authors have devoted their studies to the investigation of the solid compounds. Thus, they have shown that distinct electrochemical responses can be obtained from *cis*-$Cr(CO)_2(dpe)_2$ and *trans*-$Cr(CO)_2(dpe)_2$ (dpe = $Ph_2PCH_2CH_2PPh_2$), mechanically attached to a carbon electrode, and immersed in aqueous electrolytes. The redox processes are reported to be similar to those observed in organic solutions (see Fig. 5.42), the electron transfer through the crystal being supposedly made possible by electron hopping via self-exchange and cross redox reactions with the rate being dependent on the state of the electrode–compound–solution interface and the surface charge [B28].

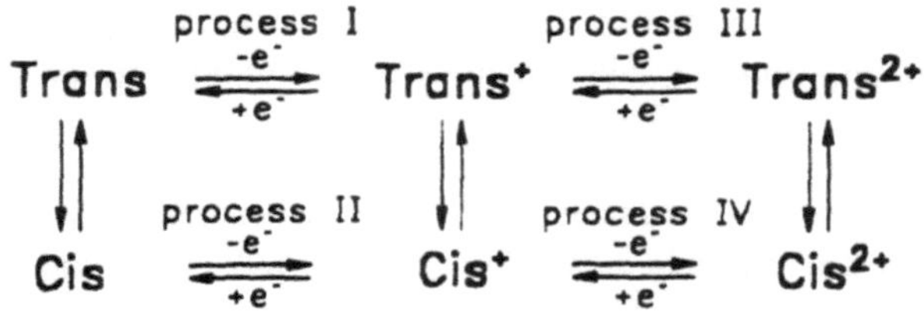

Fig. 5.41 Redox and isomerization reactions of the $[Cr(CO)_2(dpe)_2]^{0/+/2+}$ system [B31]

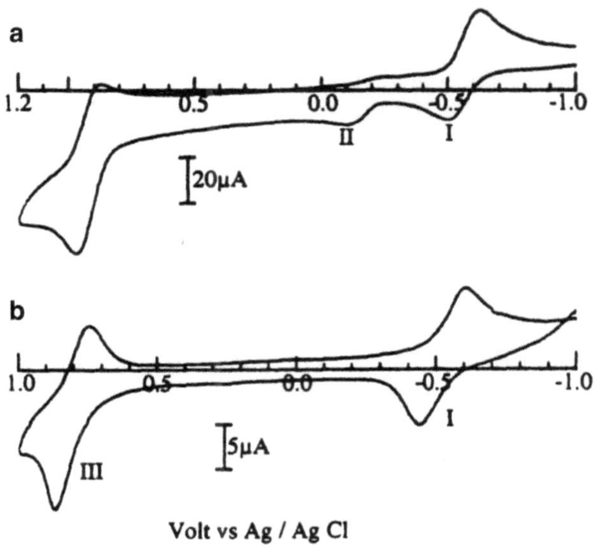

Fig. 5.42 Cyclic voltammograms obtained at 20 °C at (**a**) a platinum disk electrode in dichloromethane (0.1 M Bu_4NClO_4) for a mixture of *cis*-$Cr(CO)_2(dpe)_2$ and *trans*-$[Cr(CO)_2(dpe)_2]^+$ and (**b**) in aqueous (0.1 mol L^{-1} $NaClO_4$) media for solid *trans*-$Cr(CO)_2(dpe)_2$ mechanically attached to a polished basal plane pyrolytic graphite electrode (scan rate 50 mV s^{-1}) [B28]. Copyright ©1993 with permission from American Chemical Society

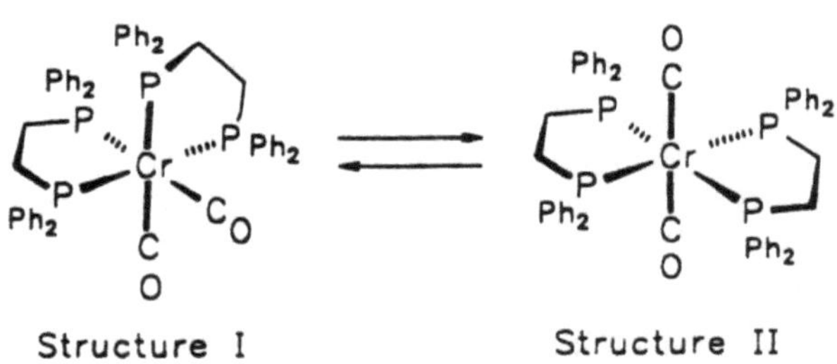

Fig. 5.43 Structures of *cis*-$Cr(CO)_2(dpe)_2$ (I) and *trans*-$Cr(CO)_2(dpe)_2$ (II) [B31]. Copyright ©1994 with permission from American Chemical Society

The solid-state isomerization of *cis*-$[Cr(CO)_2(dpe)_2]^+$ to *trans*-$[Cr(CO)_2(dpe)_2]^+$ (Fig. 5.43) has been found to be much slower than in solution, with the effect that the otherwise thermodynamically unstable *cis*-$[Cr(CO)_2(dpe)_2]^+$ could—for the first time—be spectroscopically identified (see also Sect. 5.5).

The oxidation of *trans*-$Cr(CO)_2(dpe)_2$ is accompanied by a slow uptake of anions from the aqueous electrolyte solution. By help of electrochemical quartz crystal measurements, it has been shown that the anions are transferred into the solid phase in their unhydrated form [B53]. An interesting phenomenon is that, although the expulsion of the anions during reduction is rapid, this process does not proceed through the entire microcrystal. Fewer anions are released during the reduction scan than were transferred into the solid during oxidation. The conclusion, drawn from the experiments, is that the reduction process predominately expels the anions that are situated relatively close to the solid–solution interface.

In solar energy conversion devices, separation of charge in semiconductors occurs upon excitation of an electron from a valence to a conduction band,[7] and this is translated into current flow in an external circuit. In voltammetric studies on solids, charge separation in the crystalline materials usually is generated via application of an external electrical field rather than photolytically. However, the voltammetry of microparticles can also be utilized to probe solid compounds for their photoelectrochemical properties. Thus, Eklund and Bond have examined carbonyl compounds with respect to potential photocatalyzed processes. Based on the photoactivity of *fac*-$Mn(CO)_3(\eta^2\text{-}Ph_2PCH_2PPh_2)Cl$ in organic solvents,[8] they have studied the behavior of the immobilized solid compound in an aqueous environment [B128]. Two possible reaction pathways were subject of discussion: (1) Like in the case of semiconductor electrodes, charge separation occurs within the microcrystalline environment upon radiation, resulting in a current flow and concurrent charge balance through diffusion of ions into the crystalline material. (2) The material attached to the electrode could undergo a transformation to form a material within the solid environment that can be oxidized or reduced at a different potential from that of the starting material and hence give rise to a photocatalytic reaction.

As it is depicted in Fig. 5.44a, b, the oxidation of immobilized microparticles of *fac*-$Mn(CO)_3(\eta^2\text{-}Ph_2PCH_2PPh_2)Cl$ is considerably enhanced upon radiation. Using a special photoelectrochemical cell (see Fig. 4.29), this photoprocess has been studied with the help of the electrochemical quartz crystal microbalance, EQCM. It could be demonstrated that the photocatalyzed processes at the immobilized *fac*-$Mn(CO)_3(\eta^2\text{-}Ph_2PCH_2PPh_2)Cl$ particles are accompanied by an anion uptake from the surrounding electrolyte solution (Fig. 5.45).

From the combination of voltammetric, EQCM, ex situ electron spin resonance spectroscopy, and chemical doping experiments, two possible reaction pathways have been proposed: (1) formation of a charge transfer excited state of the Mn(I) species, which is then rapidly oxidized to a *fac*-Mn(II) cationic species, that can then isomerize to the *mer*$^+$ form and (2) photoisomerization of *fac*-$Mn(CO)_3(\eta^2\text{-}Ph_2PCH_2PPh_2)Cl$ to the *mer* form, which is then directly photooxidized to the *mer*$^+$ cationic form.

---

[7] Pleskov YV (1990) Solar energy conversion, a photoelectrochemical approach. Springer, Berlin

[8] Compton RG, Barghout R, Eklund JC, Fisher AC, Bond AM, Colton R (1993) J Phys Chem 97:1661–1664

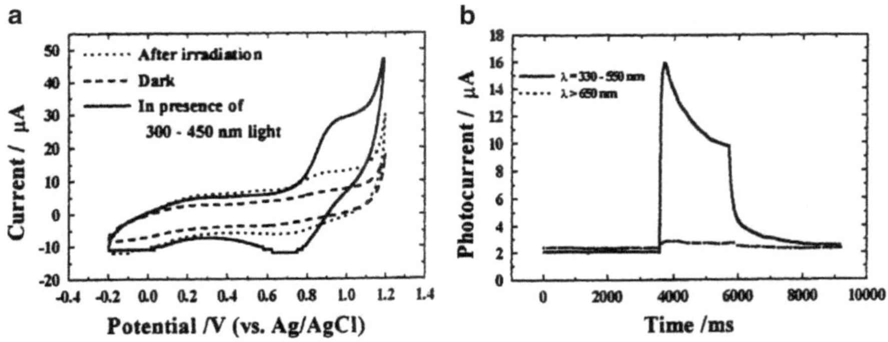

Fig. 5.44 (**a**) Cyclic voltammograms obtained for the oxidation of *fac*-$Mn(CO)_3(\eta^2\text{-}Ph_2PCH_2PPh_2)Cl$ attached to a pyrolytic graphite electrode and placed in contact with 0.1 mol L^{-1} NaCl electrolyte (scan rate 100 mV s^{-1}, irradiation intensity 10 mW cm^{-2}). (**b**) Phototransient response obtained for the oxidation of *fac*-$Mn(CO)_3(\eta^2\text{-}Ph_2PCH_2PPh_2)Cl$ attached to a graphite electrode and placed in contact with 0.1 mol L^{-1} $NaClO_4$ [B128]. Copyright ©1999 with permission from American Chemical Society

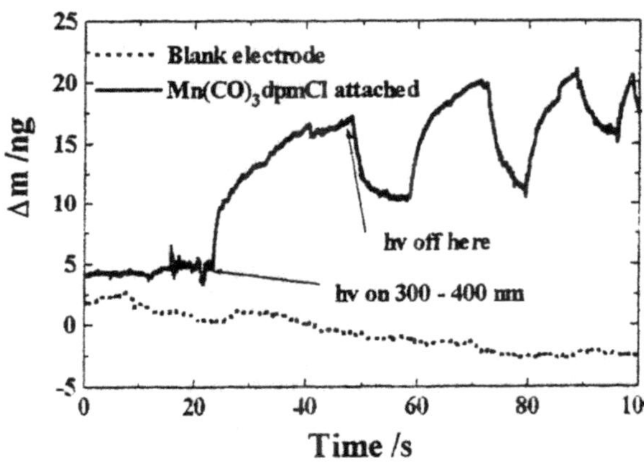

Fig. 5.45 EQCM data obtained when solid *fac*-$Mn(CO)_3(\eta^2\text{-}Ph_2PCH_2PPh_2)Cl$, attached to a gold-coated quartz crystal and placed in contact with aqueous 0.1 mol L^{-1} NaCl electrolyte, is irradiated periodically with 300–450 nm light [B128]. Copyright ©1999 with permission from American Chemical Society

Another highly interesting group of organometallic compounds to be studied by voltammetry is the class of metallocenes and their derivatives. The simplest and most studied member of this class is ferrocene (Fc). Ferrocene itself is sparingly soluble in water; its oxidation product, however, is very soluble. This, together with the well-defined voltammetry, based on the redox couple Fc/Fc^+, is an ideal basis for the study of the mechanisms of the electrochemistry of immobilized microparticles. In one of the first studies, the voltammetry of ferrocene was studied and compared to other sparingly soluble metal complexes and salts [B14]. In each case, the electrochemical perturbation generates at least one solution soluble species. For this kind of process, a reaction scheme, Fig. 5.46, has been proposed, showing how

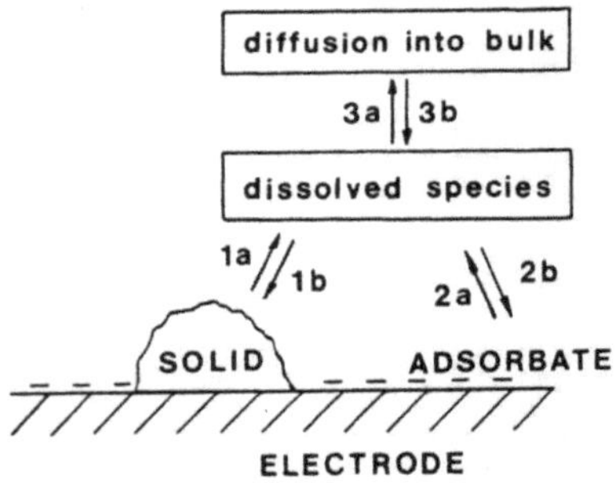

Fig. 5.46 Reaction scheme applying for the electrochemistry of ferrocene and other sparingly soluble solid compounds [B14]. Copyright ©1991 with permission from American Chemical Society

the electrochemistry is interfered by dissolution and adsorption processes of the involved redox species.

Being much more hydrophobic, decamethylferrocene (dmfc) exhibits a considerably lower solubility in water. This compound can be considered virtually insoluble, and even its oxidized form, $dmfc^+$, is insoluble in water. As for its relative, ferrocene, the voltammetry of the decamethyl compound is based on a one-electron oxidation step (Reaction VIII).

$$Fe\left(\eta^5\text{-}C_5(CH_3)_5\right)_2 \rightleftarrows \left[Fe\left(\eta^5\text{-}C_5(CH_3)_5\right)_2\right]^+ + e^- \qquad \text{(VIII)}$$

For charge compensation, the oxidation is accompanied by the uptake of anions [X^-, Reaction (VII)] from the electrolyte solution.

$$Fe\left(\eta^5\text{-}C_5(CH_3)_5\right)_{2\,\text{(solid)}} + X^-_{\text{(solution)}} \rightleftarrows \left[Fe\left(\eta^5\text{-}C_5(CH_3)_5\right)_2\right]X_{\text{(solid)}} + e^- \qquad \text{(IX)}$$

The nature of the anion not only has an effect on the shape of the voltammetric signal, it also influences the redox potential of the oxidation process. Thus, the presence of hydrophobic ions in the electrolyte solution leads to an oxidation at lower potentials, whereas the presence of hydrophilic ions, like fluoride, shifts the redox potential of Reaction (IX) toward more positive potentials. It can be assumed that the Gibbs energy of transfer of the anions from the aqueous into the organic phase is responsible for this phenomenon (see Chap. 6 for more detail). In contrast to the oxidation of ferrocene, Reaction (IX) represents a solid-to-solid transformation. As depicted in Fig. 5.47, it is a kinetically controlled process, and its rate strongly depends on temperature. The increase of the peak currents with temperature is much more pronounced than in solution-phase voltammetry.

The voltammetric study of mechanically immobilized organic or organometallic solids is of considerable importance for compounds that are insoluble in water and in nonaqueous solvents, as it is the case for many metal phthalocyanines. For example, in a series of studies, Kucernak and coworkers have thoroughly investigated the electrochemical behavior of platinum phthalocyanine [B138, B167, B185, B190]—here, the voltammetry of microcrystals offers the simplest and yet a reproducible way of depositing the target compound onto an electrode surface.

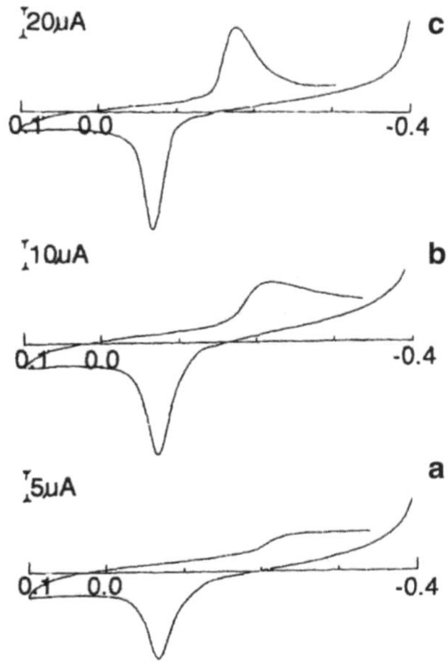

Fig. 5.47 Cyclic voltammograms obtained at a scan rate of 10 mV s^{-1} for the oxidation of microcrystalline decamethylferrocene attached mechanically to a basal plane pyrolytic graphite electrode and placed in aqueous medium (0.1 mol L^{-1} KPF_6) at temperatures of (**a**) 22 °C, (**b**) 52 °C, and (**c**) 82 °C [B29]. Copyright ©1994 with permission from Elsevier

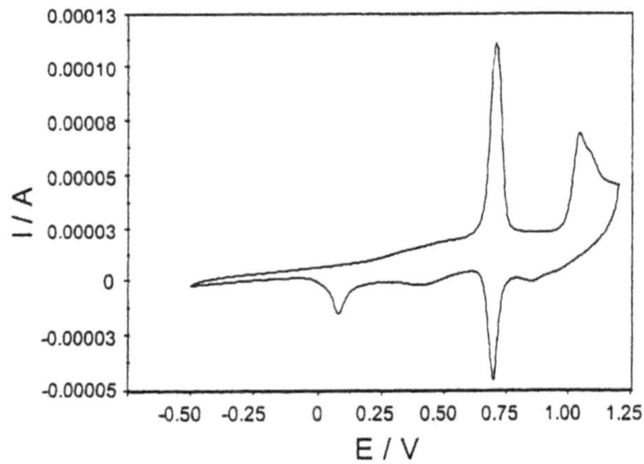

Fig. 5.48 Cyclic voltammogram at 5 mV s^{-1} for platinum phthalocyanine microcrystals deposited onto a glassy carbon electrode in acetonitrile (0.1 mol L^{-1} $N(Bu)_4ClO_4$) [B138]. Copyright ©2000 with permission from Elsevier

The electrochemical doping process of platinum phthalocyanine microcrystalline films in acetonitrile electrolyte was monitored using electrochemical impedance spectroscopy (EIS), the impedance behavior being close to that obtained for conducting polymers where, as expected, the kinetic parameters of the electrochemical doping process depended strongly on the doping potential. At low doping levels, the rate of the first electrochemical step is slow and determined by the conductivity of the microcrystalline film, but, once the film conductivity increases, the electrochemical reaction is accelerated considerably [B167]. Well-defined voltammetric responses are obtained for microcrystals in aqueous and in nonaqueous media (Fig. 5.48). The authors suppose that the oxidation, as for many metal phthalocyanine complexes, takes place reversibly on the phthalocyanine ring and is accompanied by the uptake of anions from the electrolyte solution according to Reaction (X):

$$\mathrm{PtPc}_{(\mathrm{solid})} + n\mathrm{ClO}_{4\ (\mathrm{solution})}^{-} \rightleftarrows \mathrm{PtPc}^{n+}\left(\mathrm{ClO}_{4}^{-}\right)_{n(\mathrm{solid})} + ne^{-} \qquad (\mathrm{X})$$

The voltammetry of microparticles is well suitable as a tool to the screening of compounds for desired properties like their electrocatalytic activity toward the oxidation or reduction of a target compound or for their suitability as electron mediators in biochemical sensors. An example for such a study is the comparative investigation of the electrochemical properties of osmium bipyridyl complexes [B251]. In contrast to many other solid-state electrochemical reactions, the voltammetry of immobilized osmium trisdimetoxybipyridyl, [Os(OMe-bpy)$_3$](PF$_6$)$_2$, is comparable to its solution-phase voltammetry (see Fig. 5.49). The almost ideally reversible redox-switching behavior in the solid-state favors this osmium complex as an electron mediator in biosensors. Similar observations have also been made for deposited microcrystals of other osmium bipyridyl complexes [B134, B155, B252]; in one case, even unusually fast electron and anion transport processes have been observed [B139].

The extent with which an electrochemical reaction proceeds from the triple-phase junction solid–electrode–electrolyte solution into the bulk of an organic or organometallic microcrystal depends on many factors. First, the structure of the solid has to be "electrochemically open," i.e., it should enable a comparably fast electron hopping between the molecules within the crystal and the diffusion of charge-balancing ions into the crystal. This condition is fulfilled for many

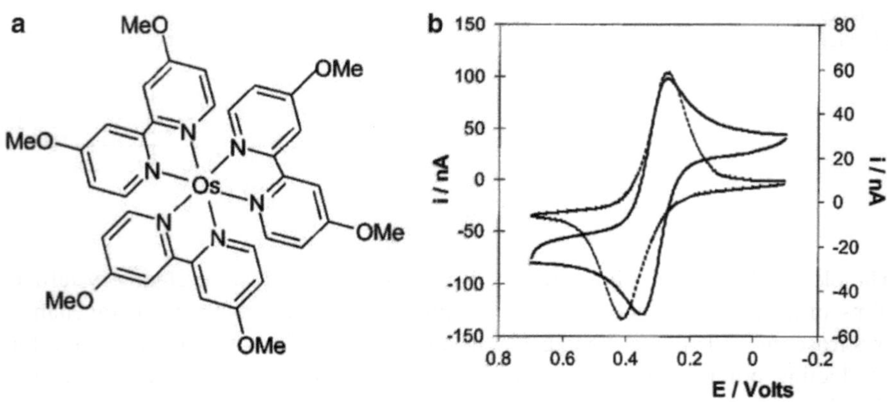

Fig. 5.49 (**a**) Structure of osmium trisdimetoxybipyridyl, [Os(OMe-bpy)$_3$](PF$_6$)$_2$ (OMe = 4,4′-dimethoxy; bpy = 2,2′-bipyridyl). (**b**) Cyclic voltammograms for a 0.8 mM solution of [Os(OMe-bpy)$_3$]$^{2+}$ dissolved in acetonitrile (*solid line*) and as a solid deposit (*dashed line*). In both cases, the supporting electrolyte is 0.1 mol L^{-1} HClO$_4$, the working electrode is a 25 μm platinum electrode, and the scan rate is 0.1 V s^{-1} [B251]. Copyright ©2002 with permission from American Chemical Society

complexes, as, for instance, for the above-described metallocenes and chromium carbonyl compounds. In these cases, a fast and exhaustive oxidation or reduction of the deposited solid can be achieved. In other cases, however, as in the case of highly charged complexes like $[Co(mtas)_2]^{n+}$ (mtas = bis(2-(dimethylarsino)phenyl)-methylarsine), the voltammetric reaction remains confined to the surface of the microcrystal. Here, the high charge of the complex ($n = 2–3$) leads to ion pair formation with the anions that are transferred into the crystal upon oxidation and thus to an increasing resistivity toward the ion migration within the crystal [B45].

As a peculiar phenomenon, it has been reported that the voltammetric conversion of immobilized organometallic compounds can lead to a reversible extraction of metals from their complexes. Thus, Marken et al. reported on electrochemically driven reversible solid-state metal exchange processes in polynuclear copper complexes [B260]. These complexes (see the example of the tetranuclear compound, Fig. 5.50), consisting of a "Robinson-type" ligand scaffold, can accommodate up to four copper ions that are bridged to each other, as shown in Fig. 5.50. Upon reduction of the immobilized complexes, the copper ions stepwise leave the complex and are reduced at the electrode surface to form a copper deposit. In the first step, two copper ions are extracted from the complex, leaving the di-copper complex behind. In the second step, the water-insoluble ligand is formed, which remains at the electrode surface (see the schematic drawing in Fig. 5.51). Upon oxidation, the formed free copper ions are reincorporated into the ligand scaffold. Roznyatovskaya et al. reported that the demetallation of dinuclear copper complex with Robinson-type macrocyclic ligand in contact with acetate buffer was quasi-reversible despite the irreversible voltammetric pattern [B446].

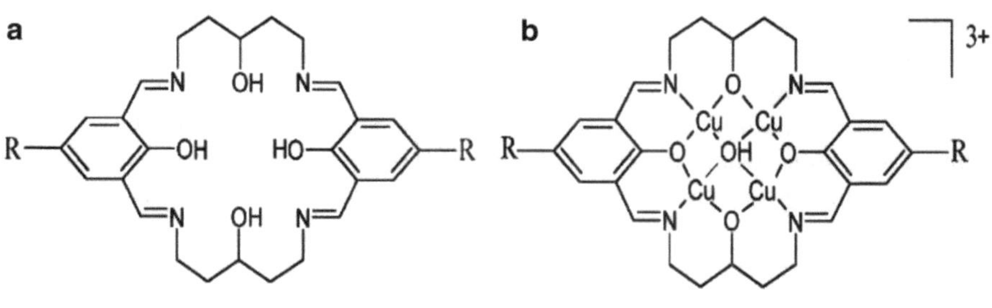

Fig. 5.50 Structures of (**a**) the expanded "Robinson-type" macrocyclic ligand H_4L, (**b**) the tetranuclear complex $[Cu_4(L)(OH)]^{3+}$ [B260]. Copyright ©2003 with permission from Springer

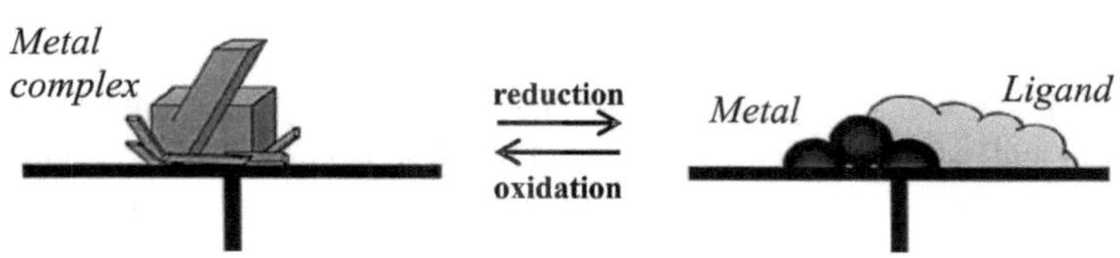

Fig. 5.51 Schematic representation of the reversible solid-state conversion of metal complex to metal deposit and free ligand [B260]. Copyright ©2003 with permission from Springer

Bond et al. [B294, B306, B308] have demonstrated that the response of microparticles of several organometallic compounds using ionic liquids as electrolyte is indistinguishable from that of the dissolved species. A rapid dissolution of the solids upon electrochemical oxidation of the neutral compounds is held responsible for this observation.

Further, it has been found that small concentrations of organic solvents in the electrolyte solution can induce a catalytic action in the insertion electrochemical reactions of microparticles [B295], causing a co-insertion and lattice widening.

The occurrence of electrochemically driven ion-insertion processes can be illustrated by scanning electron microscopy (SEM) images of a microparticulate deposits of heterometallic Au(I)-Cu(I) $[\{Au_3Cu_2(C_2\ R)_6\}Au_3(PPh_2C_6H_4PPh_2)_3](PF_6)_2$ alkynyl cluster complexes [B360, B361]. In contact with aqueous electrolytes, such complexes display an essentially reversible solid-state oxidation process at potentials around +0.5 V which can be described in terms of ferrocenyl oxidation coupled with the ingress of anions from the electrolyte into the solid lattice. The examination by SEM of microparticulate deposits on graphite electrodes before and after application of a constant potential sufficiently positive to promote the aforementioned oxidation process revealed that no significant morphological changes occurred in the crystals, except for the appearance of holes and apparent partial decapping, as can be seen in Fig. 5.52.

Quantitative scanning electron microscopy–energy dispersive X-ray analysis (SEM/EDX), shown in Fig. 5.53, confirmed the insertion of halide ions. The EDX spectra of original crystals show intense peaks for Fe, Au, Cu, P, O, and F, in agreement with their elemental composition. After electrolysis in 0.10 mol L^{-1} NaCl at +0.75 V, such peaks retain their relative heights but accompanied by the signal for Cl, clearly suggesting that anion insertion effectively occurred.

Recent studies on metal complexes [B362] and metal phthalocyanines [B363, B364, B365] have been reported. In this context, the electrochemistry of biometallic compounds has received increasing attention. This is the case of hemoglobin films, where the Fe(III)-heme/Fe(II)-heme couple provides a characteristic voltammetric signature in contact with aqueous electrolytes. Such signature has been used to detect the presence of proteinaceous binders in Taiwanese polychromed architectural decoration and studying its aging for dating purposes [B350].

A recent application of the voltammetry of immobilized microparticles in the biomedical area is the screening of molecules with potential antimalarial activity through heme-binding mechanism [B366]. This activity is patterned as a result of peak splitting appearing for Fe(III)-heme aqueous solutions in contact with microparticulate deposits of the potentially active antimalarials on graphite electrodes, as illustrated in Fig. 5.54. Remarkably, this assay does not require prior incubation and can be applied to different Fe(III) substrates, namely, hemoglobin, whole blood, and erythrocytes, as well as hemolyzed erythrocytes and blood samples [B366].

In this context, the solid-state electrochemistry of highly ion-permeable solids formed by metal ions packed with organic ligands, the so-called metal-organic

Fig. 5.52 Scanning electron microscopy (SEM) images of a microparticulate deposit of $[\{Au_3Cu_2(C_2C_6H_4Fc)_6\}Au_3(PPh_2C_6H_4PPh_2)_3](PF_6)_2$ on graphite (**a**) before and (**b**) after application of a constant potential step at +0.75 V in 0.10 mol L^{-1} NaCl during 15 min [B361]. Copyright ©2010 with permission from Springer

frameworks (MOFs) or coordination polymers (see footnote 3), has also been studied. These coordination polymers have received considerable attention because of their high porosity, large surface area, structural variety, and considerable permeability to guest molecules, with applications in hydrogen storage, catalysis, and sensing.[9,10 11,12] Resulting from the association of transition metal ions to polyfunctional organic ligands, the MOFs can even be prepared electrochemically.[13] However, relatively few reports on their electrochemical characterization

---

[9] Tomic EA (1965) Thermal stability of coordination polymers. J Appl Polym Sci 9:3745–3752

[10] Kitagawa S, Kitaura R, Noro S (2004) Functional porous coordination polymers. Angew Chem Int Ed 43:2334–2375

[11] Rowsell LC, Yaghi OM (2004) Micropor Mesopor Mat 78:3–14

[12] Janiak C (2003) Dalton Trans 2781–2804

[13] Mueller U, Schubert M, Teich F, Puetter H, Schierle-Arndt K, Pastré J (2006) J Mater Chem 16:626–636

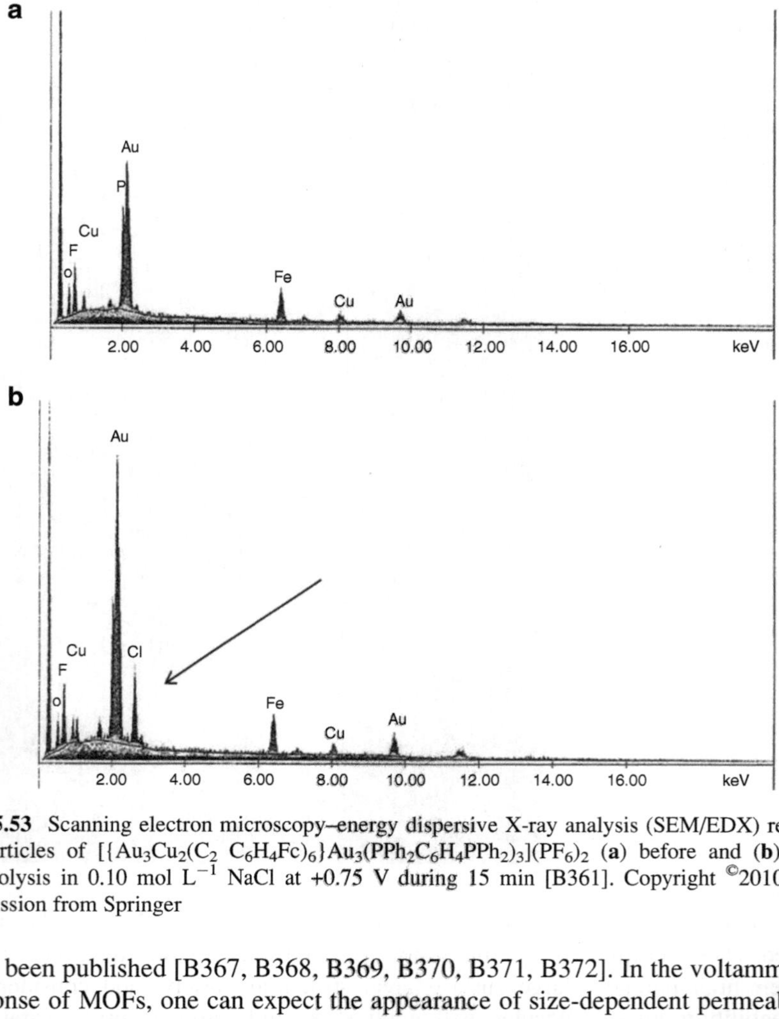

Fig. 5.53 Scanning electron microscopy–energy dispersive X-ray analysis (SEM/EDX) records of particles of $[\{Au_3Cu_2(C_2\ C_6H_4Fc)_6\}Au_3(PPh_2C_6H_4PPh_2)_3](PF_6)_2$ (**a**) before and (**b**) after electrolysis in 0.10 mol L^{-1} NaCl at +0.75 V during 15 min [B361]. Copyright ©2010 with permission from Springer

have been published [B367, B368, B369, B370, B371, B372]. In the voltammetric response of MOFs, one can expect the appearance of size-dependent permeability to electrolyte ions [B367, B368]. In the case of iron phytate deposits on tin-doped indium oxide (ITO) electrodes, well-defined voltammetric responses consistent with an immobilized Fe(III/II) redox system in aqueous and ethanolic electrolyte solutions have been observed by Marken et al.[B370]. The Fe(III/II) redox system is reversible and cation insertion/expulsion occurs fast on the time scale of voltammetric experiments even for more bulky NBu_4^+ cations and in ethanolic solution, but it appears that structural changes accompany the redox process and limit propagation [B370]. In the cases of Cu(II) and Zn(II)-terephtalates, where size-dependent effects on ionic permeability have been detected, the reductive electrochemistry is complicated by the possibility of yielding metal deposits [B367, B368].

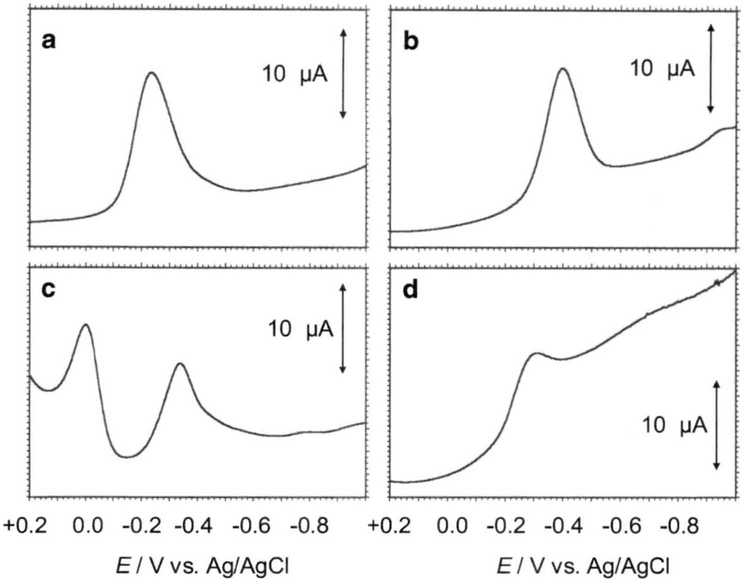

Fig. 5.54 Square-wave voltammograms for (**a**) Fe(III)-heme film on glassy carbon electrode immersed into 0.25 mol L^{-1} aqueous phosphate buffer (pH 7.4), (**b**) unmodified electrode immersed into 3 μmol L^{-1} solution of Fe(III)-heme in 0.25 mol L^{-1} aqueous phosphate buffer (pH 7.4), (**c**) 4-aminophenol-modified GCEs, (**d**) hydroquinone-modified GCEs in contact with the above solution. Potential scan initiated at +0.25 V in the negative direction. Potential step increment 4 mV, square-wave amplitude 25 mV, frequency 5 Hz [B366]. Copyright ©2013 with permission from American Chemical Society

5.4.2 *Voltammetry of Organic Compounds*

The voltammetry of immobilized microparticles can be exploited for very different purposes, be it to address mechanistic questions of electrochemical or chemical reactions, to access physical or chemical properties of a target compound, to screen compounds for a desired electrochemical property or behavior (catalysis), or to yield qualitative or quantitative information on a compound. The latter, the qualitative and quantitative determination of organic compounds is nowadays generally a domain for chromatographic and spectroscopic techniques. However, voltammetry, and especially the voltammetry of immobilized microparticles, can be highly useful as a tool for a quick and cheap analysis, without the need of any sample extraction steps. Thus, for example, Komorsky-Lovrić et al. have used the voltammetry of immobilized microparticles for the detection of trace amounts of cocaine powder on contaminated paper money/bills (see Fig. 5.55). Utilizing square-wave voltammetry, the authors determined the detection limit for cocaine to be 0.3 μg cm^{-2} surface concentration.

Although due to the limitations caused by the mechanical immobilization of a sample onto an electrode surface the analytical information on a target compound is

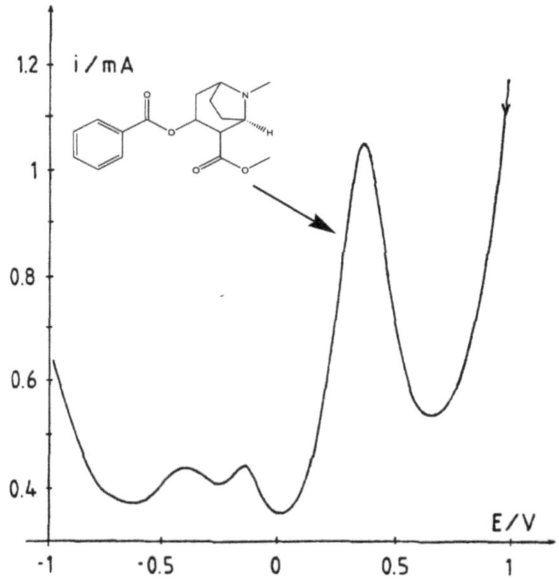

Fig. 5.55 Square-wave voltammogram of traces of cocaine powder which was mechanically transferred to a paraffin-impregnated graphite electrode from the surface of contaminated paper money (bills). The frequency was 250 Hz, amplitude 100 mV, step increment 2 mV, and starting potential 1 V [B109]. Copyright ©1999 with permission from Wiley-VCH

primarily of qualitative nature, it has been shown for the screening of pesticides on the surface of cucumber and lettuce that by following identical electrode preparation procedures, also semiquantitative information can be obtained [B59]. This fast screening provides information on the presence or absence of a certain pesticide and can thus save labor and costs for unnecessary analysis by, e.g., chromatography.

So far, more emphasis than on the analytical use has been placed on the application of the voltammetry of microparticles for studying the redox properties of organic solids and the mechanisms of their electrochemical reactions. These reactions can proceed under dissolution if the reaction product is soluble or as a solid-to-solid transformation if the product is insoluble in the electrolyte environment. Depending on the properties of the organic microcrystals, the latter reaction type can either be purely surface confined or even confined to the three-phase junction, or it can lead to an exhaustive conversion of the microcrystals. For a given compound, the extent of the reaction depends on a number of properties. First of all, solid-to-solid conversions generally require a charge compensating transfer of ions between the solid and the adjacent electrolyte solution to take place. In order to allow ions to migrate through the molecular solid, the structure of the solid must be "open," i.e., it must provide space to accommodate ions without binding them too strongly (as ion pairs). An example is the reduction of 7,7,8,8-tetracyanoquinodimethane, TCNQ, the reduction of which is accompanied by the uptake of cations from the surrounding electrolyte solution, into its layered structure, Fig. 5.56, following Reaction (XI):

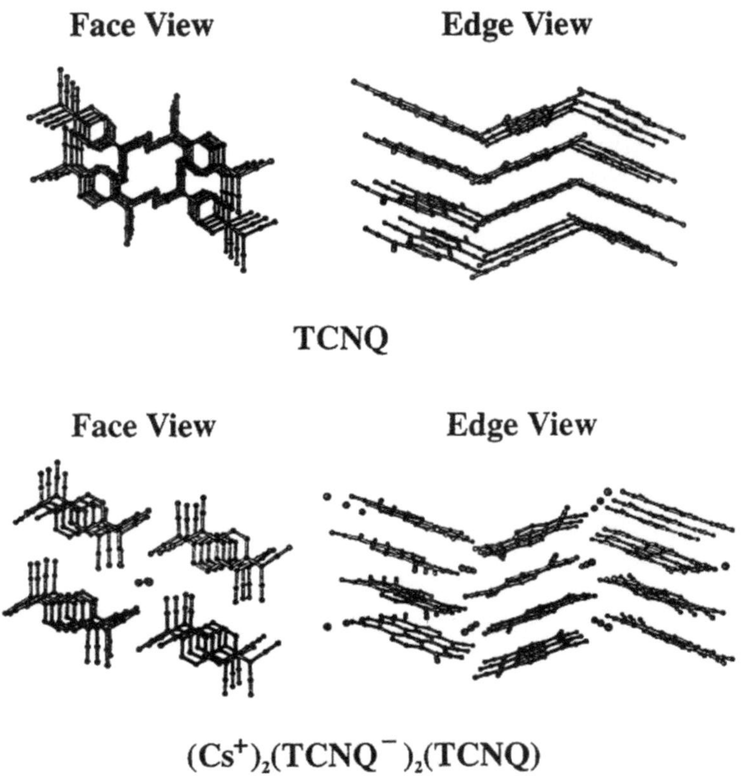

Fig. 5.56 Perspective view of the skeletal structure of TCNQ (*upper structure*) and $(Cs^+)_2$ $(TCNQ^-)_2$ (TCNQ) (*below*). Hydrogen atoms are omitted for clarity [B101]. Copyright ©1998 with permission from Royal Society of Chemistry

$$x\mathrm{TCNQ}_{(\mathrm{solid})} + ye^- + y\mathrm{M}^+_{(\mathrm{aq})} \rightleftarrows (\mathrm{M}^+)_y(\mathrm{TCNQ}^-)_y(\mathrm{TCNQ})_{(x-y)(\mathrm{solid})} \qquad \text{(XI)}$$

Of course, the electron transfer within the crystal, in most cases via electron hopping, must be assured, too.

Very often, the structure of the end product of an electrochemically induced solid-to-solid transformation is strongly deviating from that of the starting material, as shown in Fig. 5.56 for the example of TCNQ and its reduction product [B101], and the structural changes depend on the nature of the intercalating cation. The growing divergence between the oxidized and the reduced form leads to an immiscibility between starting and end product. The effect on the voltammetric response can be seen in Fig. 5.57. Clearly visible is the unusually large separation between the reduction peak and the oxidation peak. This extremely high peak-to-peak separation is caused by a large overpotential of reduction and oxidation, which the authors interpret in terms of a nucleation and growth mechanism (for the thermodynamic interpretation, see Sect. 5.11), arising from the fact that the

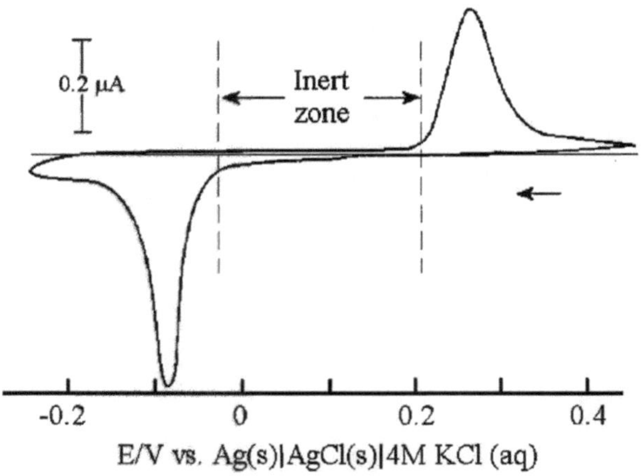

Fig. 5.57 Schematic diagram of the voltammetric response for the surface-attached $TCNQ^{0/-}$ system when interconversion of TCNQ and the reduced salt occurs by a nucleation–growth mechanism [B56]. Copyright ©1996 with permission from Royal Society of Chemistry

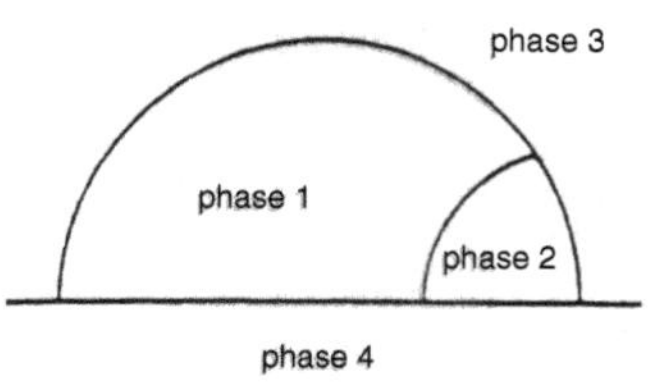

Fig. 5.58 Schematic representation of nucleation in the four-phase system electrode (phase 4)/solids (phase 1 and 2)/solution (phase 3) [B56]. Copyright ©1996 with permission from Royal Society of Chemistry

electrochemical solid-to-solid conversion does not take place via the formation of a continuous redox solid solution phase, but requires the formation of a separate product phase (as schematically drawn in Fig. 5.58) [B56, B296]. The formation of this phase, from the kinetic point of view, obeys the laws of nucleation and growth.

The described effect of the immiscibility of the corresponding redox pairs does not seem to be an exotic behavior. It has been described for other systems too, as, for instance, for the oxidation of microcrystalline tetrathiafulvalene, TTF [B54, B255], which takes place coupled with the intercalation of anions into the TTF lattice.

Evidence for nucleation and growth has also been found for the voltammetric reduction and reoxidation of the Buckminster fullerene C_{60}. The large peak-to-peak separation (Fig. 5.59a) and the typical chronoamperometric curves, showing a "rising" current transient rather than a monotonic decay (Fig. 5.59b), are characteristic and support the proposed reaction mechanism [B119].

As shown in Fig. 5.49, fullerenes exhibit a well-defined voltammetry. Depending on the applied reduction potential, they undergo a number of reduction steps, as shown in Fig. 5.60 for the example of the $(C_{60})_{1.5}(CTV)$ complex (CTV = cyclotriveratrylene). It can be derived from the voltammetric curves e–a (Fig. 5.60) that the products of the first two reduction steps are still insoluble in the

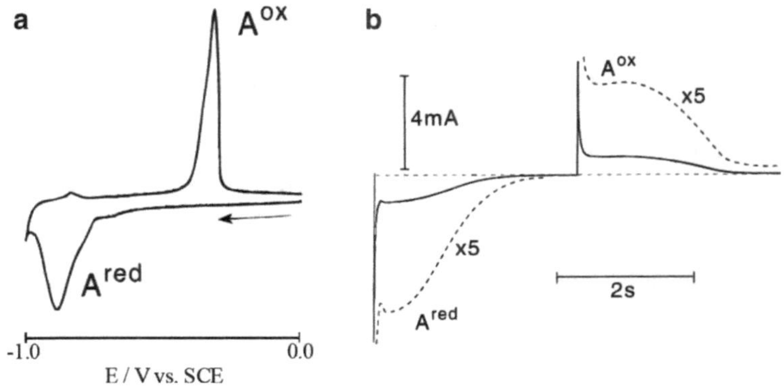

Fig. 5.59 (**a**) Cyclic voltammograms obtained for the reduction and reoxidation of solid C_{60} mechanically attached to a basal plane pyrolytic graphite electrode immersed in acetonitrile/ 0.1 mol L^{-1} NBu_4PF_6. scan rate 0.02 V s^{-1}, $J = 20 \pm 2$ °C. (**b**) Chronoamperogram for the reduction and reoxidation of solid C_{60} mechanically attached to a basal plane pyrolytic graphite electrode and immersed in acetonitrile/0.1 mol L^{-1} NBu_4PF_6. Potential steps are from 0 to −0.85 V for reduction and −0.85 V to −0.205 V versus SCE for reoxidation after ten initial potential cycles between 0 and −1.0 V versus SCE [B119]. Copyright ©1999 with permission from American Chemical Society

Fig. 5.60 Cyclic voltammograms obtained over different potential ranges at a scan rate of 0.100 V s^{-1} for reduction of microcrystalline $(C_{60})_{1.5}$(CTV) complex adhered to a 0.10-cm diameter glassy carbon electrode placed in CH_3CN (0.10 mol L^{-1} Bu_4NClO_4). (**a**) 0.0 to −3.0 V, (**b**) 0 to −2.35 V, (**c**) 0 to −2 V, (**d**) 0 to −1.55 V, and (**e**) 0 V to −1.2 V versus Ag/Ag$^+$ [B170]. Copyright ©2001 with permission from American Chemical Society

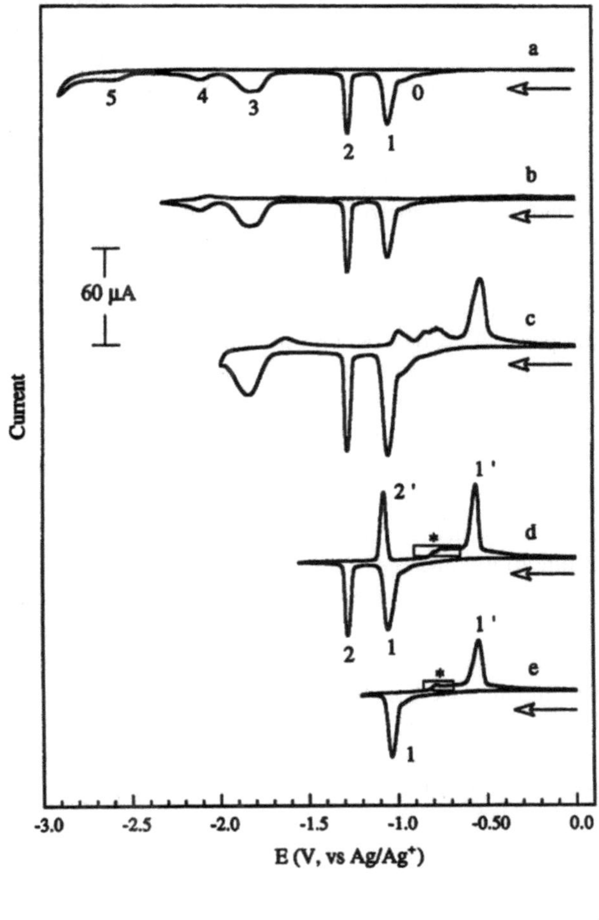

electrolyte solution; the reoxidation of the compound is thus possible. However, potential sweeps toward more negative potentials lead to the formation of soluble reaction products, visible in the irreversibility of the subsequent reduction.

Organic chemistry, and thus also organic electrochemistry, can generally be much more complex than in the above examples. Thus, the electrochemical oxidation or reduction of an organic compound can be followed by a chemical reaction of the reaction products. These EC reactions are a research field on their own, and as demonstrated for the example of electrochemically driven polymerization reactions at the mechanically immobilized microparticles [B211, B218, B272], the voltammetry of microparticles is most suited to interrogate them.

The surface identification of carbon particles with organic compounds has been shown to provide interesting new materials for the solution analysis of sulfide ions and pH [B305, B309, B310] when these particles are immobilized on electrode surfaces. Similar results can be achieved by co-immobilization of carbon nanotubes and pH-sensing compounds like 9,10-phenanthraquinone [B315].

The voltammetry of microparticles of organic compounds is also of application in pharmacology. Komorsky-Lovrić et al. have studied the electrochemistry of different pharmacologically active compounds including cocaine [B109], local anesthetics and antithusics [B373], 5-aminosalicylic acid, ciprofloxacin, and azithromycin [B374], and others [B375]. Applications include the screening of pesticides [B59] and different families of adulterant chemicals illegally added to commercial herbal phytotherapeutic formulations, so that anorexics, benzodiazepinic anxiolytics, antidepressants, diuretics, and hypoglycemics can be discerned electrochemically [B323]. In all cases, characteristic voltammetric signals were recorded on solid micro- or nanosamples attached to graphite electrodes immersed into aqueous electrolytes. Komorsky-Lovrić and Novak have applied the voltammetry of microparticles to estimate of antioxidative properties of vegetables [B376, B377]. Other studies deal with the properties of pharmacologically active substances [B378] and the interaction of dsDNa with 3,4-dibenzyl-2,5-dimethylthiophene-S-oxide using a glassy carbon electrode modified with a thick thiophene-*S*-oxide + dsDNA mixture [B379] and screening of different DNA forms [B426]. Bioelectroanalysis of pharmaceutical compounds has been recently reviewed [B380, B381, B382].

Figure 5.61 compares the square-wave voltammetric responses for lorazepam (anxiolytic) and furosemide (diuretic) in contact with aqueous acetate buffer. Well-defined, drug-characteristic voltammetric signals are recorded that, interestingly, are retained in the voltammograms recorded for mixtures of those compounds. This is due to the microparticulate nature of the sample, where separate grains of each one of the drugs exist. As a result, the observed voltammetry corresponds to the sum of the signals recorded for each one of the compounds separately, a feature that, in general, does not occur in solution electrochemistry, where cross-reactions can accompany electrochemical runs.

Electrochemical measurements can be applied to monitor the reactivity of intermediates involved in biochemical processes and can provide insight into pharmacological activity of compounds. Generation [B383, B384, B385] of free

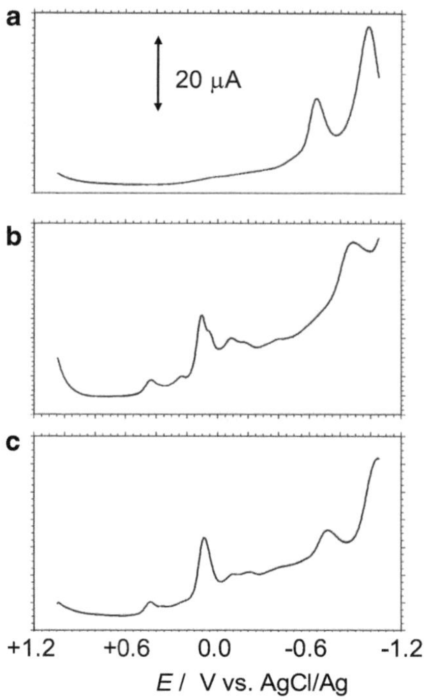

Fig. 5.61 Square-wave voltammograms for PIGEs modified with (**a**) lorazepam, (**b**) furosemide, and (**c**) a 50 % (w/w) of lorazepam plus furosemide, immersed into 0.25 M aqueous acetate buffer. Potential scan initiated at +1.05 V in the negative direction. Potential step increment 4 mV, square-wave amplitude 25 mV, frequency 5 Hz [B323]. Copyright ©2013 with permission from Elsevier

hydroxyl radicals at electrode surfaces and quantification [B386] of the hydroxyl radical scavenging activity of medicinal plant extracts are illustrative examples. Similarly, the voltammetry of immobilized particles scene can be extended to the study of films of proteins [B387], lymphocytes [B388], enzymes [B389], and mitochondria [B390].

5.4.3 Voltammetry of Organic Pigments and Hybrid Materials

The voltammetry of microparticles has been applied for the identification of organic pigments in pictorial samples. These include several families of natural colorants, namely, anthraquinonic and naphtoquinonic [B265], flavonoid [B302, B321, B322, B391, B392], curcumoid [B393], and indigoid [B72, B126, B394, B395] dyes. Identification of dye and dye mixtures in ancient textile samples has been reported [B391]. Dye screening is complicated, however, by the variety of compounds with similar chemical composition and, therefore, similar electrochemistry, so that electroanalytical procedures involving sequential use of different electrolytes [B321] and application of reductive/oxidative potential steps [B321, B322] can be used. Embedding dyes into carbon paste electrodes has also been reported

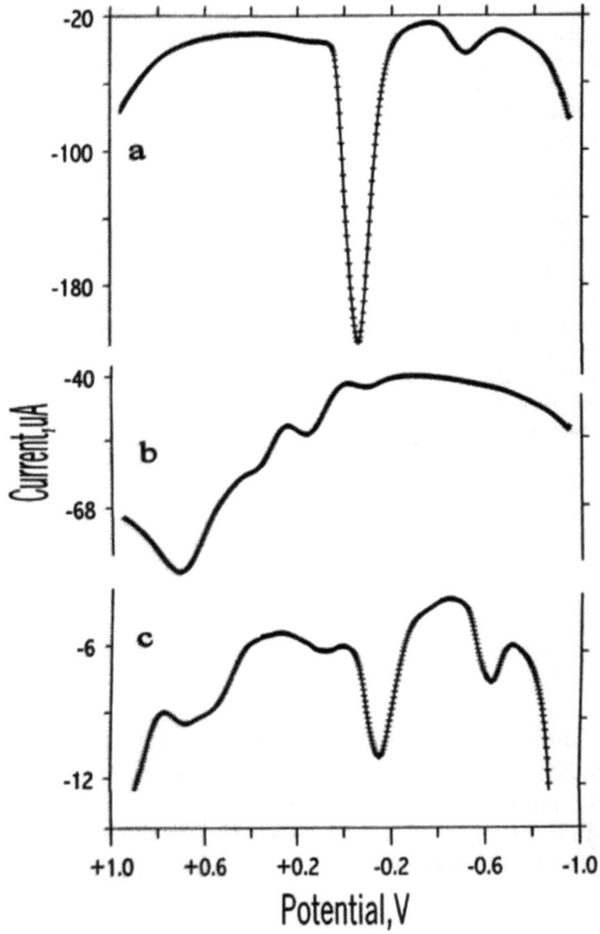

Fig. 5.62 Square-wave voltammograms of (**a**) carmine, (**b**) Brazilwood, and (**c**) sample from a cope found in the Church of Castellfort (Valencia Region, Spain) attributed to the fifteenth-century production of textile workshops in Morella immersed into 0.25 mol L^{-1} HAc + 0.25 mol L^{-1} NaAc aqueous solution. Potential scan initiated at −0.85 V in the positive direction. Potential step increment 4 mV square-wave amplitude 25 mV, frequency 15 Hz [B391]. Copyright ©2005 with permission from Elsevier

[B396, B397]. In general, the electrochemistry of dye particles was studied in contact with aqueous electrolytes, so that reduction/oxidation processes can be described in terms of the entrance/issue of protons into/from the organic solid from/to the electrolyte, as confirmed by chronoamperometric [B319] and coupled voltammetric/AFM data [B320]. Figure 5.62 shows an example of application of the voltammetry of microparticles for identifying the dyestuffs used in a medieval brocade piece found in the Castellfort church (Valencia Region, Spain). Clearly, the voltammogram of the textile sample in contact with aqueous acetate buffer resembles that resulting from the superposition of carmine and Brazilwood [B391].

An interesting group of pigmenting materials is formed upon attachment of organic dyes to inorganic hosts. Such systems, generically termed as organic–inorganic hybrid materials, stabilize the organic guest molecules upon anchorage to silicates and aluminosilicates [B398, B399]. The paradigmatic example is the

Maya blue, a historical pigment used by the Mayas and other Mesoamerican peoples in pre-Columbian times, characterized by its enormous chemical stability and peculiar brightness and variable blue-greenish hue [B400]. The absence of historical sources describing the recipe for the preparation of the pigment motivated controversies about its fabrication and its chemical composition. Although today it is recognized that the Maya blue results from the attachment of indigo to palygorskite, a phyllosilicate clay, there is no agreement with regard to the location of indigo molecules on the clay framework and the nature of that attachment and the origin of this peculiar hue and chromatic variability, in principle attributed to the bathochromic shift of indigo absorption bands due to the attachment of the dye to the clay or to light dispersion by often accompanying iron metal and iron oxide nanoparticles, as recently reviewed [B400]. The simplest way to prepare the pigment is the crushing of indigo and palygorskite mixtures (ca. 1 % wt of indigo) and heating it at temperatures between 100 and 180 °C, so that zeolitic water molecules are partially removed from the clay.

The voltammetry of the Maya blue in contact with aqueous electrolytes is close to that of indigo, with distinctive signals corresponding to the indigo oxidation to dehydroindigo and the reduction of indigo to leucoindigo (see Fig. 5.58) [B401, B402, B403]. The analysis of the voltammetric signals suggested to Doménech-Carbó et al. that a substantial proportion of the dye was actually in its oxidized form, dehydroindigo [B401]. This idea, supported by multispectral [B401, B402], nuclear magnetic resonance [B403], and chromatographic/mass spectrometry [B432, B433] data, provides a straightforward mechanism for justifying the peculiar chromatic properties of the Maya blue: the dehydroindigo (yellow)/indigo (blue) ratio determines the more or less greenish hue of the resulting material. Thermochemical parameters for the association of these dyes to the clay support, calculated from solid-state voltammetric data, were in agreement with the above idea and suggested that the ancient Mayas could modulate the color of the pigment by varying the temperature of the thermal treatment, a possibility consistent with voltammetric data for genuine samples of Maya blue from different archaeological sites [B404, B405]. These studies lead to the possibility of defining an evolutionary scheme for the Maya blue, whose method of preparation could surely experience local and time variations during the Maya epoch [B401, B402, B403, B404, B405]. The most recent electrochemical studies indicate that similar "Maya yellow" and "Maya green" [B406] and green plasters [B407] were used by this ancient people. Figure 5.63 compares the square-wave voltammograms for a commercial indigo paste and a genuine Maya blue sample from Chichén Itza, both showing the characteristic indigo peaks in contact with aqueous acetate buffer. In this figure, the voltammogram for a synthetic wet mixture of indoxyl (1 % w/w) plus palygorskite wet mixture is also shown. Indoxyl is one of the precursors of indigo in plants [B394] and in contact with palygorskite yields a system with a high proportion of dehydroindigo and other secondary dyes. This last voltammetric response is close to that recorded for glassy carbon electrodes modified with the extract of a green pellet from La Blanca archaeological site (Late classic period), after extractive

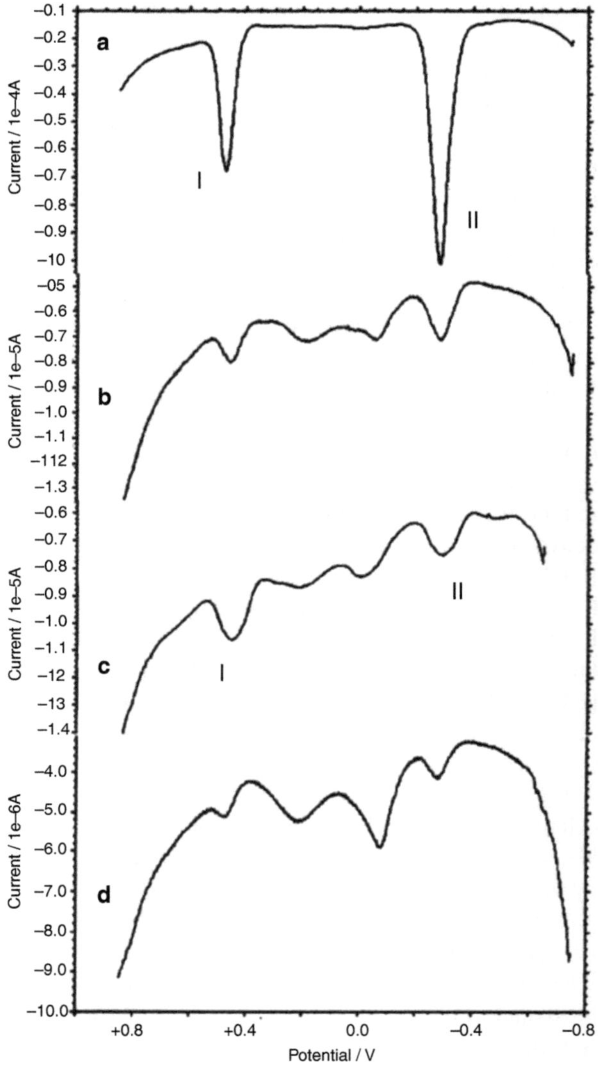

Fig. 5.63 Square-wave voltammograms of (**a**) indigo microparticles and samples from (**b**) Calakmul, substructure II-C, Late Preclassical period; (**c**) Dzibilnocac-2, Late Classical period; and (**d**) Mayapán-2, Early Postclassical period. Electrolyte 0.50 mol L^{-1} acetate buffer, pH 4.85. Potential scan initiated at −750 mV in the positive direction. Potential step increment 4 mV, square-wave amplitude 25 mV, frequency 5 Hz [B401]. Copyright ©2006 with permission from American Chemical Society

contact with DMSO overnight, where dehydroindigo is identified as the main organic component (Fig. 5.64).

Analysis of peak currents of voltammetric curves permits an in-depth analysis of the distribution of the dye molecules and its oxidation state [B399, B403, B408, B409] in the palygorskite grains, as well as a study of the kinetics of the solid-state reaction [B410], using a methodology previously developed for analyzing the kinetics of the tetragonal to monoclinic shift of zirconia [B411]. The central idea is that for diffusion-controlled electrochemical processes, the voltammetric response is representative of the composition of the region reached during the advance of the diffusion layer. Roughly, this can be approached by the time τ characteristic of the experiment (given by the magnitude of the potential scan rate in cyclic voltammetry or the frequency in square-wave voltammetry) and the diffusion coefficient (D) attributable to the species whose motion is rate-determining. The apparent depth of the diffusion region, $x_{\text{difus}}^{\text{app}}$, can be estimated as $x_{\text{difus}}^{\text{app}} = (2D\tau)^{1/2}$, so that the variation of the voltammetric parameters with the potential scan rate can be converted into a variation of such parameters with the depth, provided that the diffusion coefficient is known, for instance, from

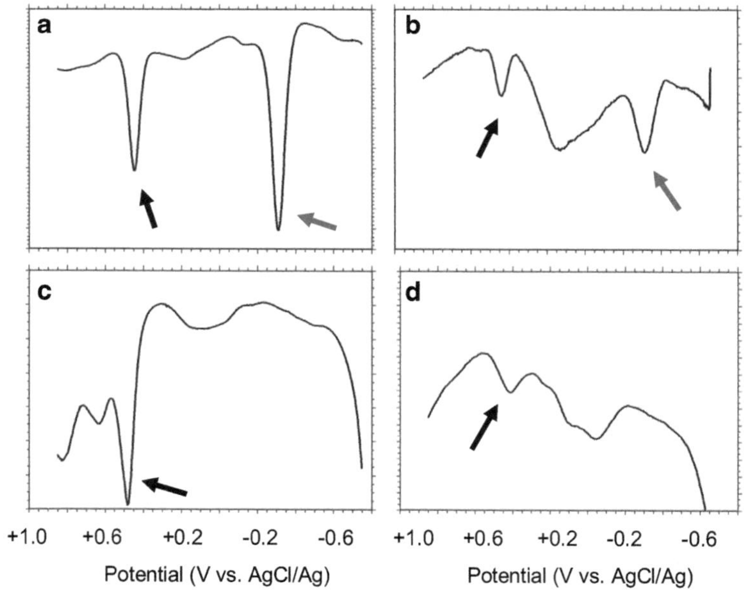

Fig. 5.64 Square-wave voltammograms for (**a**) commercial indigo paste, (**b**) Maya blue sample from Chichén Itza, (**c**) indoxyl (1 % w/w) plus palygorskite wet mixture attached to paraffin-impregnated graphite electrodes, and (**d**) glassy carbon electrodes modified with the extract of a green pellet from La Blanca archaeological site in contact with DMSO overnight. Electrolyte 0.50 mol L^{-1} HAc/NaAc, pH 4.85. Potential scan initiated at −0.75 V versus AgCl/Ag in the positive direction, potential step increment 4 mV, square-wave amplitude 25 mV, frequency 5 Hz. *Black arrow*: dehydroindigo/indigo couple; *gray arrow*: indigo/leucoindigo couple [B406]. Copyright ©2011 with permission from Wiley-VCH

chronoamperometric experiments [B319, B401, B403, B452] using the Lovrić–Scholz–Oldham modeling for the electrochemistry of ion-insertion solids [B87, B88, B107, B142].

The same methodology can be applied to other materials displaying ion-insertion properties. Micro- and mesoporous silicates and aluminosilicates incorporating electroactive organic guests have been extensively studied [B412, B413], with particular attention to their analytical applications [B414, B415, B416]. An interesting aspect to be underlined is the possibility of obtaining discernible voltammetric responses for different topological redox isomers, i.e., guest molecules or ions located in different positions and/or binding environments of the inorganic host framework [B225, B226, B246, B301, B417, B418, B419].

5.4.4 Overview of Studied Compounds

Organometallics and Metal Complexes with Organic Ligands

Compound	References
Tris(2,2′-bipyridine)ruthenium(II) hexafluorophosphate	[B134]
cis-$Ru^{(II)}(dcbpy)_2(NCS)_2$ (dcbpy = 2,2′-bipyridine-4,4′-dicarboxyl acid)	[B102]
$[\{M(bipy)_2\}\{M'(bpy)_2\}(\mu\text{-}L)](PF_6)_2$ (M, M′ = Ru, Os; L = 1,4-Diydroxy-2,5-bis(pyrazol-1-yl)benzene	[B139]
Osmium bis(bipyridyl)tetrazine chloride, $[Os(bpy)_2\text{-4-tet-Cl}](ClO_4)$ tet = 3,6-bis(4-pyridyl)-1,2,4,5-tetrazine	[B155]
Osmium trisdimetoxybipyridyl, $[Os(OMe\text{-}bpy)_3](PF_6)_2$ OMe = 4,4′-dimethoxy; bpy = 2,2′-bipyridyl	[B251]
Triazole bridged osmium dimer, $[Os(bpy)_2Cl\ \text{4-bpt}\ Os(bpy)_2Cl]PF_6$	[B252]
Ruthenium(III) dithiocarbamate Ruthenium(III) diphenyldithiocarbamate $[Ru(bpy)_3][PF_6]_2$	[B100] [B98] [B134]
$[Ru(bpy)_3]_3[P_2W_{18}O_{62}]$	[B303]
fac-$Mn(CO)_3(\eta^2\text{-}Ph_2PCH_2PPh_2)Cl$	[B128]
cis-$[Mn(CN)(CO)_2\{P(OPh)_3\}(dppm)]$ (dppm = $Ph_2PCH_2PPh_2$)	[B294]
cis-$[W(CO)_2(dppe)]$ (dppe = $Ph_2P(CH_2)_2PPh_2$)	[B294]
cis- *trans*-$[Re(CO)_2(P\text{-}P)_2]^+$ and *trans*-$[Re(CO)(P\text{-}P)_2X]$ P-P = diphosphine ligand, X = Cl, Br	[B90]
trans-Re(Br)(CO)X (X = $(dppe)_2$, (dppz)(dppm), (dppe),(dppm) (dppe = $Ph_2P(CH_2)_2PPh_2$, dppm = $Ph_2PCH_2PPh_2$, dppz = $(Ph_2P)_2C_6H_4$)	[B295]
cis, *trans*-$Cr(CO)_2(dppe)_2$ (dppe = $Ph_2P(CH_2)_2PPh_2$)	[B28, B31, B53, B295]
$[(C_4H_9)_4N][Cr(CO)_5I]$	[B65]
$Fe(\eta^5\text{-}C_5Ph_5)((\eta^6\text{-}C_6H_5)C_5Ph_4)$	[B106]
1,3,5-tris(3-((ferrocenylmethyl)amino)pyridiniumyl)-2,4,6-triethylbenzene Hexafluorophosphate	[B306]

(continued)

Bis(η^5-pentaphenyl-cyclopentadienyl)iron(II)-iron(III)	[B129]
$[Co(mtas)_2](X)_n$ (mtas = bis(2-(dimethylarsino)phenyl)-methylarsine; $X = BF_4^-$, $n = 3$; $X = ClO_4^-$, $n = 2, 3$; $X = BPh_4^-$, $n = 2$)	[B45]
Co and Mn phthalocyanines	[B43, B63]
Pt phthalocyanine	[B138, B167, B185, B190]
Polynuclear copper complexes $[Cu_2(H_3L)(OH)][BF_4]_2$, $[Cu_4L(OH)][NO_3]_3$ $L = O_4N_4$	[B260]
Copper(II) 5,10,15,20-tetraphenyl-21H,23H-porphyrin (H_2TPP)	[B135, B159, B178]
Azurin	[B297]
Decamethylferrocene	[B29]
Ferrocene/cobaltocene	[B14, B308]
Decaphenylferrocene	[B295]

Organics

Compound	References
Azobenzene	[B68]
Carbazole/polycarbazole	[B272]
Cocaine	[B109]
2,2-diphenyl-1-picrylhydrazyl (DPPH)	[B114]
Diphenylamine (B 197 (droplet))	[B211, B218]
Fullerenes, C_{60}	[B119, B154, B172, B263]
$C_{60} \subset L_2$ [(L = *p*-benzyl-calix[5]arene)$_2$] ·8toluene	[B156]
HQBpt (HBpt = 3,5bis(pyridine-2-yl)-1,2,4-triazole); 1,4-hydroquinone H_2Q	[B179]
Indigo	[B72, B126]
Lead and mercury dithiocarbamate	[B13, B14]
N,N′-bis(4-Cyanophenyl)-3,4,9,10-perylene-bis (dicarboximide) and *N,N′*-bis(4-Cyanophenyl)-1,4,5,8-naphthalenediimide	[B313]
Quinhydrone, acridine	[B126]
Tetraphenylviologen	[B295]
TCNQ	[B56, B101, B124, B255, B296, B299]
1,3,5-Tris[4-[(3-methylphenyl)phenylamino]phenyl] benzene	[B316]
TTF	[B54, B196, B267]
Organic dyes and pigments	[B265, B302, B321, B322, B391, B392, B393, B394, B395]

5.5 Electrochemically Initiated Chemical Reactions of Solid Particles

5.5.1 Electrochemically Initiated Isomerizations

Similar to the scenario of electrochemical reactions of species in solution, solid compounds can also undergo chemical follow-up reactions. Such systems can be conveniently studied using particles of the solid compound immobilized on an electrode surface. This is very interesting because a comparison can be made between the behavior of a species dissolved and in the solid state. As a first example, we shall discuss an isomerization that was well known to occur when iron(II) hexacyanochromate(III), $KFe[Cr(CN)_6]$, is heated to temperatures around 100 °C. Under this condition, the cyanide ions turn around to form chromium(III) hexacyanoferrate(II), $KCr[Fe(CN)_6]$. This isomerization takes place at room temperature when microparticles of $KFe[Cr(CN)_6]$ are immobilized on a graphite electrode and are subjected to potential cycling in a range where the hexacyanochromate unit is switched between the chromium(III) and chromium (II) states [B33]. Figure 5.65 depicts the cyclic voltammograms and it is obvious how the hexacyanochromate system decreases simultaneous to the growing of the hexacyanoferrate system. The rate constant of the isomerization was 4×10^{-9} mol s^{-1}. The isomerization can be easily understood considering the general inertness of chromium(III) ions against substitution and, on the other

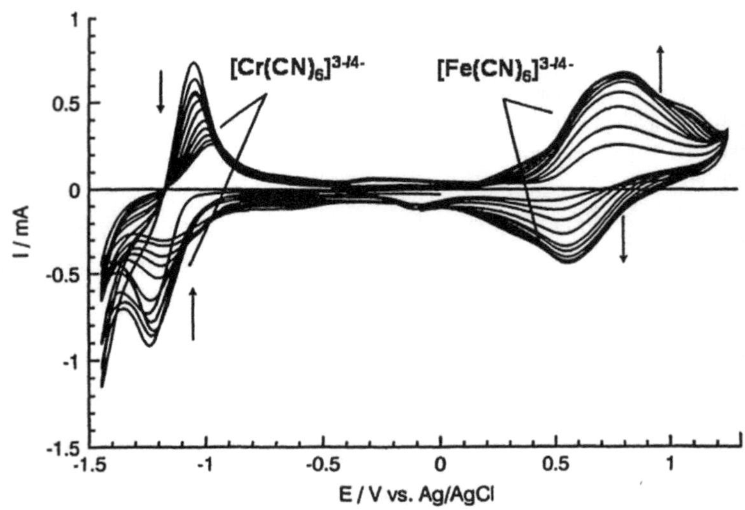

Fig. 5.65 Subsequent cyclic voltammograms of iron(II) hexacyanochromate(III) mechanically immobilized on a PIGE. The isomerization to chromium(III) hexacyanoferrate(II) can be deduced from the simultaneous decrease of the hexacyanochromate system accompanied by the growing of the hexacyanoferrate system. Electrolyte 0.1 mol L^{-1} KCl, scan rate 0.1 V s^{-1} [B33]. Copyright ©1995 with permission from American Chemical Society

hand, a high lability of the ligands of chromium(II) ions. Thus, the cyanide ions are labilized in the hexacyanochromate(II) and allowed to turn around, i.e., coordinating on the carbon end to the iron ions. A similar isomerization was observed in the case of iron(II) hexacyanomanganate(III), although in this case the compound could not be synthesized without some amount of the compound having already isomerized, i.e., with some amount of manganese(II) hexacyanoferrate(III) already present when the compound was subjected to potential cycling. Figure 5.66 shows very well how the two voltammetric systems of hexacyanomanganate (III)/(II) and hexacyanomanganate(II)/(I) decrease continuously while the hexacyanoferrate (III)/(II) system grows simultaneously [B103].

Whether the hexacyanomanganate ion is stable in its reduced or its oxidized state depends both on the nitrogen-coordinated metal ions and on the voltage range, i.e., what redox transitions are performed. Figure 5.67 depicts subsequent cyclic voltammograms of three hexacyanomanganates, (a) chromium (II) hexacyanomanganate(III), (b) iron(II) hexacyanomanganate(III), and (c) manganese(II) hexacyanomanganate(III), all measured in the potential range from +0.1 V to −1.55 V versus Ag/AgCl. Comparing Fig. 5.67b with Fig. 6.51 reveals that the iron(II) hexacyanoferrate(III) is rather stable when the electrode is not polarized to potentials where iron(II) is oxidized to iron(III). Figure 5.67a shows that the chromium(II) hexacyanomanganate is even completely stable, whereas the manganese(II) hexacyanomanganate(III) signal quickly deteriorates (Fig. 5.67c). However, when only the system of $[Mn(CN)_6]^{4-/5-}$ is measured in the limited potential range from −0.7 to −1.55 V versus Ag/AgCl, then it remains completely stable (Fig. 5.67d).

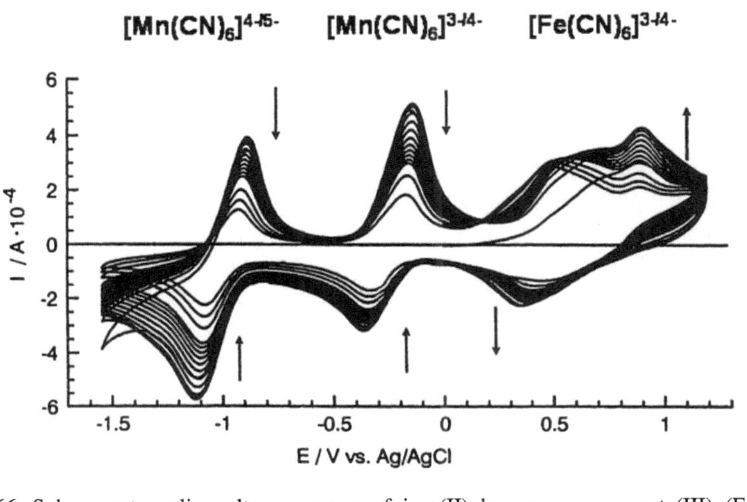

Fig. 5.66 Subsequent cyclic voltammograms of iron(II) hexacyanomanganate(III) (Fe *hcm*) mechanically immobilized on a PIGE. The isomerization of Fe *hcm* to manganese (II) hexacyanoferrate (Mn *hcm*) can be deduced from the simultaneous decrease of the hexacyanomanganate signals accompanied by the growing of the hexacyanoferrate signal. Electrolyte 0.1 mol L^{-1} KCl, scan rate 0.1 V s^{-1} [B103]

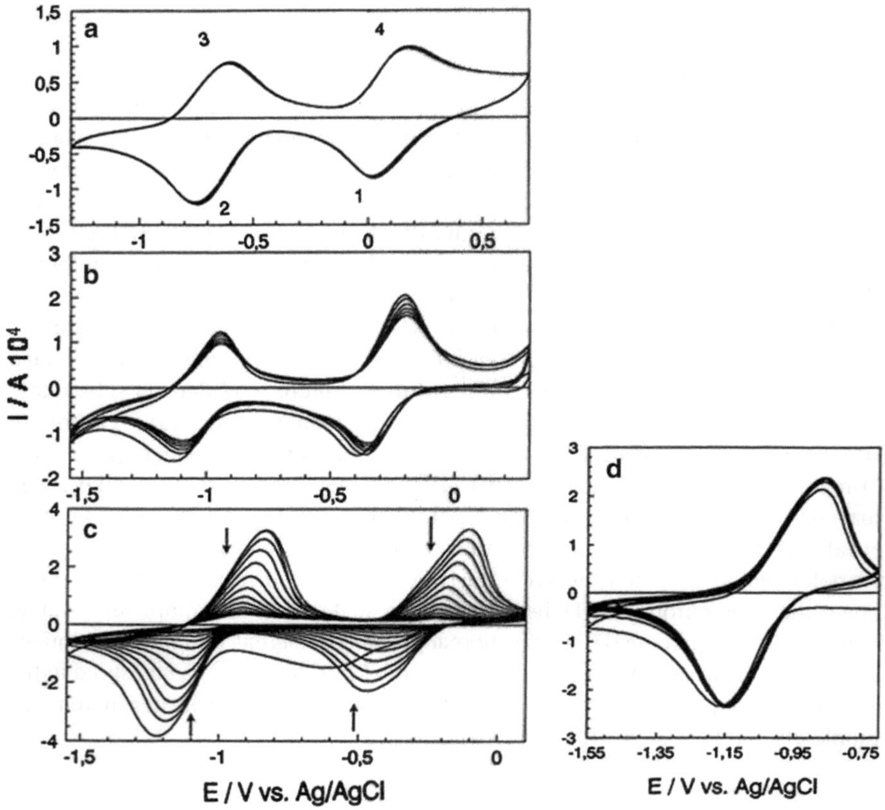

Fig. 5.67 Subsequent cyclic voltammograms of immobilized samples of (**a**) chromium (II) hexacyanomanganate(III), (**b**) iron(II) hexacyanomanganate(III), (**c**) manganese (II) hexacyanomanganate(III), (**d** as **c**) but in a limited potential range [B103]

Figures 5.65, 5.66 and 5.67 are good examples for showing that the voltammetry of immobilized particles yields curves that allow following chemical reactions in the solid state in a straightforward way, very similar to spectroscopic techniques.

Bond et al. [B75] undertook a thorough study of the electrochemical oxidation of the solid organometallic compounds *cis, mer-* and *trans*-$Mn(CO)_2(\eta^3\text{-}P_2P')Br$ ($P_2P' = \{Ph_2P(CH_2)_2\}_2\text{-}PPh$), *cis,mer*-$Mn(CO)_2(\eta^3\text{-}P_3P')Br$ ($P_3P' = \{Ph_2PCH_2\}_3P$), and binuclear *cis,fac*-$\{Mn(CO)_2(\eta^2\text{-dpe})Br\}_2(\mu\text{-dpe})$ (dpe $= Ph_2P(CH_2)_2PPh_2$) and of the electrochemical reduction of *trans*-$[Mn(CO)_2(\eta^3\text{-}P_2P')Br]BF_4$, *trans*-$[Mn(CO)_2(\eta^3\text{-}P_3P')Br]BF_4$, and *trans*-$[\{Mn(CO)_2(\eta^2\text{-dpe})Br\}_2(\mu\text{-dpe})](BF_4)_2$. Aqueous electrolyte solutions have been used in these studies. A major result of this investigation was that isomerizations that were known to occur in the case of organic solutions of these compounds upon electrochemical transformations also occur in the solid state, however, at a markedly slower rate. An example is the following isomerization prompted by oxidation of *cis, mer*- $Mn(CO)_2(\eta^3\text{-}P_2P')Br$:

Electrochemical oxidation using a chloride-containing electrolyte:

$$\textit{cis,mer-}\ \mathrm{Mn(CO)_2(\eta^3\text{-}P_2P')Br_{(solid)}} + \mathrm{Cl^-} \rightleftarrows$$
$$\textit{cis,mer-}\ \mathrm{[Mn(CO)_2(\eta^3\text{-}P_2P')Br]Cl_{(solid)}} + \mathrm{e^-}$$

Slow isomerization:

$$\textit{cis,mer-}\mathrm{[Mn(CO)_2(\eta^3\text{-}P_2P')Br]Cl_{(solid)}} \rightarrow \textit{trans-}\mathrm{[Mn(CO)_2(\eta^3\text{-}P_2P')Br]Cl_{(solid)}}$$

In the case of the solid binuclear complex *trans*-$[\{Mn(CO)_2(\eta^2\text{-dpe})Br\}_2(\mu\text{-dpe})](BF_4)_2$, a single reversible two-electron reduction was observed (cf. Fig. 5.68a), whereas in a dichloromethane solution, two one-electron reductions appear. The occurrence of a reversible single two-electron oxidation was attributed to a decreased communication of the metal centers in the solid compound, probably due to less flexibility. When the electrolyte contains chloride ions, an ion exchange reaction precedes the reduction:

Ion exchange:

$$\textit{trans-}\mathrm{[\{Mn(CO)_2(\eta^2\text{-dpe})Br\}_2(\mu\text{-dpe})](BF_4)_{2(solid)}} + 2\mathrm{Cl^-} \rightleftarrows$$
$$\textit{trans-}\mathrm{[\{Mn(CO)_2(\eta^2\text{-dpe})Br\}_2(\mu\text{-dpe})]Cl_{2(solid)}} + 2\mathrm{BF_4^-}$$

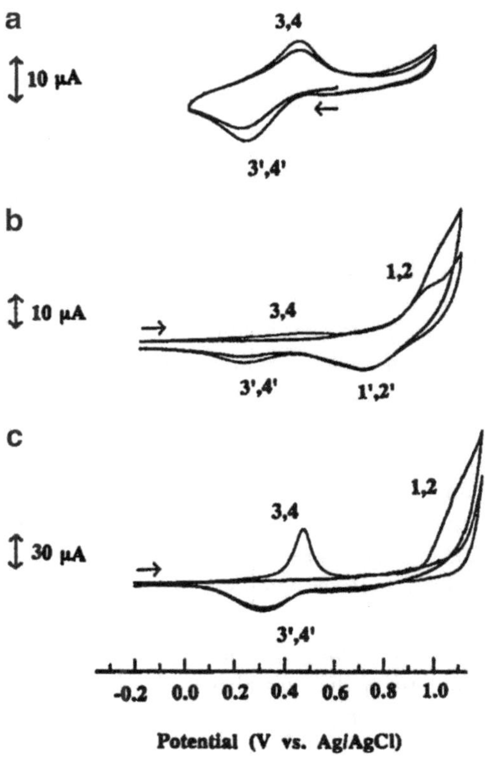

Fig. 5.68 Cyclic voltammograms (scan rate, 200 mV s^{-1}) of solid complexes mechanically attached to graphite electrodes and placed in aqueous solution (0.1 M NaCl): (**a**) *trans*-$[\{Mn(CO)_2(\eta^2\text{-dpe})Br\}_2(\eta\text{-dpe})](BF_4)_2$ at 20 °C, (**b**) *cis,fac*-$\{Mn(CO)_2(\eta^2\text{-dpe})Br\}_2(\mu\text{-dpe})$ at 20 °C, and (**c**) *cis,fac*-$\{Mn(CO)_2(\eta^2\text{-dpe})Br\}_2(\mu\text{-dpe})$ at 50 °C [B75]. Copyright ©1997 with permission from American Chemical Society

Electrochemical reduction/oxidation:

$$trans\text{-}\left[\left\{\mathrm{Mn(CO)_2(\eta^2\text{-}dpe)Br}\right\}_2(\mu\text{-dpe})\right]\mathrm{Cl}_{2(\mathrm{solid})} + 2e^- \rightleftarrows$$
$$trans\text{-}\left\{\mathrm{Mn(CO)_2(\eta^2\text{-}dpe)Br}\right\}_2(\mu\text{-dpe})_{(\mathrm{solid})} + 2\mathrm{Cl}^-$$

5.5.2 Electrochemically Initiated Substitutions

Solid Metal Hexacyanoferrates The simplest case of an electrochemically initiated substitution is when a solid contains the cations C^+, the electrolyte solution contains the cations C'^+, and the solid compound can be oxidized accompanied by a release of C^+. When the oxidized form is reduced, the cations C'^+ will be inserted as the released C^+ ions have diffused into the solution, provided the time suffices or the solution is stirred. In this way, it is easy to pump *out* one sort of cations and to pump *in* another one. The same holds true for anions if their transfer accompanies electrochemical reactions. As an example for cation exchange may serve the electrochemical transformation of potassium containing Prussian blue into sodium-containing Prussian blue:

Pumping out potassium ions during oxidation of the low-spin iron(II) ions:

$$\mathrm{K}\left\{\mathrm{Fe}\left[\mathrm{Fe(CN)_6}\right]\right\}_{(\mathrm{solid})} \rightarrow \left\{\mathrm{Fe}\left[\mathrm{Fe(CN)_6}\right]\right\}_{(\mathrm{solid})} + \mathrm{K}^+ + e^-$$

Pumping in sodium ions during reduction of the low-spin iron(III) ions:

$$\left\{\mathrm{Fe}\left[\mathrm{Fe(CN)_6}\right]\right\}_{(\mathrm{solid})} + e^- + \mathrm{Na}^+ \rightarrow \mathrm{Na}\left\{\mathrm{Fe}\left[\mathrm{Fe(CN)_6}\right]\right\}_{(\mathrm{solid})}$$

These reactions have to be carefully taken into account whenever a solid compound is studied that is capable of undergoing chemically reversible insertion electrochemical reactions. Often, the compound will contain ions from synthesis that differ from those present in the electrolyte solution used for the electrochemical measurements. In such a case, the first scan will considerably differ from the following and one should make sure to equilibrate the solid by extended oxidation–reduction cycles. Complete repeated oxidation and reduction and stirring of the solution usually ensures that the final composition of the solid is in equilibrium with the solution composition. This is especially important when scan rate dependencies are recorded. For fast scan rates, the layer of converted solid is, of course, much thinner than for slow scan rates. If the solid was not equilibrated before the scan rate dependence is measured, it could happen that at each scan rate, a layer of the solid is studied that possesses a different composition compared to the previous (i.e., faster) scan rate. Peak potentials derived from such measurements would be slightly incorrect.

Electrochemically initiated substitutions are not confined to the exchange of counterions; they can also occur with ions more strongly bound in the framework of the solid compound. Very early it was observed that the high-spin iron ions of

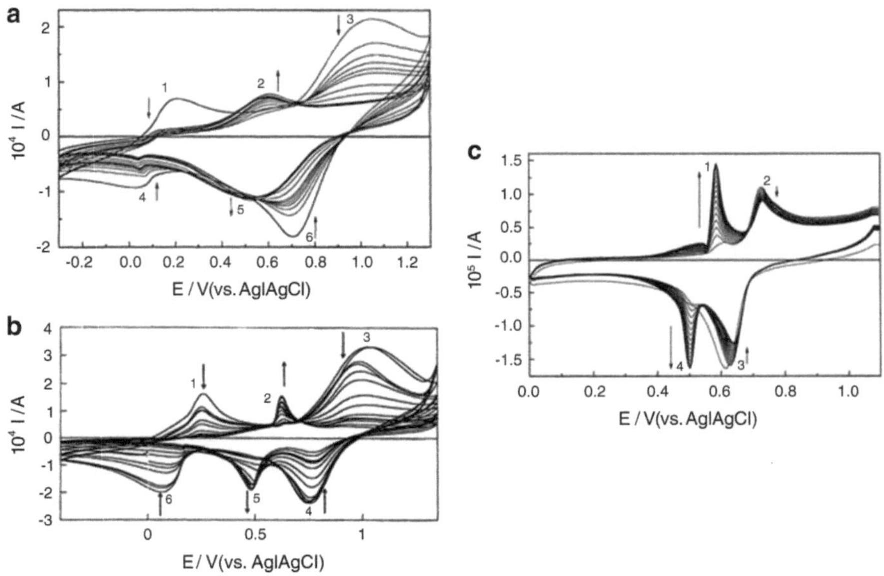

Fig. 5.69 Cyclic voltammograms recorded during the conversion of (**a**) Prussian blue to nickel hexacyanoferrate, (**b**) Prussian blue to cadmium hexacyanoferrate, and (**c**) silver hexacyanoferrate to cadmium hexacyanoferrate. Electrolyte solutions: (**a**) 0.1 mol L^{-1} nickel nitrate and (**b** and **c**) 0.1 mol L^{-1} cadmium nitrate. The scan rate was 0.1 V s^{-1} [B55]. Copyright ©1996 with permission from Elsevier

Prussian blue can be substituted by cadmium ions when Prussian blue is cyclically oxidized and reduced in a Cd^{2+} ions containing electrolyte [B35]. Later it was shown that analogous substitutions lead to the formation of nickel hexacyanoferrate when Prussian blue is cycled in a Ni^{2+}-containing solution and to the formation of cadmium hexacyanoferrate when silver hexacyanoferrate is cycled in a Cd^{2+}-containing electrolyte [B55]. Figure 5.69 shows cyclic voltammograms recorded during the electrochemically initiated substitution. It could be shown that the reaction leads not to a simple mixture of the mother and daughter hexacyanoferrate, but it proceeds via bilayered structure. This can be deduced from a scan rate dependence of the ratio of the signals of the mother and daughter hexacyanoferrate as shown in Fig. 5.70 for the cadmium and iron hexacyanoferrate systems. In this figure, the ratio of peak currents of signals 4 and 5, Fig. 5.69b, is depicted compared with the voltammograms of a simple powder mixture of the same two hexacyanoferrates (inset of Fig. 5.70). In case of a powder mixture of cadmium and iron hexacyanoferrate, both compounds are equally accessible for the electrochemical reactions and both signals increase in the same way upon increasing the scan rate. However, when the daughter hexacyanoferrate forms a layer on the mother hexacyanoferrate that must be crossed by the inserting ions before the mother hexacyanoferrate can react, then the ratio of the peak currents must show a strong scan rate dependence as depicted in Fig. 5.70. A very thorough analysis of this

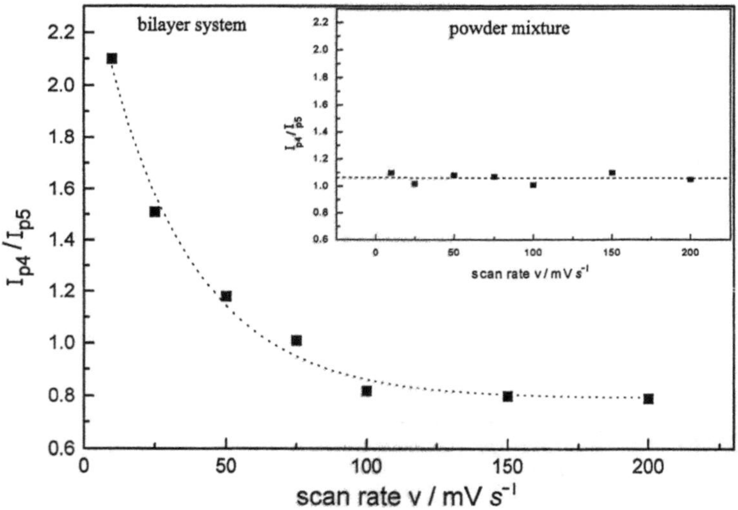

Fig. 5.70 Plots of the ratio of peak currents i_{p4}/i_{p5}, i.e., ratio of cathodic peak current of iron hexacyanoferrate to cathodic peak current of cadmium hexacyanoferrate, as a function of scan rate. The *inset* shows the same dependence for the case that a powder mixture of these hexacyanoferrates was immobilized [B55]. Copyright ©1996 with permission from Elsevier

system [B173] using cyclic voltammetry, impedance spectroscopy, and scanning electron microscopy in conjunction with energy dispersive X-ray detection (ex situ) confirmed the following reaction mechanism. Prussian blue can be oxidized to Berlin green as follows:

$$\{KFe_N^{3+}Fe_C^{2+}(CN)_6\} \rightleftarrows \{Fe_N^{3+}Fe_C^{3+}(CN)_6\} + K^+ + e^- \quad (XII)$$

or it can be reduced to Berlin white according to

$$\{KFe_N^{3+}Fe_C^{2+}(CN)_6\} + K^+ + e^- \rightleftarrows \{K_2Fe_N^{2+}Fe_C^{2+}(CN)_6\} \quad (XIII)$$

If Prussian blue, or either of its redox congeners, is merely soaked in a solution containing Cd^{2+} ions, some of these ions may replace the K^+ in the cages, but no substantial reaction occurs [B35]. However, if cyclic voltammetry is carried out on Berlin white microcrystals $\{K_2Fe_N^{2+}Fe_C^{2+}(CN)_6\}$ adhered to graphite in a solution containing both K^+ and Cd^{2+} ions, then an uptake of cadmium ions into the lattice of the metal hexacyanoferrate occurs gradually. It is postulated that the cycling "pumps" Cd^{2+} (and K^+) ions in and out of the cages, permitting the substitution reaction

$$\{Fe_N^{3+}Fe_C^{3+}(CN)_6\} + Cd^{2+} \rightarrow \{Cd_N^{2+}Fe_C^{3+}(CN)_6\}^- + Fe^{3+} \quad (XIV)$$

to occur. The twin-process voltammogram described by Reactions (XII) and (XIII) is gradually replaced by a single-process cyclic voltammogram corresponding to the reaction

$$\{Cd_N^{2+}Fe_C^{3+}(CN)_6\}^- + e^- \rightleftarrows \{Cd_N^{2+}Fe_C^{2+}(CN)_6\}^{2-} \qquad (XV)$$

The charge compensating cations are not given in Reactions (XIV) and (XV) because their identity is not exactly known. Depending on time and location, K^+, Cd^{2+}, and Fe^{3+} are possible candidates. Of course, Reactions (XII), (XIII), and (XV) will also take place under insertion of Cd^{2+} into and out of the solid. Even though it is the nitrogen-bound iron atom that undergoes substitution in Reaction (XIV), the carbon-bound iron atom is also a factor in the process. It is essential that this atom is in the +3 state—then the conceivable reaction

$$\{Fe_N^{3+}Fe_C^{2+}(CN)_6\} + Cd^{2+} \rightarrow \{Cd_N^{2+}Fe_C^{2+}(CN)_6\}^{2-} + Fe^{3+} \qquad (XVI)$$

does *not* significantly take place, as evidenced by the observation that cadmium substitution does not occur if the voltammetric cycling is restricted to the range involving only Berlin green and Prussian blue. On the other hand, the reaction cannot be as simple as Reaction (XIV) suggests, because the cages of Berlin green are unoccupied, so $\{Cd_N^{2+}Fe_C^{3+}(CN)_6\}^-$ cannot be formed by a pure substitution reaction as implied by Reaction (XIV). Instead, it may be postulated that substitution and oxidation occur concurrently during the second half of the oxidative scan:

$$\{Fe_N^{3+}Fe_C^{2+}(CN)_6\}^- + Cd^{2+} \rightarrow \{Cd_N^{2+}Fe_C^{3+}(CN)_6\}^- + e^- + Fe^{3+} \qquad (XVII)$$

The substitution reaction continues during the reduction of low-spin Fe(III) as long as the latter is still present, that is,

$$Fe_N^{3+}Fe_C^{3+}(CN)_6 + e^- + Cd^{2+} \rightarrow \{Cd_N^{2+}Fe_C^{2+}(CN)_6\}^{2-} + Fe^{3+} \qquad (XVIII)$$

One can sum both substitution possibilities defining the following formal exchange reaction with a rate constant k_f:

$$2Fe_N^{3+}Fe_C^{3+}(CN)_6 + 3Cd^{2+} \xrightarrow{k_f} Cd^{2+}\left(Cd_N^{2+}Fe_C^{3+}(CN)_6\right)_2 + 2Fe^{3+} \qquad (IXX)$$

Figure 5.71 schematically shows how the bilayered system is formed and how cyclic voltammetry and EDX mirror the proceeding of the electrochemically induced transformation. See also Sect. 5.11 for the results of theoretical simulations of this substitution reaction.

Solid Porphyrins When solid microparticles of 5,10,15,20-tetraphenyl-21H,23H-porphyrin (H_2TPP) immobilized on an electrode surface are subjected to reduction–oxidation cycles in a copper(II)-containing electrolyte solution, the incorporation of

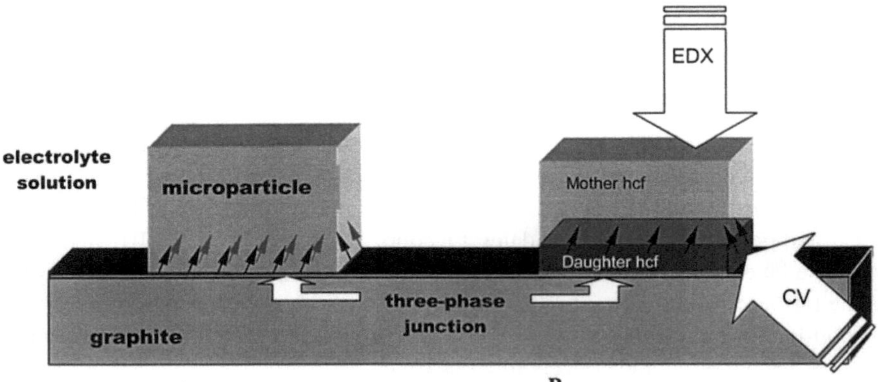

Fig. 5.71 Schematic view of the situation when a microparticle of a mother hexacyanoferrate is transformed into a daughter hexacyanoferrate. The scheme visualizes the different perspectives from which cyclic voltammetry (CV) and energy dispersive X-ray analysis sees the reaction. With EDX it takes much longer until the product can be detected, whereas with CV the mother compound will only be seen when slow scan rates are used and the diffusion layer expands within the particles up to the mother compound [B173]. Copyright ©2001 with permission from Elsevier

Cu^{2+} ions has been observed [B135, B159, B178]. The following mechanism has been proposed:

1. The solid H_2TPP is oxidized at 0.95 V to solid TPP:

$$\{H_2TPP\}_{solid} \rightleftarrows \{TPP\}_{solid} + 2e^- + 2H^+ \quad (XX)$$

2. The solid TPP is able to incorporate copper(II) ions into the porphyrin ring, presumably accompanied by an association of chloride ions:

$$\{TPP\}_{solid} + Cu^{2+} + 2Cl^- \rightleftarrows \{TPPCu^{2+}(Cl^-)_2\}_{solid} \quad (XXI)$$

3. The porphyrin skeleton of $\{TPPCu^{2+}(Cl^-)_2\}_{solid}$ is reduced at potentials negative to 0.95 V according to

$$\{TPPCu^{2+}(Cl^-)_2\}_{solid} + 2e^- \rightleftarrows \{TPP^{2-}Cu^{2+}\}_{solid} + 2Cl^- \quad (XXII)$$

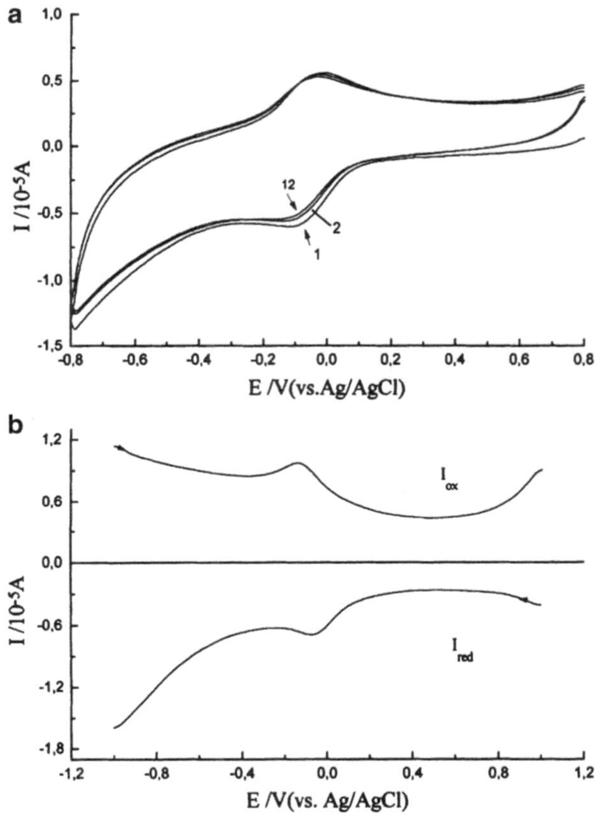

Fig. 5.72 (**a**) Cyclic voltammograms of an electrode with immobilized H_2TPP after copper(II) ion incorporation (0.1 mol L^{-1} KCl, pH 5.5, scan rate 0.1 V s^{-1}). The 1st, 2nd, and 12th cycles are shown. (**b**) Square-wave voltammograms of the same electrode [B135]. Copyright ©2000 with permission from World Sc. Publishing Co

4. The solid compound $\{TPP^{2-}Cu^{2+}\}_{solid}$ undergoes a reversible one-electron reduction–oxidation accompanied by potassium ion transfer at −0.07 V according to (see Fig. 5.72)

$$\left\{TPP^{2-}Cu^{2+}\right\}_{solid} + e^- + K^+ \rightleftarrows \left\{TPP^{2-}Cu^+K^+\right\}_{solid} \qquad \text{(XXIII)}$$

While the electrochemical system (XXIII) has been unambiguously identified, it was not possible to detect the incorporated copper ions with scanning electron microscopy coupled EDX. This can be explained by a very small conversion zone at the three-phase junction so that the electron beam does not access it. The small currents support this idea, suggesting that only a very small fraction of the porphyrin compound is incorporating the copper ions.

5.6 Assessing Structure–Property Relationships

It is the intention of this section to show how systematic studies of certain groups of compounds can be performed with the help of the voltammetry of immobilized particles in order to derive relationships between their structures and their electrochemical properties. Here, we shall discuss some relations between the structure of compounds and their thermodynamic properties.

Any chemically synthesized compounds that are insoluble and electrochemically active can be studied by means of the voltammetry of immobilized particles. For the characterization of these compounds, a wealth of analytical tools is available, ranging from elemental analysis to the most sophisticated techniques for structure analysis. This is by far not the case for compounds that are precipitated by electrochemical means as thin films on electrodes. In the latter case, the means for analysis are rather limited and the data of these analyses are less reliable. There are also much fewer possibilities to deliberately vary the composition or structure of the synthesized compound.

5.6.1 *Solid Metal Hexacyanometalates*

A first example of a systematic study was the determination of the formal potentials of many metal hexacyanometalates and the search for a correlation of these data with properties of the nitrogen-coordinated metal ions [B41]. Figure 5.73 depicts

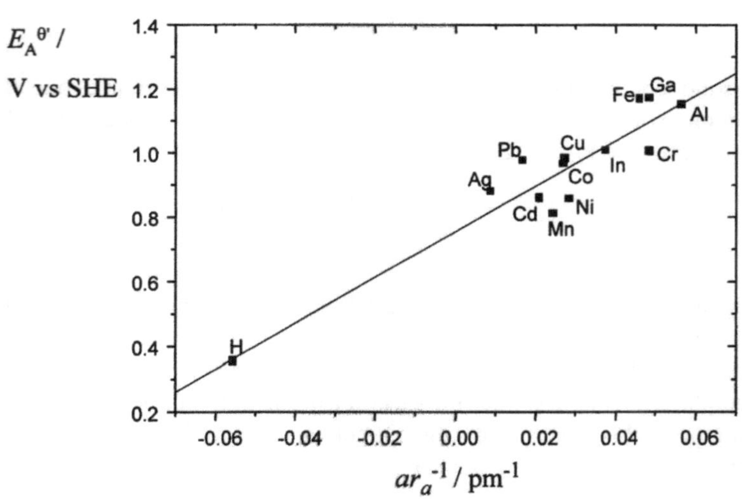

Fig. 5.73 Correlation between the formal potentials of the hexacyanoferrate units of the solid hexacyanoferrates and the ion potential (charge/radius) of the nitrogen-coordinated metal ions [B41, B209]. Copyright ©1995 with permission from Wiley-VCH

the correlation of the formal potentials of the following reactions with the ion potential of the metal ions Me^{+}, Me^{2+}, and Me^{3+}:

$$\left\{K_3Me^{I}\left[Fe(CN)_6\right]\right\} \rightleftarrows \left\{K_2Me^{I}\left[Fe(CN)_6\right]\right\} + K^{+} + e^{-} \tag{XXIV}$$

($Me^{I} = Ag$)

$$\left\{K_2Me^{II}\left[Fe(CN)_6\right]\right\} \rightleftarrows \left\{KMe^{II}\left[Fe(CN)_6\right]\right\} + K^{+} + e^{-} \tag{XXV}$$

($Me^{II} =$ Cd, Zn, Pb, Mn, Ni, Co)

$$\left\{KMe^{III}\left[Fe(CN)_6\right]\right\} \rightleftarrows \left\{Me^{III}\left[Fe(CN)_6\right]\right\} + K^{+} + e^{-} \tag{XXVI}$$

($Me^{III} =$ Ga, In, Al, Fe, Cr, Co).

All these reactions are typical solid-state insertion electrochemical processes with coupled electron and ion transfers. In Fig. 5.73 included is also the standard potential of the hexacyanoferrate ions in aqueous solution. Very interestingly, there is a rather good correlation between these data, and this is the first example that one and the same correlation bridges the formal potentials of an ion in different solids *and* in aqueous solution. It is reasonable to assume that the protons of the water are the nearest neighbors of the nitrogen atoms of the cyanide ions surrounding the iron, as the metal ions $Me^{I,\,(II),(III)}$ ions are the nearest neighbors in the solid. On the basis of elementary thermodynamics, it was later possible to derive the following equation for the formal potentials [B163, B209]:

$$E_{\mathrm{A}}^{\theta'} = -\frac{\kappa}{zF} + \frac{\xi}{zF}\frac{a}{r_{\mathrm{Me}}} \tag{5.14}$$

(κ and ξ are constants, a is the charge of the nitrogen-coordinated metal ions $Me^{I,\,(II),(III)}$, and r_{Me} their radius.)

As can be seen from Fig. 5.73, the experimental data follow this equation. Other examples for systematic studies of the electrochemical and structure data of solid hexacyanometalates concern the dependence of the formal potential of the hexacyanoferrate unit and the ion potential of the inserting metal ions and the correlation between the formal potentials of the hexacyanometalate ions and the lattice constants (the latter correlation includes hexacyanochromates and hexacyanomanganates) [B41, B163, B209].

It has been known for a long time that the formal potential of the hexacyanoferrate system in solid compounds is affected by the nature of the inserting cations. On the basis of simple Born–Haber cycles, the following equation was derived for the effect of the ion potential of the inserting cations on the formal potential of the hexacyanoferrate system [B209]:

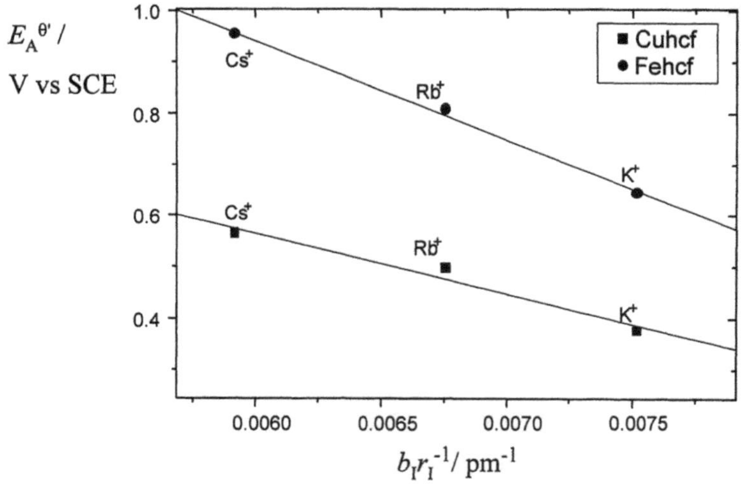

Fig. 5.74 Correlation between the formal potentials of the hexacyanoferrate units of the solid hexacyanoferrates and the ion potential (charge/radius) of the inserting metal ions [B41, B209]. Copyright ©2002 with permission from Elsevier

$$E_{\mathrm{A}}^{\theta'} \approx -\frac{const_3}{zF} - \frac{\Phi\, b_{\mathrm{I}}}{zF\, r_{\mathrm{I}}} \tag{5.15}$$

(const$_3$ and Φ are constants, and b_{I} is the charge of the inserting metal ions Me^{I} and r its radius.)

Figure 5.74 gives the experimental proof of the validity of Eq. (5.15). There exists also a relationship between the lattice constant of solid metal hexacyanometalates and the formal potential. This dependence can be experimentally found when the nitrogen-coordinated metal ion and the inserting metal ion is always the same, but the nature of the carbon-coordinated metal ions is varied, i.e., for hexacyanoferrate, hexacyanocobaltate, hexacyanomanganate, etc. [B209]. The equation that has been derived is

$$E_{\mathrm{A}}^{\theta'} \approx -\frac{2\Psi}{zF}\left[\frac{b_{\mathrm{C}} - b_{\mathrm{B}}}{L}\right] - \frac{\omega_{\mathrm{I}}}{zF} \tag{5.16}$$

(Ψ and ω_{I} are constants, (b_{C}–b_{B}) is the difference of charges of the oxidized and reduced forms of the hexacyanometalate unit, i.e., it is 1, and L is the lattice constant of the solid hexacyanometalate.)

Figure 5.75 shows that the experimental data are in accordance with this equation.

The results given in Figs. 5.73, 5.74 and 5.75 show that the technique of immobilizing microparticles of solid compounds on solid electrodes easily allows studying series of related compounds with the aim of understanding how the thermodynamic properties change with certain structure parameters.

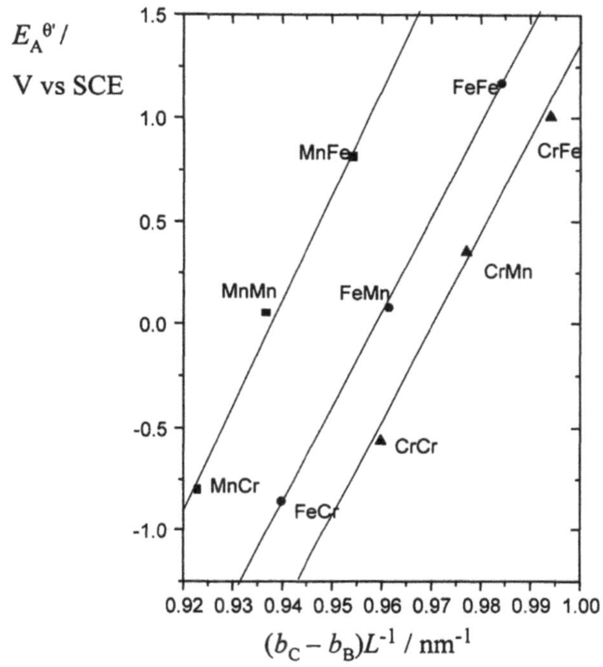

Fig. 5.75 Correlation between the formal potentials of the hexacyanometalate units of the solid hexacyanometalates and the inverse lattice constant L of the solid compounds [B41, B209]. Copyright ©2002 with permission from Elsevier

5.6.2 The Stability Constants of Metal Dithiocarbamate Complexes and Their Structure

The electrochemistry of solutions of metal dithiocarbamates in organic solvents has been intensively studied by Alan Bond and his group [14], [15], [16]. On this sound basis, it was possible to perform a systematic study of the electrochemistry of these highly water-insoluble compounds as immobilized particles on electrodes [B13, B14]. All studied mercury(II) and lead(II) dithiocarbamate complexes displayed one redox system in cyclic voltammetry that was characterized by the following features: the response decreased during subsequent cycling, the mid-peak potentials converged against a constant value in subsequent cycles, and the anodic to cathodic peak separation was in most cases below 100 mV. Figures 5.76 and 5.77 show one example for a lead complex and one for a mercury complex.

Eleven differently substituted dithiocarbamate complexes of lead(II) and 17 differently substituted dithiocarbamate complexes of mercury(II) have been studied. For some of them, the overall stability constants in water have been known. With

[14] Bond AM, Colton R, Dillon, ML, Moir, JE, Page DR (1984) Inorg Chem 23:2883–2891

[15] Bond AM, Colton R, Hollenkamp AF, Hoskins BF, McGregor K (1987) J Am Chem Soc 109:1969–1980

[16] Bond AM, Colton R, Hollenkamp AF (1990) Inorg Chem 29:1991–1995

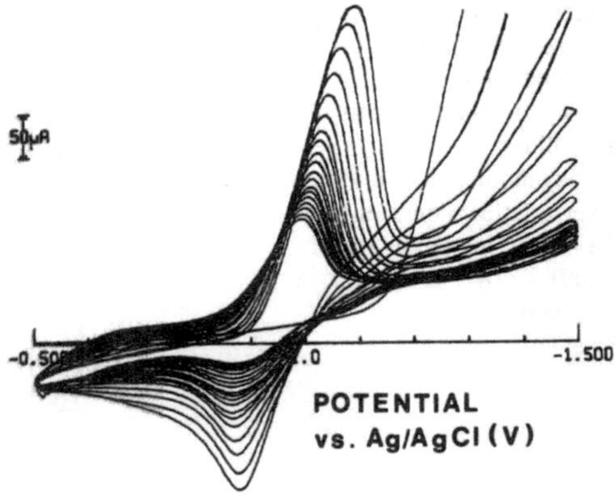

Fig. 5.76 First twenty cyclic voltammograms for reduction of $Pb(i\text{-}Pr_2dtc)_2$ mechanically attached to a PIGE with aqueous 0.1 mol L^{-1} KNO_3 as electrolyte: scan rate 100 mV s^{-1} [B13]. Copyright ©1991 with permission from American Chemical Society

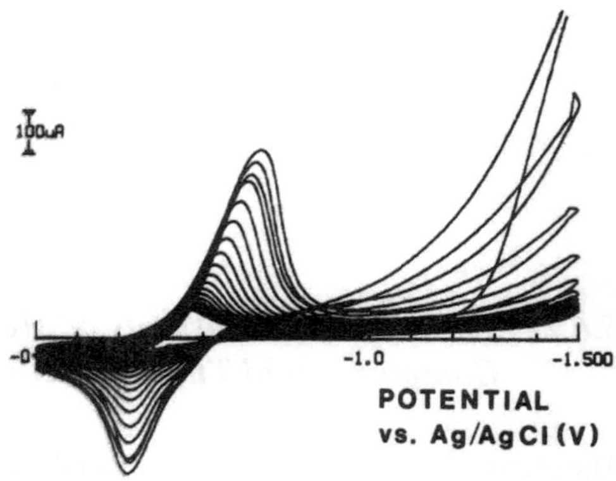

Fig. 5.77 Multiple cyclic voltammograms for reduction of $Hg(Et_2dtc)_2$ mechanically attached to a PIGE with aqueous 0.1 mol L^{-1} KNO_3 as electrolyte: scan rate 100 mV s^{-1} [B13]. Copyright ©1991 with permission from American Chemical Society

the help of these data, it was found that the mid-peak potentials of the cyclic voltammograms are determined by these stability constants. Figure 5.78 illustrates the situation on the electrode surface. The solid metal complex is partially reduced to metal and the dissolved ligand ions. This first reaction can be formulated as follows:

$$\left\{\text{Me}(\text{ligand})_2\right\}_s + 2e^- \rightarrow \{\text{Me}\}_s + 2(\text{ligand})^- \qquad \text{(XXVII)}$$

In the following cyclic polarizations, the dissolution–precipitation equilibrium seems not established, perhaps because of a retardation of nucleation. The voltammograms are completely governed by the following electrochemical system involving dissolved complex molecules and ligands:

$$\left\{\text{Me}(\text{ligand})_2\right\}_{\text{dissolved}} + 2e^- \rightleftarrows \{\text{Me}\}_s + 2(\text{ligand})^- \qquad \text{(XXVIII)}$$

The lead complex of *n*-hexyl-dithiocarbamate crystallizes very slowly, and so

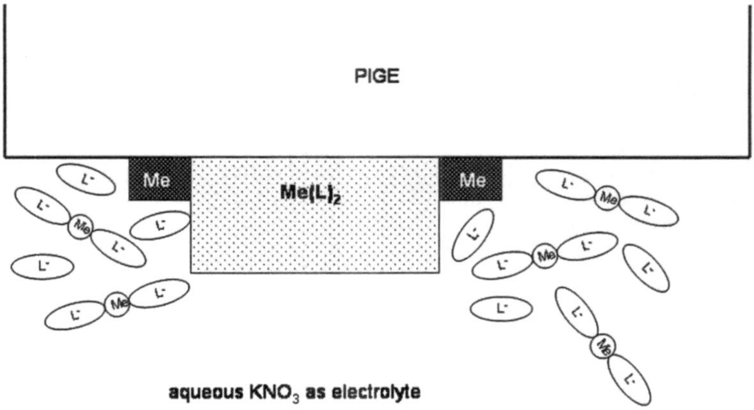

Fig. 5.78 Situation at an electrode surface with immobilized $Pb(i\text{-}Pr_2dtc)_2$ that is partially reduced to lead metal and dissolved ligands (the *ovals*)

several experiments have been performed with attached microdroplets of this compound. Interestingly, the electrochemical response was exactly the same as that of the crystalline compound. These experiments with the oil droplets of this lead compound were obviously the first experiments with attached droplets that further developed so vividly (see Chap. 6).

For the redox system (XXVIII), the Nernst equation reads as follows:

$$E = E^{\ominus}_{Pb^{2+}/Pb} + \frac{RT}{2F}\ln\beta_2^{-1} + \frac{RT}{2F}\ln\frac{a_{Me(ligand)_2}}{a^2_{ligand^-}} \tag{5.17}$$

The formal potential $E^{\ominus'}_A$ of the system is

$$E^{\ominus'}_A = E^{\ominus}_{Pb^{2+}/Pb} + \frac{RT}{2F}\ln\beta_2^{-1} + \frac{RT}{2F}\ln\frac{\gamma_{Me(ligand)_2}}{\gamma^2_{ligand^-}} \tag{5.18}$$

Assuming that the activity coefficients are unity, or at least constant at a certain ionic strength, the overall stability constants can be calculated. The intriguing result of this study was that for both series of metal dithiocarbamate complexes, the log β values are linear functions of the molecular weight of the complexes (cf. Fig. 5.79). This is further supported by the fact that the stability constants of the lead complexes of *n*-propyl- and *i*-propyl-dithiocarbamate have exactly the same values, as have the *i*-butyl- and *n*-butyl-substituted complexes. The cyclohexyl- and *n*-hexyl-substituted complexes have almost the same stability constants. Following this observation, a search for similar series of metal complexes revealed that there are several other examples for such dependences, although, the number of complexes was much smaller in each series. A prerequisite for these dependencies seems to be that the chelating region of the ligand is electronically not, or only marginally,

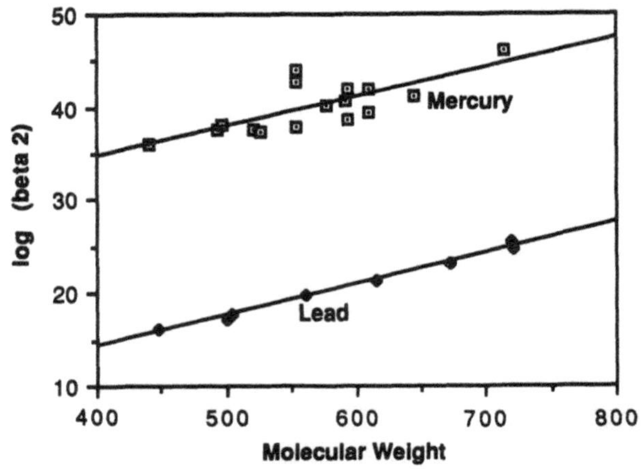

Fig. 5.79 Plot of $\log\beta_2$ values of mercury and lead dithiocarbamate complexes versus their molecular weight. The electrolyte solution was 0.1 mol L^{-1} KNO_3 [B13]. Copyright ©1991 with permission from American Chemical Society

affected by the substituents of the ligand that increase the molecular weight. A quite similar dependence was also observed for the literature data of pK_a values of carboxylic acids on their molecular weight. Until now, there has not been given a theoretical explanation for these correlations. The study of the $\log\beta$ values is presented here to show how electrochemical measurements of immobilized particles can yield interesting new insight also in cases where dissolution–precipitation processes are coupled to redox and complex equilibria.

5.7 Assessing Kinetic Properties of Electrochemical Dissolution Reactions

When solid particles immobilized on an electrode undergo an electrochemical dissolution reaction, the electrochemical measurements should provide information on the size and shape of the particles and on the electrochemical and chemical steps of the reaction. T. Grygar was the first who realized the potential of electrochemical measurements with immobilized particles and who contributed most of the material which is presented in this chapter [B37, B40, B50]. The methodology of these studies was developed using iron(III) oxides and hydroxyoxides which undergo reactions as the following:

$$Fe^{III}OOH_{(s)} + 3H^+ + e^- \rightleftarrows Fe^{2+} + 2H_2O \qquad \text{(IXXX)}$$

Reaction (XXXII) is a proton-assisted reductive dissolution of the solid phase FeOOH yielding dissolved iron(II) ions. Grygar applied the general kinetic Eq. (5.12) for reaction rate J to the special case of electrochemical dissolution of immobilized microparticles of oxides:

$$J = -\frac{dn}{dt} = -kN_0G(c)F\left(\frac{N_t}{N_0}\right) \tag{5.19}$$

where k is the rate constant, N_0 is the total number of moles of compound to be dissolved, N_t is the number of moles of compound dissolved until the time t, and $F(N_t/N_0)$ characterizes the dependence of reactivity on the size and shape distribution of the dissolving particles. $G(c)$ is a function that depends on the composition of the solution phase. If diffusion is controlling the dissolution, this function is simply $G = c_s - c$ with c_s being the concentration of the saturated solution and c being the actual concentration. In the case of reductive dissolution, this function depends on the concentration of all reactants and the adsorption capacity of the particles. When all these factors can be kept constant, it may be included in k. In that case, the reaction will be of pseudo-first order and the unit is s^{-1}. Eq. (5.19) expressed in electrochemical terms reads

$$I_t = kQ_0G(c)F\left(\frac{Q_t}{Q_0}\right) \tag{5.20}$$

where I_t is the current at time t and Q_t and Q_0 are the charges flown until the time t and until complete dissolution, respectively. When the solution composition is kept constant, the function $G(c)$ can be included in the rate constant, and for the function $F(N_t/N_0)$, an experimentally found expression can be used to write Eq. (5.20) as follows:

$$I_t = k'Q_0\left(\frac{Q_t}{Q_0}\right)^g \tag{5.21}$$

The exponent g has been interpreted as characterizing the heterogeneity of the powder with respect to its reactivity. Grygar has shown that the rate constant k' in Eq. (5.21) follows a potential dependence that can be written as

$$k' = k_0'\exp\left[-\frac{\alpha nF(E - E^0)}{RT}\right] \tag{5.22}$$

Equation (5.22) is limited to growing overpotentials when the subsequent reaction step, the dissolution of Fe^{2+} compounds, becomes rate limiting. Figure 5.80 depicts experimental plots of ln k' versus electrode potential for several iron hydroxyoxides and an oxide, proving that, within the used potential range, the reductive dissolution is charge transfer controlled. Table 5.4 lists rate constants and g-values as determined from chronoamperometric measurements. From this table and other publications, one can see that the rate constants of electrochemical dissolution of α-FeOOH vary, depending on the electrode potential and electrolyte solution, between 2.3×10^{-3} and 26×10^{-3} s^{-1} [B40, B50, B66]. With the help of in situ AFM, the average dissolution rate for seven microcrystals was found to be

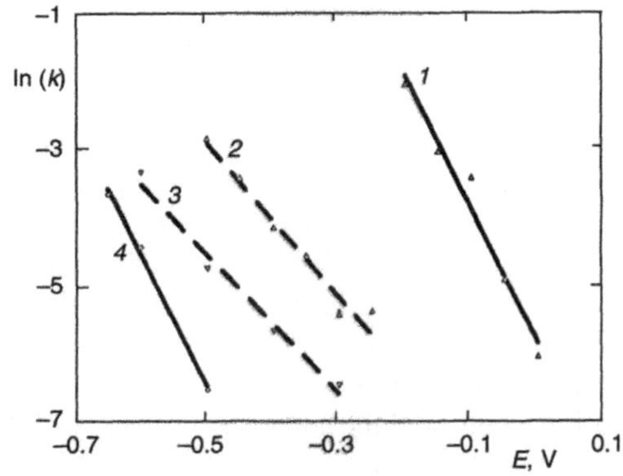

Fig. 5.80 Dependence of the logarithm of the electrochemical dissolution rate constant k on the working electrode potential E. *Curves*: 1 ferrihydrite, 0.05 M chloroacetate; 2 goethite, 0.2 mol L^{-1} chloroacetate +0.5 mol L^{-1} KCl; 3 goethite, 0.05 M chloroacetate; 4 hematite, 0.2 mol L^{-1} chloroacetate [B40]. Copyright ©1995 with permission from publisher

2.0×10^{-3} s^{-1} [B276], what is remarkably near to the data with arrays of immobilized microcrystals that have been used in Grygar's experiments. In experiments with goethites and Al-substituted goethites, Grygar found the exponent g in a range from 0.8 to 1.2 and interpreted these values as typical for dissolutions that proceed in a shape-conservative fashion [B66]. Since k and k' in Eqs. (5.20) and (5.21) depend on the specific surface area σ of the sample at the start, Eq. (5.22) can be written to describe the peak potential in differential pulse voltammetry, assuming that the DP peak is shifted from the DC peak by a constant value, in the form

$$E_{\mathrm{p,DP}} = B + \frac{RT}{\alpha nF} \ln \sigma \tag{5.23}$$

It has been shown that such dependencies hold true for the reductive electrochemical dissolution of Ba-hexaferrite, hematite, goethite, maghemite, and magnetite [B50]. This linear dependence is also obeyed by Li–Mn–O spinels as can be seen in Fig. 5.81 [B191].

The influence of the parameter g on the shape of linear sweep voltammograms is depicted in Fig. 5.82. Grygar has derived the following equation for the dependence of the peak current on g:

$$I_{\mathrm{p}}/Q_0 = \alpha nFv(RT)^{-1} g^{g/(1-g)} \tag{5.24}$$

and has proved that the following equation, which was derived earlier by Brainina for paste electrodes, holds true also for immobilized microparticles:

$$E_{\mathrm{p}} = E^0 - \frac{RT}{\alpha nF} \ln \frac{\alpha nFv}{RT} + \frac{RT}{\alpha nF} \ln \frac{k_{\mathrm{E}}}{d} \tag{5.25}$$

Table 5.4 Values of the k' and g parameters in the kinetic Eq. (5.14) for the dissolution of the phases in the 0.1 M oxalate buffer at working electrode potentials of -0.2 and −0.4 V versus SCE and the voltammetric half-wave and peak potentials in 0.1 M chloroacetate buffer, scan rate 10 mVs^{-1}

	−0.2 V		−0.4 V		$E_{½}$, V
Phase (Sample)	$k' \times 10^{-3}$, s^{-1}	g	$k' \times 10^{-3}$, s^{-1}	g	
Ferrihydrite (A)	45	1.32	>100	/	0.00
(B)	33	1.38	>100	/	−0.10
α-FeOOH					
(A)	26	1.15	22	1.36	−0.34
(B)	8.0	1.18	15	1.13	−0.49
(D)	3.8	1.18	6.9	1.28	−0.53
(E)	7.1	1.17	5.9	0.93	−0.55
(F)	4.0	1.01	5.6	1.38	−0.50
(G)	3.6	1.38	/	/	−0.47
β-FeOOH	4.1	0.88	[a]		−0.16[b]
γ-FeOOH	55	1.35	>100	/	−0.01
δ-FeOOH	43	1.31	>100	/	0.00
Fe_3O_4	65	1.98	>100	/	−0.02
α-Fe_2O_3	0.6	1.12	2.7	1.04	−0.56
γ-Fe_2O_3	63	2.62	>100	/	−0.08
Sediments: BRS			15	3.31	[c]
SLI	16	1.73	16	2.81	[c]
Fe^{3+}(aq)[d]	Reduction process obeying Cottrell's equation				−0.08[e]
					+0.30[f]

The samples designated by A, B, C, etc., were prepared under different experimental conditions [B40]

[a]Chronoamperometric curve is not monotonous

[b]Broad peak, slow current decrease after the maximum

[c]No clear maximum, slowly rising wave

[d]$c_{Fe} = 1$ mmol l^{-1}

[e]Clean electrode surface

[f]Electrode surface with deposited FeOOH

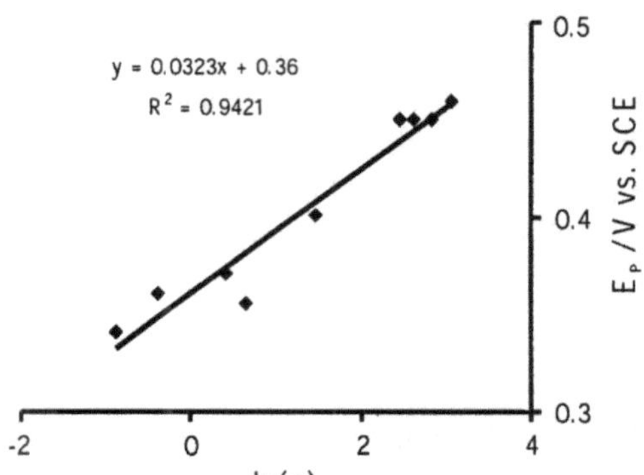

Fig. 5.81 Peak potential of reductive dissolution of Li–Mn–O spinels versus natural logarithm of specific surface area [B191]. Copyright ©2001 with permission from Springer

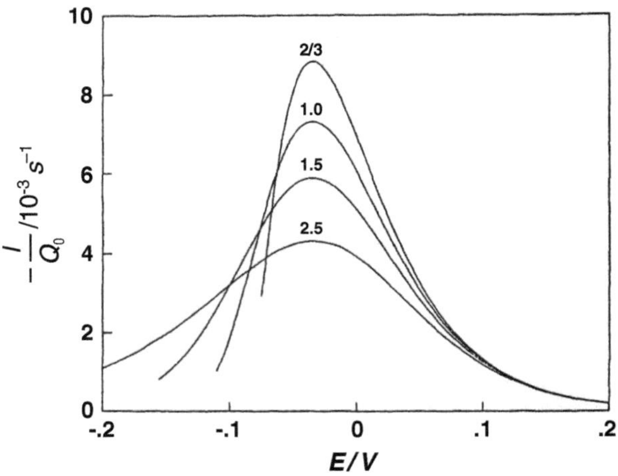

Fig. 5.82 The influence of the g parameter on the shape of the voltammetric curves of the reductive dissolution. The curves were calculated for $\alpha = 0.5$, $k_0 = 1 \times 10^{-2}\ \mathrm{s}^{-1}$, and $v = 1\ \mathrm{mV\ s}^{-1}$. The g-values are given at the curves [B52]. Copyright ©1996 with permission from Elsevier

with k_E being the rate constant of the electrode reaction in m s^{-1} at E^0 and d the particle diameter. Equation (5.24) was approximated by the following equation:

$$\left(\frac{I_\mathrm{p}}{Q_0}\right) = 0.37\frac{\alpha nF}{RT}v(1 - 0.47\ \ln g) \tag{5.26}$$

that could be observed experimentally for several oxides with only slightly different coefficients [B52]:

$$\left(\frac{I_\mathrm{p}}{Q_0}\right) = 0.2\frac{\alpha nF}{RT}v(1 - 0.5\ \ln g) \tag{5.27}$$

For some metal oxides, more elaborate models have been developed to describe the electrochemical dissolution. These models take into account an intermediate surface complexation. The general approach was to separate the potential and conversion-dependent functions that determine the overall kinetics of dissolution. Both contributions have been separated by the combination of potentiostatic and potentiodynamic measurements [B81]. In a study of the electrochemical oxidative dissolution of chromium oxides, good indications have been found for two one-electron steps preceding the dissolution [B92]. This result was obtained using a potential pulse technique with which a formal charge transfer coefficient can be determined on the basis of the following equation:

$$a_E = \frac{nF}{RT}\alpha_E = \frac{\ln(I_{\mathrm{B}+\Delta}) - \ln(I_\mathrm{B})}{\Delta} \tag{5.28}$$

The meaning of the symbols follows from Fig. 5.83. From the dependence of a_E on the electrode potential and a fitting of these curves based on kinetic models, the

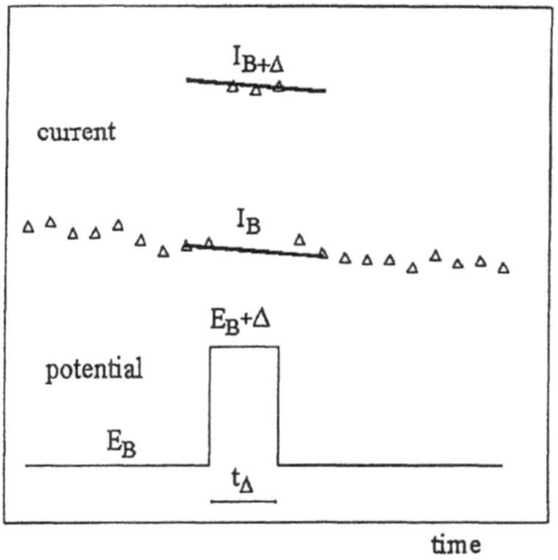

Fig. 5.83 Potential time scheme and current response for the determination of formal transfer coefficients according to [B92]. Copyright ©1998 with permission from Springer

author came to conclusions about the validity of the models. Figure 5.84 depicts these curves for some oxidic compounds. Note the two inflection points in case of the pure chromium oxides indicating two one-electron steps. Similar studies have been performed with the reductive decomposition of manganates (V). From the experiments, it was derived that the reduction is controlled by an irreversible one-electron surface reaction, whereas the reduction of MnO_2 is controlled by a reversible reaction coupled with a discussion in the solid [B105].

Several similar systems have been studied by Grygar and coworkers with respect to their reactivity under electrochemical conditions [B104, B117, B118, B125, B141, B181, B191, B194]. In [B192] the application of the discussed methodology to the study of the dissolution kinetics of metallic iron particles has been described.

5.8 The Electrochemistry of Single Particles

This monograph would be incomplete without a discussion of studies of the electrochemistry of single particles. The use of single particles is, of course, extremely appealing as such studies can, in principle, provide information that is otherwise inaccessible. In case of the AFM investigations discussed in Sect. 4.2, in fact single particles have been imaged and the course of their electrochemical transformations was visualized. This was achieved by immobilizing arrays of particles on an electrode surface and focusing the AFM on a single or just some particles. Of course, the simultaneously recorded electrochemical signals were caused by all immobilized particles.

Here we shall discuss the preparation and immobilization of a single particle on an electrode with the aim of performing the electrochemical measurement with just

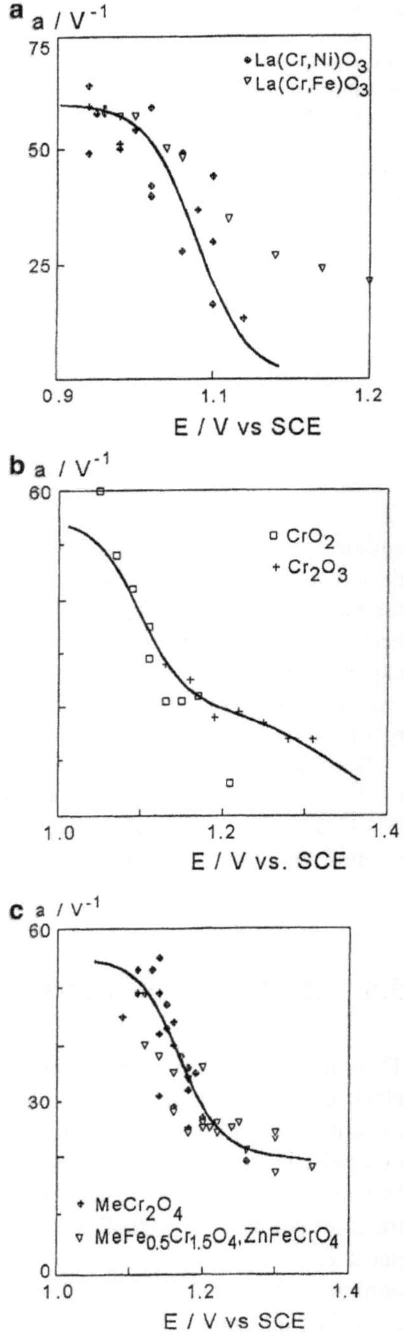

Fig. 5.84 a_E versus potential curves for Cr oxides: (**a**) perovskites, (**b**) pure Cr oxides, and (**c**) spinels. *Solid lines* are simulated based on a kinetic model [B92]. Copyright ©1998 with permission from Springer

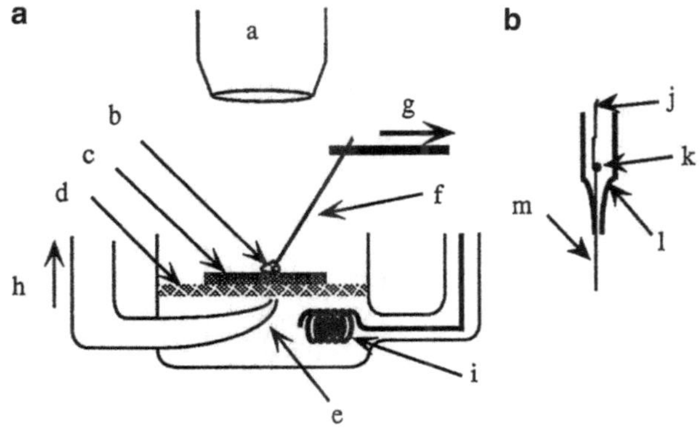

Fig. 5.85 Experimental setup for single-particle measurements: (**a**) the measurement system and (**b**) the carbon fiber microelectrode. (a) microscope, (b) particle to be studied, (c) glass separator, (d) glass frit, (e) Luggin capillary, (f) microelectrode, (g) *x*–*y*–*z* manipulator, (h) reference electrode, (i) counter electrode, (j) lead wire, (k) Ag paste, (l) glass tube, (m) carbon fiber (diameter 10 μm) [B47]. Copyright ©1995 with permission from Elsevier

one particle. This is rather difficult and such measurements will probably be always confined to specialized laboratories.

In 1990 Bursell and Björnbom published an experimental setup that allows contacting agglomerates of particles from 5 μm onwards with the help of a carbon fiber [B10]. The agglomerate was transferred with a syringe, together with electrolyte solution, onto an electrolyte-soaked separator that was in contact with a reference electrode. Uchida et al. [B47] have improved the methodology for single-particle measurements (see Fig. 5.85). These authors have used their equipment for the determination of the diffusion coefficients of hydrogen in $LaNi_5$ particles. For this purpose, they measured the current-time response for the oxidation of hydrogen after a previous loading of the particle. The response was analyzed in a short time and a long time range, as the diffusion regime was supposed to change from semi-infinite planar diffusion to finite diffusion. In the case of LaN_5 particles, the long time response yielded diffusion coefficients that were in agreement with independent NMR data; however, the short time range yielded deviating values that were attributed by the authors to a micro-roughness of the particles. In case of measurements with very smooth spherical Pd particles with a diameter of 520 μm, the authors obtained in both time domains diffusions coefficients of hydrogen that were in good agreement.

In a later study, Uchida et al. [B78] investigated the electrochemistry of single particles of $LiCoO_2$ and $LiMn_2O_4$. The size of the single particles was between 7 and 16 μm. The measured currents were in the nA range. The authors emphasize that cyclic voltammetric measurements of single particles with scan rates between 1 and 10 mV/s yield information on the cycle performance of these battery materials because the entire volume of the particle is converted.

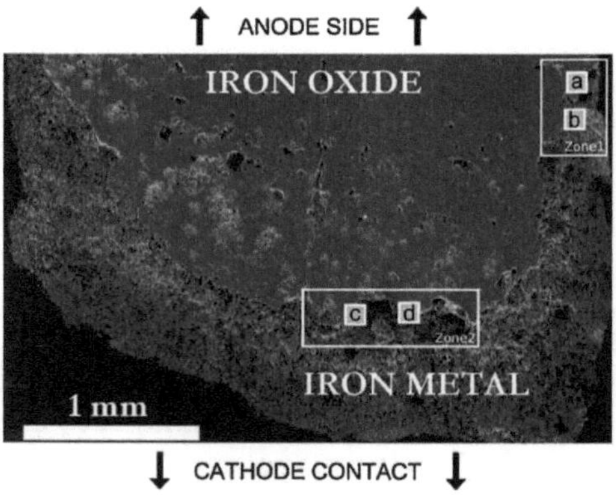

Fig. 5.86 Cross section of a hematite particle partially electrolyzed in 50 wt% NaOH. The cathode contact point is at the bottom and the anode direction is the upper side of the view. Zones 1 and 2 correspond, respectively, to the outer progressing interface and confined zones [B427]. Copyright ©2010 with permission from Elsevier

The same methodology has been applied by Perdicakis et al. [B80] to study the voltammetric behavior of silver chalcogenide particles, i.e., Ag_2Se, Ag_2Te, and Ag_2S.

The examples published so far show that the advantages of performing the electrochemical measurement with just one particle are practically the same as those performed with an array of well-separated particles immobilized on an electrode, as described in all the other parts of this chapter. Unfortunately, nobody has yet endeavored to use single-particle measurements to study the relation between the shape and the size of single particles and their electrochemical response (see also Sect. 5.11).

Recently, Allanore et al. [B427] have studied the electrochemical conversion of single hematite particles in contact with alkaline solutions at both macro- and microscopic scales, using a Ni disk electrode where hematite particles of mass between 3 mg and 1 g were attached. These authors observe the formation of a light gray layer of Fe metal which develops upward on the particle surface, from the cathode electrical contact point toward the anode direction (see Fig. 5.86). Cross-sectional examination of the particles (Fig. 5.87) indicates that the reaction occurred within the particle volume, toward its core, and is accompanied by the appearance of a porous magnetite phase at the hematite grain boundaries, as confirmed by Mössbauer spectroscopy and chemical analysis.

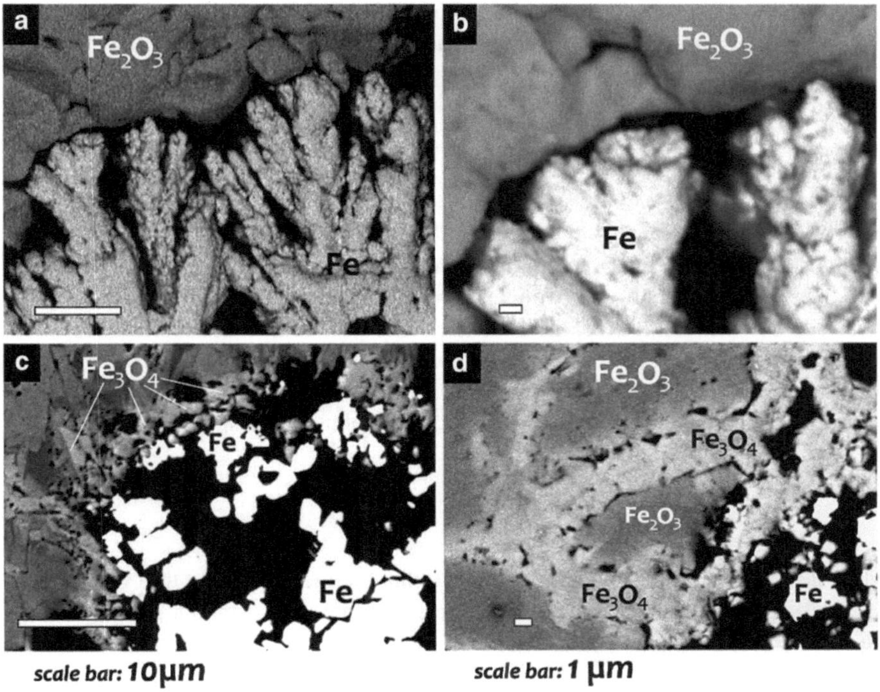

Fig. 5.87 Backscattered electron microscopy images of oxide/metal interfaces at high magnification for the partially electrolyzed hematite particle presented in Fig. 5.71. (**a** and **b**) Zone 1: the outer surface progressing interface. (**c** and **d**) Zone 2: the confined zone [B427]. Copyright ©2010 with permission from Elsevier

5.9 Chemical and Enzymatic Conversion of Immobilized Particles

5.9.1 In Situ Studies

Electrochemical techniques can also be applied for monitoring chemical or biochemical reactions of powders when the latter are immobilized on an electrode surface. First, we shall discuss the case that a chemical dissolution of metal particles is followed by monitoring the potential of the electrode [B84]. Let us consider the following dissolution of a metal M by oxygen that is present in the solution:

$$M + \frac{1}{2}O_2 + 2H^+ \rightarrow M^{2+} + H_2O \qquad \text{(XXX)}$$

Assuming that the activity of the solid metal is unity, the so-called stripping potential is defined by the equation for an electrode of the first kind. The concentration of metal ions at the electrode surface depends on the flux of O_2. It can be assumed that for a stirred solution, a steady state is established and that the bulk concentration of metal ions is negligible as well as the concentration of O_2 at the surface of the electrode. The stripping potential E_S of the diffusion-controlled metal oxidation depends on the bulk concentration of O_2 as follows:

$$E_S = E^{\theta}_{M} + \frac{RT}{2F}\ln c^*_{O_2} + \frac{RT}{2F}\ln\frac{2D_{O_2}\delta_M}{D_M\delta_{O_2}} \tag{5.29}$$

where D_{O_2}, δ_{O_2}, $c^*_{O_2}$, D_M, and δ_M are diffusion coefficients, diffusion layer thicknesses, and the bulk concentration of O_2 and the metal ions, respectively. The stripping potential does not depend on the amount of metal immobilized on the electrode surface. Of course, the formation of metal complexes in solution would shift the stripping potential. The amount of immobilized metal determines the time that is necessary to dissolve the metal in the oxidation reaction. The rate of dissolution depends on the concentration gradient of oxygen according to the equation:

$$\frac{\rho S}{M_M}\left(\frac{dl^*}{dt}\right) = -\frac{2SD_{O_2}c^*_{O_2}}{\delta_{O_2}} \tag{5.30}$$

when it is assumed in a first approximation that the metal forms a thin layer with a thickness l^* and a surface area S. M_M is the molar mass of the metal. The solution of Eq. (5.30) is as follows:

$$\tau = \frac{\rho\delta_{O_2}l^*}{2D_{O_2}M_Mc^*_{O_2}} \tag{5.31}$$

τ is the so-called transition time determined from the potential versus time recording.

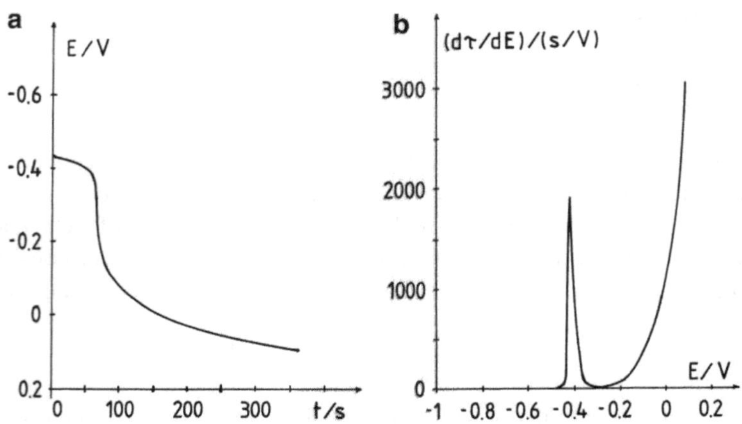

Fig. 5.88 Direct (**a**) and derivative (**b**) stripping chronopotentiogram of the dissolution of lead microparticles that have been transferred to a PIGE. The electrolyte was a 0.5 mol L^{-1} $NaClO_4$ solution with a pH = 2 ($HClO_4$). The oxygen concentration was 2.6×10^{-4} mol L^{-1}. The initial potential was −1 V [B84]. Copyright ©1998 with permission from Elsevier

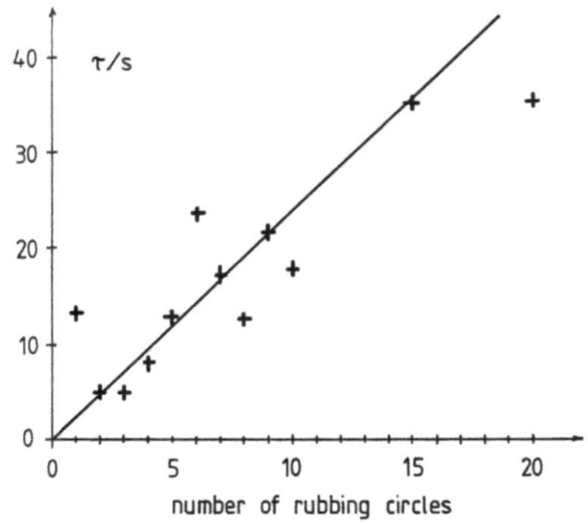

Fig. 5.89 Dependence of transition time τ on the number of rubbing circles in which the lead was transferred from the piece of lead metal to the PIGE [B84]. Copyright ©1998 with permission from Elsevier

Figure 5.88 shows the direct and derivative stripping chronopotentiogram of dissolution of lead traces that have been transferred by the abrasive technique from a piece of lead to a PIGE electrode. Figure 5.89 depicts the dependence of the transition times on the number "rubbing circles" that have been performed to transfer lead from the piece of metal onto the PIGE. The latter correlation is remarkably good. From these experiments, one can conclude that such chronopotentiometric measurements are suitable to study the amount of metals that are transferred in mechanical friction experiments. The second example of a study of a chemical conversion of immobilized microparticles concerns the enzymatic oxidation of quinhydrone [B133]. Quinhydrone microcrystals have been immobilized onto a gold electrode by evaporation of a solution in ethanol. Laccase, an enzyme isolated from cultures of the fungus *Pycnoporus cinnabarinus*, is capable of catalyzing the oxidation of hydroquinone and thus also that of quinhydrone to quinone. From the potential of the quinhydrone-modified electrode, it is easy to calculate the amount of hydroquinone. Figure 5.90 shows the dependence of the amount of hydroquinone on the electrode surface as a function of time when the electrode is subjected to a laccase solution that was oxygen saturated. This figure shows that the laccase concentration clearly determines the rate of oxidation of hydroquinone. These two examples show that chronopotentiometry is a powerful technique for monitoring chemical and biochemical reactions of microparticles that are attached to the surface of electrodes.

5.9.2 *Ex Situ Studies*

The third example concerns a chemical reaction between dissolved ions (hexacyanoferrate(II) ions) and an oxide powder (goethite) and of a colloid (Prussian blue) and an oxide (goethite) [B187]. These reactions are of some environmental interest because there are many places in industrial areas where the soil is contaminated with Prussian blue, esp. from coal gasification plants. To study these

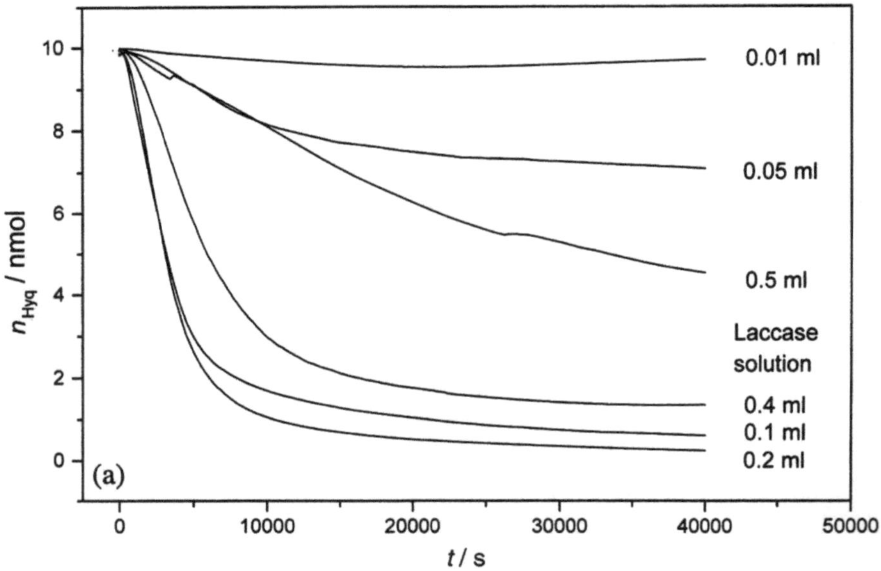

Fig. 5.90 The amount of hydroquinone at the electrode surface as a function of time of the enzymatic oxidation of immobilized quinhydrone for different additions of a laccase solution (the laccase solution had an activity of 0.012 U mL^{-1}) [B133]. Copyright ©2000 with permission from Elsevier

reactions, suspensions of goethite have been exposed to hexacyanoferrate(II) ions or Prussian blue colloidal solutions in a well-defined way. The reaction products were isolated and small amounts of the dry powders were immobilized on a PIGE. The electrochemical response of the immobilized microparticles proved that hexacyanoferrate(II) ions are sorbed on goethite. Figure 5.91 shows that the goethite particles that were exposed to a hexacyanoferrate(II) solution exhibit a signal of high-spin iron, but no signal of low-spin iron. This result is in line with spectroscopic results and indicates that the hexacyanoferrate ions are surface bound. The iron(III) ions on the goethite surface are not electroactive as they are in Prussian blue. Figure 5.92 depicts cyclic voltammograms of goethite powder that was exposed to a Prussian blue colloid. Clearly, the typical Prussian blue response is visible; however, the response decreases upon cycling. This bears witness of a rather low stability of the Prussian blue-goethite surface complex.

Recently, Sander et al. [B331] have introduced mediated electrochemical reduction and oxidation techniques in order to determine the number of electrons transferred to and from natural simples. Such techniques are based on the use of one-electron-transfer mediating compounds to facilitate electron transfer between the sample of interest and a working electrode. In this approach, an electrochemical cell is set to a constant potential value while the current is measured over time. In the presence of a mediator, known amounts of a sample are added to the electrolyte, resulting in current responses whose integration yields the number of electrons transferred. This methodology has been applied to assess the redox properties of humic substances [B428, B429] and quantify structural Fe in Fe-bearing smectite clays [B430, B431]. In this

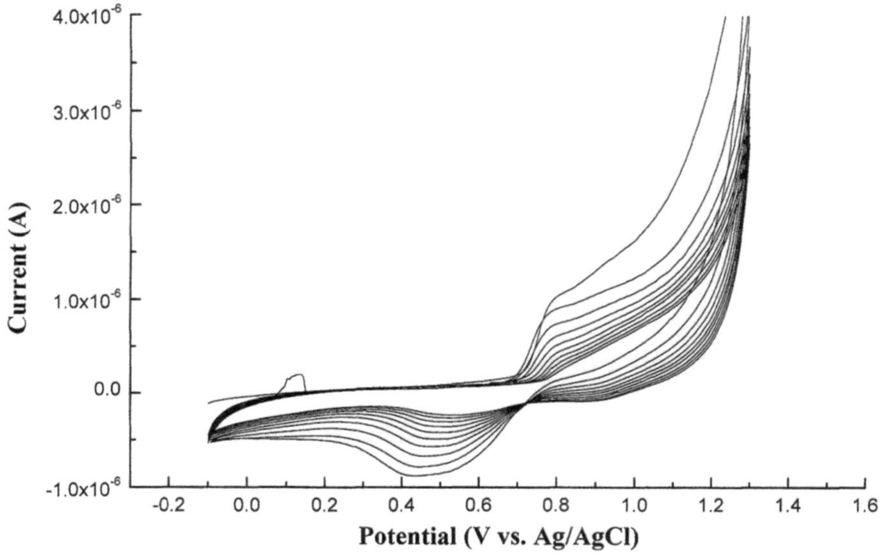

Fig. 5.91 Cyclic voltammogram of a goethite sample with a load of 6.05 g CN^- (as hexacyanoferrate)/kg goethite. Electrolyte 0.1 mol L^{-1} KCl, scan rate 0.05 V s^{-1} [B187]. Copyright ©2001 with permission from Elsevier

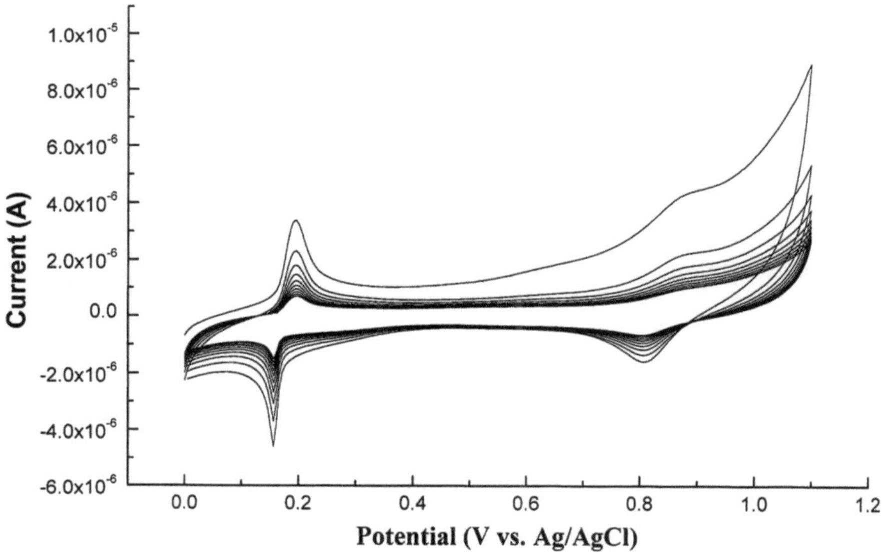

Fig. 5.92 Cyclic voltammogram of a goethite sample with a load of 36.7 g CN^- (as Prussian blue)/kg goethite. Electrolyte 0.1 mol L^{-1} KCl, scan rate 0.05 V s^{-1} [B187]. Copyright ©2001 with permission from Elsevier

last case, experiments were performed on suspensions of the clays optionally treated with dithionite in an electrochemical cell using a cylindrical vitreous carbon working electrode and diquat as a redox mediator. Mediated electrochemical oxidation/reduction techniques allow to directly quantify the electron-accepting and electron-donating capacities of smectites and other clay minerals and discriminate between total Fe and redox-active Fe contents in these minerals [B430] as well as assess redox properties due to reversible and irreversible structural changes that occur as a result of reduction and oxidation of structural Fe [B431].

5.10 The Screening of Electrocatalysts

Electrocatalysts facilitate the electrochemical reduction or oxidation of dissolved compounds on electrodes by decreasing the overpotential, i.e., the activation barrier. Voltammetry of immobilized particles lends itself for the screening of potential electrocatalysts as it allows for a very quick comparison of their effectiveness. Later, the mechanical immobilization of a chosen electrocatalyst may even be used for the modification of working electrodes for analysis. Figure 5.93 shows the results of a search for the most effective electrocatalysts for hydrazine oxidation among many metal hexacyanoferrates.

Table 5.5 gives an overview on the analytical application of electrodes modified by immobilized particles for the sake of electrocatalytic determination of dissolved substrates.

Kucernak et al. [B147, B208, B243] have used the mechanical attachment of nanostructured platinum particles to study their catalytic activity for hydrogen evolution and the electrooxidation of formic acid. They immobilized the particles on highly oriented pyrolytic graphite and polished gold electrodes. Rolison et al. [B223] assessed the electrocatalytic properties of several metal and metal oxide particulate materials by immobilizing the compounds on the surface of carbon-wax composite electrodes that consisted of 35 wt% carbon black (acetylene black) and 65 wt% wax (eicosane, beeswax, or paraffin wax).

5.11 Theoretical Aspects of the Voltammetry of Immobilized Particles

Parallel to the experimental studies, efforts have been made for a deeper understanding of the theoretical basis of electrochemical reactions of immobilized particles. At the beginning of the mathematical modeling, the initial conditions and the models have to be defined. Since crystals have finite dimensions, a finite diffusion space has to be considered leading to voltammetric features that differ from those for semi-infinite planar diffusion. For this reason, the following section starts with a voltammetric theory for a thin-layer cell.

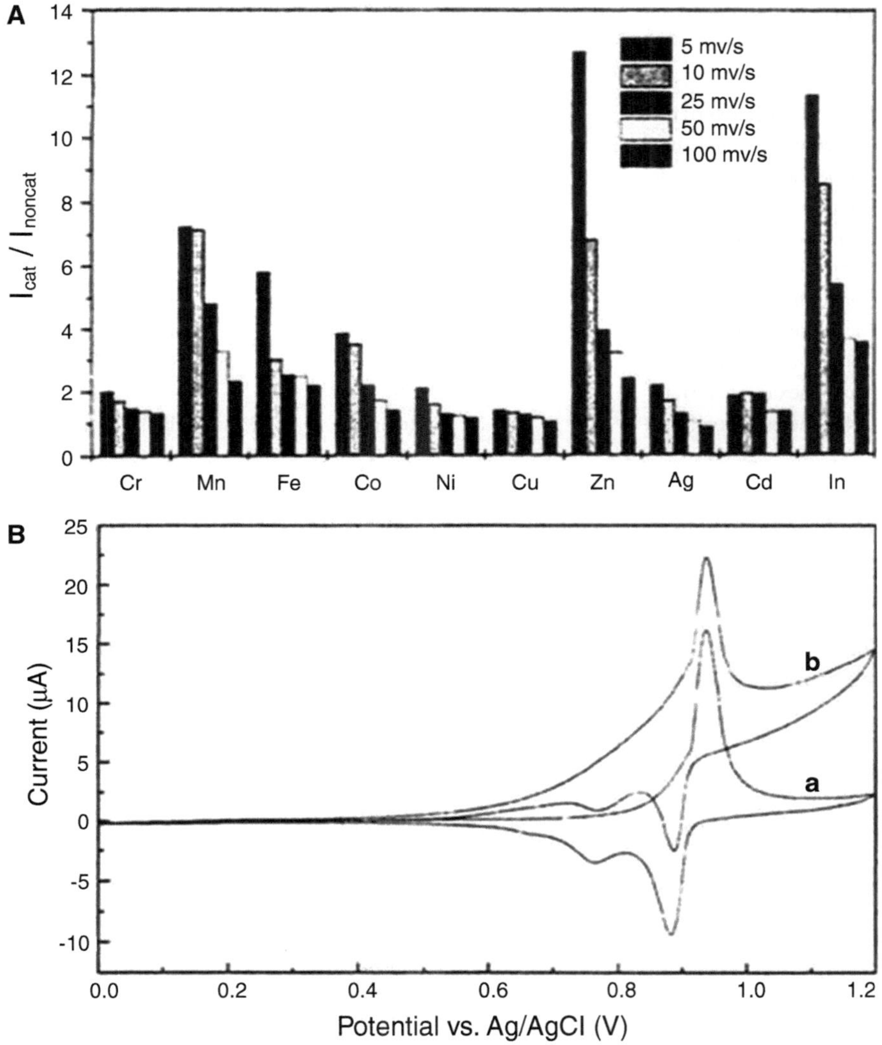

Fig. 5.93 (**A**) Comparison of the catalytic activities of metal hexacyanoferrates for the electrocatalytic oxidation of hydrazine at 700 mV versus Ag/AgCl (3 mol L^{-1} KCl). Electrolyte 0.1 mol L^{-1} KNO_3, pH 6 (phosphate buffer), 1 mmol L^{-1} hydrazine. The ratio of the catalytic to the noncatalytic current is given for different scan rates. (**B**) Cyclic voltammograms for the electrocatalytic oxidation of hydrazine (a) without and (b) with zinc hexacyanoferrate [B113]. Copyright ©1999 with permission from Wiley-VCH

5.11.1 Staircase Voltammetry with Finite Diffusion Space

Due to the digital data processing implemented in all modern voltammetric instrumentation, it was necessary to model the theoretical response of a redox system

Table 5.5 Electrocatalysts applied in the form of particles immobilized on electrodes and substrates to be determined in analyses

Electrocatalyst	Substrate	References
Metal hexacyanoferrates	Hydrazine	[B13, B202, B253]
Ruthenium(III) diphenyldithiocarbamate	Sulfhydryl compounds	[B98]
Copper hexacyanoferrate	Cysteine	[B99]
Ruthenium(III) dithiocarbamate	Ascorbic acid	[B100]
Nickel hexacyanoferrate	Thiosulfate	[B130]
Copper hexacyanoferrate	Sulfur dioxide	[B120]
Copper hexacyanoferrate	Ascorbic acid	[B115]
Nickel hexacyanoferrate	Sulfur dioxide	[B160]
Copper hexacyanoferrate	Glutathione	[B199]
Nickel hexacyanoferrate	Thiols	[B201]
Cobalt hexacyanoferrate	Glutathione	[B212]

confined to a finite diffusion space in staircase voltammetry. For this purpose, a simple redox reaction proceeding in a thin-layer cell has been considered [B64]:

$$\mathrm{A} + n\mathrm{e}^- \rightleftarrows \mathrm{B} \tag{XXXI}$$

The diffusion of the reactant and the product of this redox reaction in the space between two parallel planes is defined by the following differential equations:

$$\frac{\partial c_\mathrm{A}}{\partial t} = D\frac{\partial^2 c_\mathrm{A}}{\partial x^2};\frac{\partial c_\mathrm{B}}{\partial t} = D\frac{\partial^2 c_\mathrm{B}}{\partial x^2} \tag{5.32}$$

with the initial and boundary conditions:

$$\begin{aligned}
&t = 0, 0 \le x \le L : c_\mathrm{A} = c_\mathrm{A}*, c_\mathrm{B} = 0\\
&t > 0 : \left(\tfrac{\partial c_\mathrm{A}}{\partial x}\right)_{x=0} = \left(\tfrac{\partial c_\mathrm{B}}{\partial x}\right)_{x=0} = 0;\\
&\left(\tfrac{\partial c_\mathrm{A}}{\partial x}\right)_{x=L} = -\left(\tfrac{\partial c_\mathrm{B}}{\partial x}\right)_{x=L} = -\frac{I}{nFSD};\\
&\frac{I}{nFS} = k_s\exp(-\alpha\varphi)\left[(c_\mathrm{A})_{x=L} - \exp(\varphi)(c_\mathrm{B})_{x=L}\right];\ \varphi = \frac{nF(E - E^\ominus)}{RT}
\end{aligned} \tag{5.33}$$

Here, c_A, c_B, and c_A* are concentrations; t and x are the time and space variables, respectively; L is the distance between the parallel planes; D is the diffusion coefficient; S is the electrode surface area; I is the current; E is the applied potential; $E^\ominus$ is the standard redox potential of the couple A/B; k_s is the standard rate constant; and α is the electron transfer coefficient, while n, F, R, and T have their usual meaning. If Reaction (XXXI) is fast and reversible, the dimensionless peak currents of the cyclic staircase voltammograms are a function of the dimensionless thickness parameter ω defined as $\omega = 5L/\sqrt{Dt}$ with $t = \Delta E/v$, v being the scan rate and ΔE being the potential increment. This dependence has a sigmoidal shape as shown in

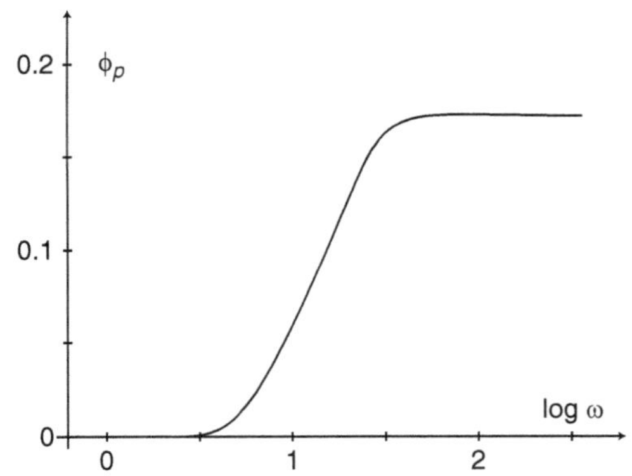

Fig. 5.94 Theoretical dependence of the dimensionless peak current Φ_p of a reversible reaction on the logarithm of the dimensionless thickness parameter ω of the cell. The potential increment was $\Delta E = 5$ mV [B64]. Copyright ©1997 with permission from Wiley-VCH

Fig. 5.77. A characteristic feature of the staircase voltammograms obtained in a cell with $\log(\omega) < 1.5$ is a decrease of the peak current followed by the vanishing of the voltammetric response for a very thin film. This is a consequence of the current sampling procedure in all staircase techniques. If all redox species inside the cell acquire the Nernst equilibrium before the sampling time, there will be no current to be measured at the end of the potential pulses, i.e., when the current is sampled. When Reaction (XXXI) is controlled by the kinetics of charge transfer, the voltammetric features are additionally controlled by the dimensionless kinetic parameter λ defined as $\lambda = k_s\sqrt{t}/\sqrt{D}$. The influence of the redox kinetics depends on the cell thickness depicted in Fig. 5.94. For films with moderate thicknesses (curves 2 and 3 in Fig. 5.95), there is a parabolic dependence of the dimensionless peak currents on λ. This feature is well known for surface-confined redox reactions as a so-called quasi-reversible maximum. It is used for determining the standard rate constant k_s. The quasi-reversible maximum, together with the narrower peak-to-peak separation, distinguishes the models of finite diffusion space from those with semi-infinite planar diffusion.

5.11.2 The Propagation of a Redox Reaction Through Microcrystals

A key issue for understanding the voltammetry of microparticles is the propagation of a redox reaction through the microcrystals. It is reasonable to assume that the Nernst equilibrium is initially established at the *three-phase junction*, where the particle, the solution, and the electrode meet. The reaction starts at this junction line, and from this locus, it may advance along the surface and into the particle.

Chronoamperometry has been considered in a first attempt to theoretically tackle the question of the propagation of a redox reaction through a microcrystal [B69]. In order to solve the problem of the coupled transport of electrons and ions, an infinite crystal attached to an infinite electrode surface has been chosen (cf. Fig. 5.96). The

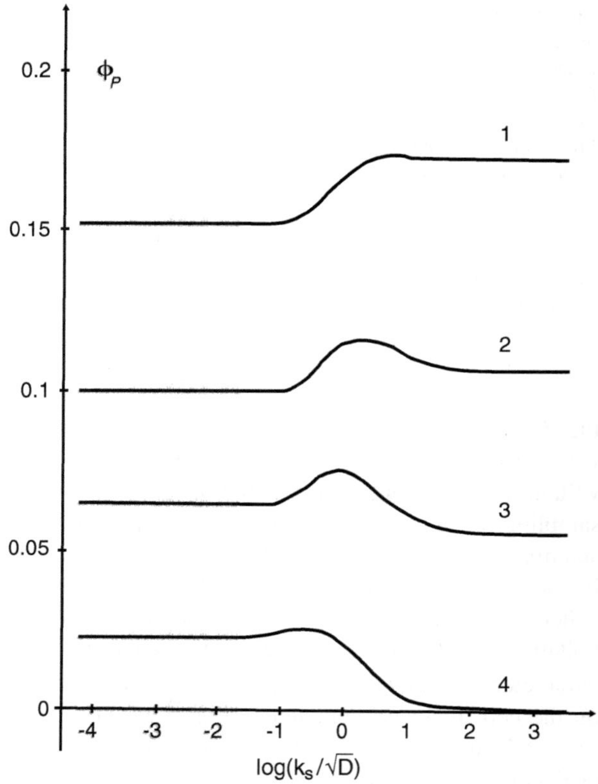

Fig. 5.95 Dependence of the dimensionless peak currents on the logarithm of the standard rate constant. $\Delta E = 5$ mV, $v = 50$ mV/s; $5LD^{-0.5}/s^{0.5} = 100$ (1), 5 (2), 3 (3), and 1 (4) [B64]. Copyright ©1997 with permission from Wiley-VCH

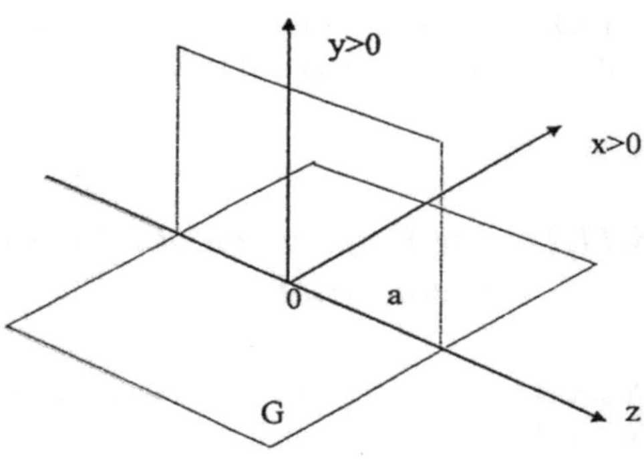

Fig. 5.96 Coordinate system with a part of the crystal–solution interface (a) and a part of the electrode graphite surface (G) [B69]. Copyright ©1997 with permission from Springer

axis z coincides with the three-phase junction. The electrode surface is located in the x–z plane, and its body lies in the space for $y < 0$. The remaining space ($y > 0$) is divided between the crystal ($x > 0$) and the solution ($x < 0$). It is assumed that the Nernst equilibrium of the insertion electrochemical reaction

$$\{C_m\mathrm{AX}\}_{\mathrm{solid}} + ne^- + n[\mathrm{C}^+]_{\mathrm{solution}} \rightleftarrows \{C_{m+n}\mathrm{BX}\}_{\mathrm{solid}} \quad \text{(XXXII)}$$

is initially established along the axis z, where both electrons and ions are immediately available. Thereafter, the current is conducted along the crystal surface. This surface redox reaction creates a gradient of electrochemical potential and defines the initial conditions for the diffusion of ions into the crystal. The fluxes of electrons and ions inside the solid particle are perpendicular to each other. The surface diffusion of the electrons is described by the differential equation of Eq. (5.34):

$$\frac{\partial \Gamma_{\mathrm{B}}}{\partial t} = D_{\mathrm{e}} \frac{\partial^2 \Gamma_{\mathrm{B}}}{\partial y^2} \tag{5.34}$$

with the following initial and boundary conditions:

$$\begin{aligned}
&t = 0, y \geq 0 : \Gamma_{\mathrm{B}} = 0, \Gamma_{\mathrm{A}} = \Gamma;\ t > 0, y \to \infty : \Gamma_{\mathrm{B}} \to 0, \Gamma_{\mathrm{A}} \to \Gamma; \\
&y \geq 0 : \Gamma_{\mathrm{A}} + \Gamma_{\mathrm{B}} = \Gamma;\ y = 0 : \Gamma_{\mathrm{A},y=0} = \Gamma_{\mathrm{B},y=0} \exp(\varphi); \\
&\varphi = \frac{nF\left(E - E_{\mathrm{c}}^{\ominus'}\right)}{RT};\ \frac{I_{\mathrm{s}}}{nFbD_{\mathrm{e}}} = -\left(\frac{\mathrm{d}\Gamma_{\mathrm{B}}}{\mathrm{d}y}\right)_{y=0}
\end{aligned} \tag{5.35}$$

where Γ_{A} and Γ_{B} are the surface concentrations of ions A and B and Γ is the total surface concentration. D_{e} is a diffusion coefficient of the electrons, I_{s} is the surface component of the current, and b is the length of the contact between the electrode, the crystal, and the solution. In chronoamperometry the solution for the *surface current* is given by

$$I_{\mathrm{s}} = nFb\Gamma[1 + \exp(\varphi)]^{-1} \sqrt{\frac{D_{\mathrm{e}}}{\pi t}} \tag{5.36}$$

Equation (5.29) implies that the *surface current* depends on the diffusion coefficient of the electrons and it decreases with the square root of time. Within the solid particle, the diffusions of electrons (i.e., here the propagation of electrons by hopping) and ions are coupled. The advancement of electrons is slower than at the surface since they must wait for ions (supposed that ion diffusion is slower than electron self-exchange). The diffusion of ions is described by the following differential equation:

$$\frac{\partial c_{\mathrm{B}}}{\partial t} = D_{\mathrm{C}} \frac{\partial^2 c_{\mathrm{B}}}{\partial x^2} \tag{5.37}$$

which was solved for the following initial and boundary conditions:

$$\begin{aligned}&t = 0, x \geq 0 : c_{\mathrm{B}} = 0, c_{\mathrm{A}} = \rho_{\mathrm{d}};\\&t > 0, x \to \infty : c_{\mathrm{B}} \to 0, c_{\mathrm{A}} \to \rho_{\mathrm{d}};\ x \geq 0 : c_{\mathrm{A}} + c_{\mathrm{B}} = \rho_{\mathrm{d}};\\&y = 0 : x = 0 : (c_{\mathrm{A}})_{x=0,y=0} = (c_{\mathrm{B}})_{x=0,y=0}\exp(\varphi);\\&x = 0 : (c_{\mathrm{B}})_{x=0} = \rho_{\mathrm{d}}[1 + \exp(\varphi)]^{-1}\mathrm{erfc}\left[0.5y(D_{\mathrm{e}}t)^{-0.5}\right]\\&\frac{\mathrm{d}I_{\mathrm{c}}}{\mathrm{d}S} = -nFD_{\mathrm{C}}\left(\frac{\partial c_{\mathrm{B}}}{\partial x}\right)_{x=0};\ \mathrm{d}S = b\mathrm{d}y\end{aligned} \tag{5.38}$$

Here, D_{C} is the diffusion coefficient of the ions, I_{c} is a volume component of the current, while $S = by$ is the surface area of the crystal which is exposed to the solution and ρ_{d} is a common density (mol cm^{-3}). The concentration of the product B and the current within the crystal are obtained as follows:

$$c_{\mathrm{B}} = \frac{\rho_{\mathrm{d}}}{1 + \exp(\varphi)}\mathrm{erfc}\left(\left[0.5x(D_{\mathrm{C}}t)^{-0.5}\right] + \left[0.5y(D_{\mathrm{e}}t)^{-0.5}\right]\right) \tag{5.39}$$

$$\frac{\mathrm{d}I_{\mathrm{c}}}{\mathrm{d}S} = \frac{nF\rho_{\mathrm{d}}\sqrt{D_{\mathrm{C}}}}{1 + \exp(\varphi)}\frac{1}{\sqrt{\pi t}}\exp\left(-\frac{y^2}{4D_{\mathrm{e}}t}\right) \tag{5.40}$$

Equation (5.39) shows that the product of Reaction (XXXII) appears primarily near the two crystal surfaces: the one which faces the solution ($x = 0$) and the one which is in contact with the electrode ($y = 0$). In both planes, the concentration of the product decreases with the distance from the three-phase boundary, i.e., it propagates through the body of the crystal. Equation (5.33) shows that the current density is highest at the contact line b, where $y = 0$, and decreases along the crystal surface. Equations (5.34) and (5.37) have also been solved for conditions of single scan and cyclic voltammetry. In both cases, the solutions give a steady-state current because of the infinite volume of the crystal. Since Eq. (5.39) implies that Reaction (XXXII) propagates through the crystal, a term of the *advancing front* has been introduced to depict this propagation. In the case that both the surface and the bulk (volume) reaction proceed at comparable rates, the reaction front expands from the three-phase boundary as a space with exponential borderlines in the x and y directions. Generally, the total current is the sum of the surface and the volume current. The reaction will be surface confined when the diffusion of the ions inside the crystal is hindered, and the surface current will be negligible when the volume reaction is dominating. *In any case, the three-phase junction is the starting point for the reaction front independent of the geometry of the particle and its conductivity.* Qualitatively, the same results have been obtained by using a rigorous model for the coupled diffusions of electrons and ions in two directions by using the lattice-gas concept without interactions [B107]. By that model, it has been shown that the surface and the volume currents depend on Wagner's thermodynamic factor (W) of electrons and ions. The Wagner factor is defined as $W = 1 + \partial \ln(\gamma)/\partial \ln(c)$ where γ is the activity coefficient and c is the concentration. The transport in direction perpendicular to the electrode surface depends on the Wagner factor of electrons, while the transport parallel to the electrode surface depends on the Wagner factor of ions. The final mathematical solutions of both models under conditions of chronoamperometry, single scan, and cyclic voltammetry are given elsewhere [B69, B107] and can be applied directly in any appropriate mathematical program.

5.11.3 The Effect of the Electrolyte Concentration on the Voltammetric Response of Insertion Electrodes

In the previous model of insertion electrochemical reactions with coupled ion and electron transfer, the solution side has been neglected. However, under certain conditions that may not be justified. Therefore, the following model has been developed: a conductive cylinder of the solid compound is embedded in the surface of the electrode in a way that only a circular surface of the cylinder is exposed to the solution. This surface is positioned at $x = 0$, x being positive inside the crystal and negative inside the solution. Assuming the insertion electrode to be of the type [B87]:

$$\{\mathrm{Ox}\}_{\mathrm{solid}} + \mathrm{e}^- + \mathrm{C}^+{}_{\mathrm{solution}} \rightleftarrows \{\mathrm{Cred}\}_{\mathrm{solid}} \quad \text{(XXXIII)}$$

in which cations can diffuse through the crystal surface along the longitudinal axis of the cylinder, the mass transport can be described by planar semi-infinite diffusion of cations in two media, the solution ($x \leq 0$) and the crystal ($x \geq 0$). This will be valid only if the longitudinal axis in the crystal is longer than the diffusion layer in the particle. By solving the differential equation

$$\frac{\partial c}{\partial t} = D\frac{\partial^2 c}{\partial x^2} \quad (5.41)$$

where $D = D_{\mathrm{aq}}$ and $c = [\mathrm{C}^+]$ if $x \leq 0$ and $D = D_{\mathrm{ss}}$ and $c = [\mathrm{C}_{\mathrm{red}}]$ if $x \geq 0$. D_{aq} and D_{ss} are the diffusion coefficients of the cations in aqueous solution and in the crystal, respectively. The starting and the boundary conditions for Eq. (5.41) are

$$\begin{aligned}
&t = 0, x \geq 0 : [\mathrm{Cred}] = 0, [\mathrm{Ox}] = \rho;\ x \leq 0 : [\mathrm{C}^+] = [\mathrm{C}^+]*\\
&t > 0, x \to -\infty : [\mathrm{C}^+] \to [\mathrm{C}^+]*;\ x \geq 0 : [\mathrm{Cred}] + [\mathrm{Ox}] = \rho;\\
&x \to \infty : [\mathrm{Cred}] \to 0;\ [\mathrm{Ox}] \to \rho;\ x = 0 : \frac{I}{D_{\mathrm{ss}}FS} = -\frac{\partial[\mathrm{Cred}]}{\partial x};\\
&\frac{I}{D_{aq}FS} = -\frac{\partial[\mathrm{C}^+]}{\partial x};\ [\mathrm{Ox}] = [\mathrm{red}^-]\exp\left[\frac{F(E - Eo)}{RT}\right];\ K = \frac{[\mathrm{Cred}]}{[\mathrm{red}^-][\mathrm{C}^+]}
\end{aligned} \quad (5.42)$$

where $[\mathrm{C}^+]*$ is the concentration of cations in the bulk of the solution and the meaning of the other symbols is as in the previous models. K is the equilibrium constant for $\{\mathrm{red}^-\}_{\mathrm{solid}} + \mathrm{C}^+ \rightleftarrows \{\mathrm{Cred}\}_{\mathrm{solid}}$. In the mathematical solution of the

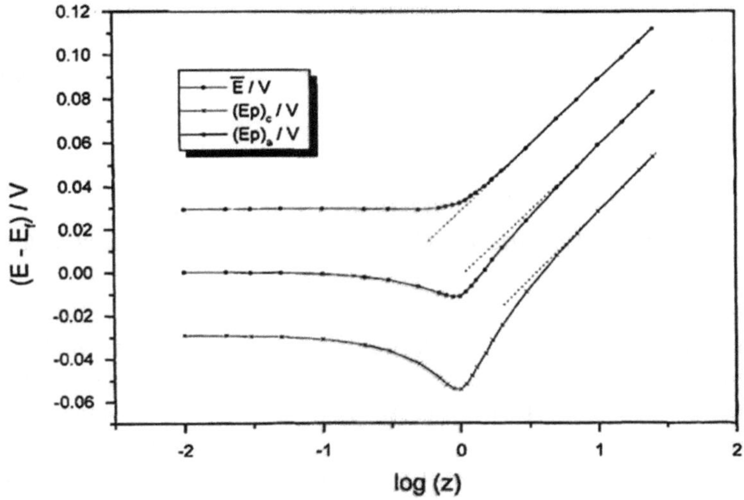

Fig. 5.97 Dependence of the anodic and cathodic peak potentials $(E_p)_a$ and $(E_p)_c$, respectively, and their median $\overline{E}$ on the logarithm of the mass transfer parameter z [B87]. Copyright ©1998 with permission from Springer

differential Eq. (5.34), two important parameters containing the term of $[C^+]^*$ appear, i.e., the formal potential $E_c^{\ominus'}$ and the mass transfer parameter z, defined as

$$E_c^{\ominus'} = E_{\mathrm{Ox/red}}^{\ominus} + \frac{RT}{F}\ln K + \frac{RT}{F}\ln\,[C^+]* \tag{5.43}$$

$$z = \sqrt{\frac{D_{ss}\ \rho}{D_{aq}[C^+]*}} \tag{5.44}$$

Cyclic voltammograms obtained from this model depend on the mass transfer parameter z. The dependence of cathodic and anodic peak potentials and of their median on log(z) are depicted in Fig. 5.97, while in Fig. 5.98 is given the dependence of the cathodic and anodic peak currents as a function of the same parameter.

For $z < 0.1$, the voltammograms are independent of the mass transfer in solution. In that case, the median of the peaks is equal to the formal potential, i.e.,

$$\overline{E} = E_{\mathrm{Ox/red}}^{\ominus} + \frac{\mathrm{RT}}{F}\ln K + \frac{\mathrm{RT}}{F}\ln\,[C^+]* \tag{5.45}$$

Under experimental conditions, such a situation exists if the concentration of the cations in the aqueous solution exceeds 1 mol L^{-1}. If $z > 10$, the median depends linearly on log(z) with a slope of 59 mV and the intercept is equal to

$$E - E_c^{\ominus'} = 2.3\frac{RT}{F}\log z \tag{5.46}$$

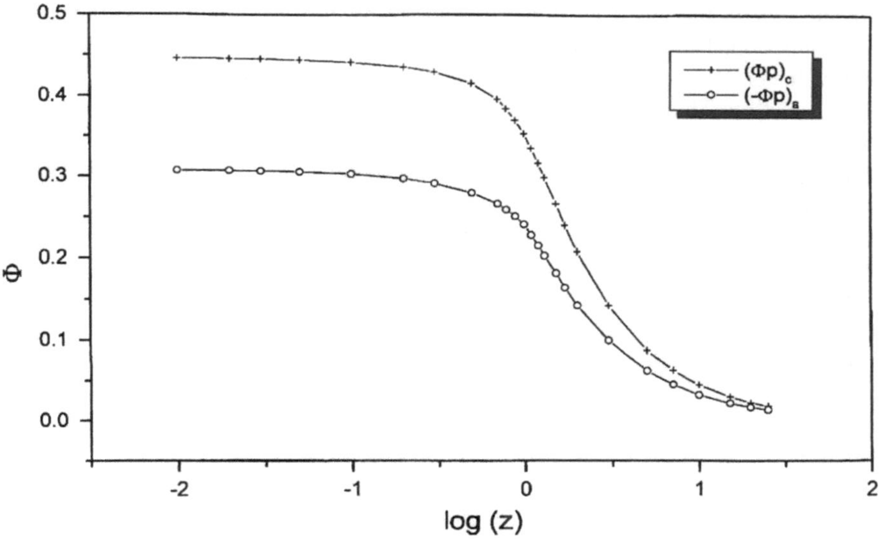

Fig. 5.98 Dependence of the absolute values of dimensionless anodic peak currents $(\Phi_p)_a$ and cathodic peak currents $(\Phi_p)_c$ on the logarithm of the mass transfer parameter z [B87]. Copyright ©1998 with permission from Springer

Then, the median does not depend on $[C^+]^*$ and is given by

$$\overline{E} = E^{\ominus}_{\mathrm{Ox/red}} + \frac{RT}{F}\ln K + \frac{RT}{F}\ln\rho + \frac{RT}{2F}\ln\frac{D_{ss}}{D_{aq}} \tag{5.47}$$

Equation (5.47) has been obtained by combining the Eqs. (5.43), (5.44), (5.45), and (5.46). However, if $z > 10$, the voltammograms vanish. The dependencies given in Fig. 5.81 can be used for the determination of D_{ss} [B95]. The described theory has been successfully applied to decide whether the diffusion of OH^- in solution or in a RuO_2-PVC film electrode controls the rate of the electrode process.[17]

5.11.4 *Solid-State Voltammetry at a Three-Phase Junction*

Oldham has mathematically modeled the processes occurring at microparticle-modified electrodes [B88]. He developed a model to address such important issues, as the possible magnitude of currents and the shape of voltammograms that can be

[17] Dharuman V (2001) A study of reaction mechanisms at RuO_2-PVC and RuO_2-NAFION film electrodes: electrooxidation of glucose, catechol and L-dopa. PhD thesis, University of Madras, Madras

expected when the electrochemical reaction is *exclusively* confined to the three-phase junction. The author has not considered the intriguing and important question of how the cations spread inside the crystal, once they have entered it via the three-phase junction. This spreading has been assumed to be so rapid that their concentration is always uniform throughout the crystal. This is more a limiting case than a realistic description of real experiments. It is, however, of interest to determine the extent to which diffusion can supply ions to or remove them from a three-phase junction. The key question is whether the bottleneck three-phase junction will allow obtaining measurable voltammetric responses.

For this, the insertion electrochemical Reaction (XXXII) has been considered. The diffusion has been supposed to occur under cylindrical conditions, and therefore the cylindrical version of the differential equation of Fick's second law

$$\frac{\partial^2 c}{\partial r^2} + \frac{1}{r}\frac{\partial c}{\partial r} = \frac{1}{D}\frac{\partial c}{\partial t} \tag{5.48}$$

has been solved for the two cases: (a) assuming a constant concentration of cations at the three-phase junction and (b) assuming a variable concentration of cations at the three-phase junction line. In both cases, qualitatively similar results have been obtained.

The current density i obtained by solving Eq. (5.48) has the form

$$i = \frac{-\pi FD\left(c^{\mathrm{b}} - c^{\mathrm{j}}\right)}{\ln\left\{\frac{\eta\sqrt{Dt}}{a}\right\}} \tag{5.49}$$

where c^{j} is the concentration of the cations at the three-phase junction line, c^{b} is the bulk concentration of cations in the solution, a is the distance of the three-phase junction at which a diffusing species has reached the three-phase junction (a has a magnitude of atomic dimensions), and η is defined as $\eta = 2\exp(-\gamma) = 1.123\ldots$, where γ is the Euler constant, $\gamma = 0.57722$. The meaning of other symbols is as usual. The weak inverse logarithmic dependence of the current density on time, for all periods of voltammetric significance, implies that the current is almost steady-state throughout any practicable experiment. In order to apply Eq. (5.49) to voltammetric purposes, one needs to estimate the length of the three-phase junction. If a total volume of the electroactive solid is V and if it consists of isolated, uniformly sized cubes of edge length L, each cube resting on the electrode via one of its faces, then the number of the cubes will be VL^{-3}, while the total length of the three-phase junction is $4VL^{-2}$. By multiplying this term with Eq. (5.49), one obtains the following equation for the current as a function of the length of the cube edges:

$$I = \frac{-2\pi FDV\left(c^{\mathrm{b}} - c^{\mathrm{j}}\right)}{L^2\ln\left\{\frac{\eta\sqrt{Dt}}{a}\right\}} \tag{5.50}$$

Table 5.6 Dependence of the current I on the edge length L of the crystal cubes

L/μm	#	$-I$	t_{min}/s
464	1	1.2 nA	3.4×10^6
215	10	5.6 nA	7.4×10^5
100	100	26 nA	1.6×10^5
46	10^3	0.12 μA	3.4×10^4
22	10^4	0.56 μA	7.4×10^3
10	10^5	2.6 μA	1.6×10^3
4.6	10^6	12 μA	340
2.2	10^7	56 μA	74
1.0	10^8	0.26 mA	16
0.46	10^9	1.2 mA	3.4
0.22	10^{10}	5.6 mA	0.74
0.10	10^{11}	26 mA	0.16

L is the edge length of cubes, # denotes the number of cubes, I is the overall steady-state current, and t_{min} is the time necessary for total reduction [B88]

This equation shows that after subdivision of the solid crystal, the three-phase junction length is a predominant parameter determining the magnitude of the total current. Using the constant values for the other parameters in Eq. (5.50), the values of the current estimated for different values of L are given in Table 5.6. The column headed by "#" represents the number of the cubes according to this model. Table 5.6 confirms that significant currents can flow into an adherent electroactive solid via the three-phase junction line, especially when the crystals are sufficiently small and numerous.

The parameter t_{min} in Table 5.6 is very important since it represents the minimum time in which total reduction can be achieved. Typical times corresponding to voltammetric experiments are encountered for L being a few micrometers.

Applying the solution of Eq. (5.48) under conditions of cyclic voltammetry, the scan rate is the major parameter that determines the shape of the cyclic voltammograms. Two types of theoretical voltammograms obtained with slow and fast scan rates are depicted in Figs. 5.99 and 5.100, respectively. Their shape considerably differs from what is common in solution electrochemistry. At the slowest scan rates (see Fig. 5.99), the cyclic voltammograms have inversion symmetry about the origin and resemble their counterparts for reversible thin-layer voltammetry. Another similarity to thin-layer cyclic voltammetry is in having peak heights proportional to the scan rate. These similarities are expected since in both cases there is a limited amount of material, which can exhaustively be reduced. In the voltammograms obtained at rather low scan rates, the current is close to zero at the right-hand side of each plot, since at these positive potentials the solid is totally oxidized. When the dimensionless scan rate is as large as 10, the total reduction is delayed until after reversal (cf. Fig. 5.100). At faster scan rates, the

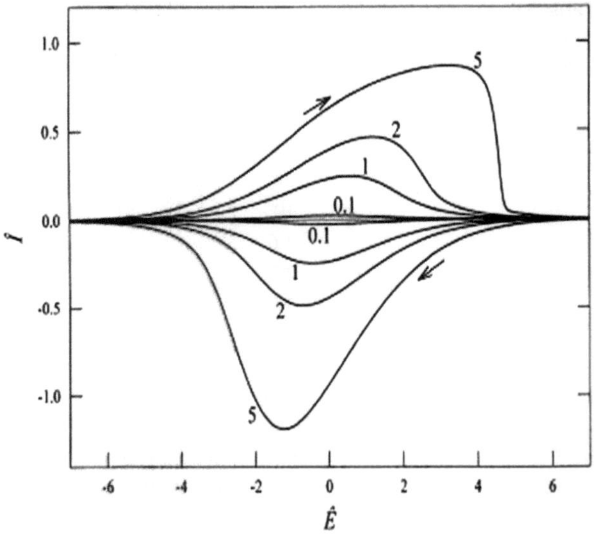

Fig. 5.99 Cyclic voltammograms obtained for several rather slow scan rates [B88]. Copyright ©1998 with permission from Springer

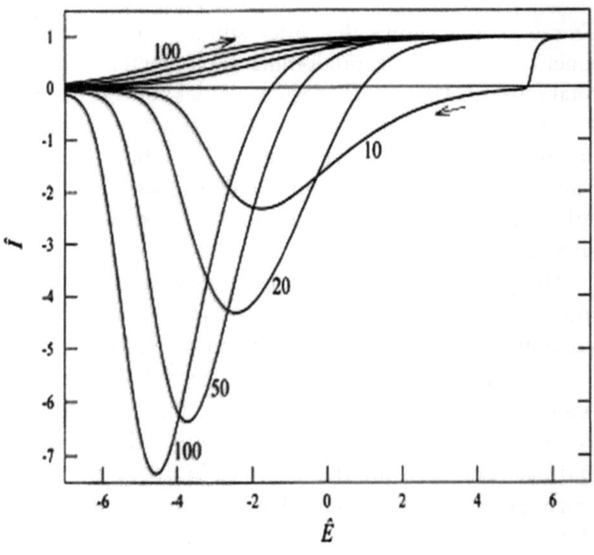

Fig. 5.100 Cyclic voltammograms obtained for scan rates one range of magnitude faster than that presented in Fig. 5.82 [B88]. Copyright ©1998 with permission from Springer

total reduction is never approached under CV conditions. Considerably higher anodic currents are flowing, but for shorter periods. The reason for that is that the junction concentration c^{j} can decrease to zero upon reduction, but can exceed c^{b} manyfold during anodization. This is equivalent with the idea that the ions can escape from the crystal through the three-phase junction much faster than they can enter the crystal from solution. The model has been solved also for the conditions of chronoamperometry (at constant potential). The main result obtained in this study is that the diffusion to a three-phase junction can provide an adequate supply of ions to permit classical voltammetric techniques to be applied within the time and current ranges traditionally used. The supply of ions to the three-phase junction seems to be in a quasi-steady state during the electrochemical experiments. If the voltammetry is reversible, the response is predicted to be very dependent on the degree of subdivision of the crystals. Depending on the crystal size and scan rate, cyclic voltammograms may mimic solution-phase voltammograms of classical thin-layer experiments or typical stripping experiments. This model, despite of its limitations, clearly demonstrates that the role of the three-phase junction must not be overlooked in modeling the voltammetry of solids.

5.11.5 The Effect of Crystal Size and Shape on Electrochemical Signals

The electrochemical reduction of a microcrystal attached to an electrode has been theoretically described by the help of two- and three-dimensional models [B142]. An electrochemically reversible reduction of redox centers in a crystal followed by insertion of cations from the adjacent aqueous phase can be described by Reaction (XXXIII) [B69]. The authors have considered a model where the simultaneous uptake of electrons and cations starts at the three-phase junction line. They have utilized two- and three-dimensional models to simulate the diffusion of cations and electrons and the current flow in response to the applied potential. A scheme of the two-dimensional matrix used in the simulations is given in Fig. 5.101. The total flux of electrons and cations is a function of the concentration gradients between each box (k, m) and its two neighbor boxes. Both fluxes follow Fick's first diffusion law [Eq. (5.51)], where ΔN represents the change of the

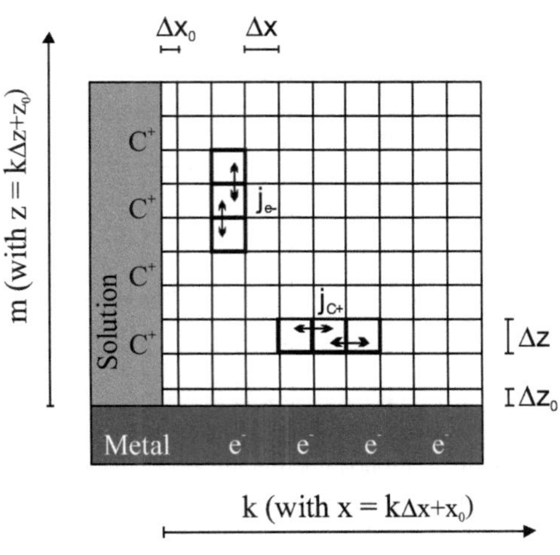

Fig. 5.101 Two-dimensional matrix of discrete boxes with $\Delta x \Delta z$ being the size of the bulk boxes, $\Delta x_0 \Delta z$ and $\Delta x \Delta z_0$ being the size of the surface boxes, and $\Delta x_0 \Delta z_0$ representing the three-phase junction [B142]. Copyright ©2000 with permission from Springer

$$j_{C^+} = \frac{1}{Y\Delta z}\frac{\Delta N}{\Delta t} \tag{5.51}$$

$$j_{e^-} = \frac{1}{Y\Delta x}\frac{\Delta N}{\Delta t} \tag{5.52}$$

moles of cations or electrons and Y is the length of the infinite three-phase junction, defined similarly as given in Fig. 5.101. Equations (5.51) and (5.52) imply that both currents, i.e., the surface (due to the electron diffusion) and bulk (or volume) currents (due to the cations diffusion) are considered separately in this model. The two- and three-dimensional models have been solved for the conditions of chronoamperometry by considering a semi-infinite diffusion space as well as a finite diffusion space. The main feature of the chronoamperometric plot is for the two-dimensional model and semi-infinite diffusion that the total current approaches a steady-state value (see Fig. 5.102). The time dependence of the total current is

$$I(t) = \frac{FY}{V_m}\left\{\frac{1}{1+\exp(-\varphi)}\right\}\left[\left(\frac{\Delta x_0\sqrt{D_e}+\Delta z_0\sqrt{D_{C^+}}}{2\sqrt{\pi t}}\right)+\sqrt{D_{C^+}D_e}\right] \tag{5.53}$$

Here, V_m is the molar volume, while φ is the dimensionless potential of the Nernstian equation representing the thermodynamic equilibrium of Reaction (XXXVI). D_e and D_{C^+} are the diffusion coefficients of electrons and cations inside the crystal. The increase of the bulk current in Fig. 5.102 is ascribed to the growth of the reaction zone as shown in Fig. 5.103a.

The semi-infinite diffusion model applied to the three-dimensional model is more realistic since for real cuboid crystals the three-phase junctions of these objects include the edges of the rectangular base. The consequence of this is that the cations in the real three-dimensional system can enter the crystal from different

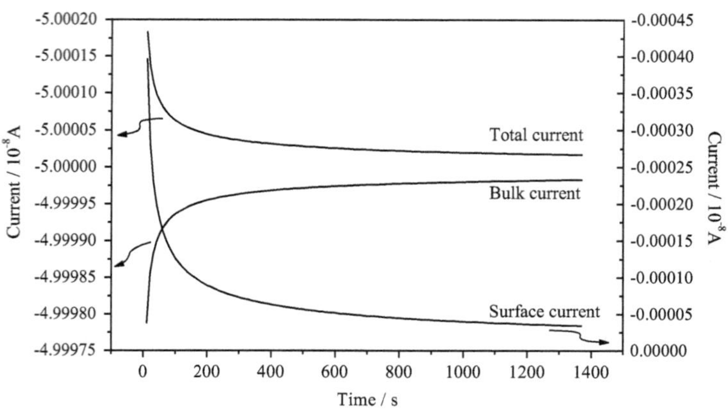

Fig. 5.102 Chronoamperogram for a potential step of a value 200 mV more negative than the formal potential of Reaction (XXXIII). $D_e = D_{C^+} = 5 \times 10^{-8}\,\text{cm}^2\,\text{s}^{-1}$: $\Delta x_0 = \Delta z_0 = 10^{-4}\,\Delta x$ [B142]. Copyright ©2000 with permission from Springer

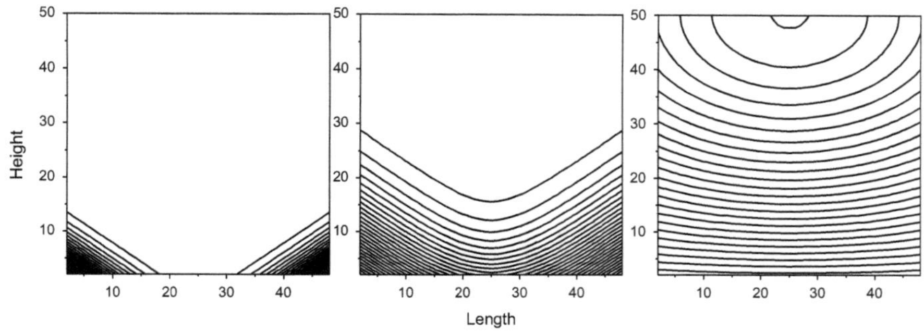

Fig. 5.103 Concentration profiles of a partially converted two-dimensional microcrystal at different times after a potential step (**a**) 3 s, (**b**) 10 s, and (**c**) 40 s. $D_e = 5 \times 10^{-9}\,cm^2\,s^{-1}$; $D_{C^+} = 10^{-8}\,cm^2\,s^{-1}$ [B142]. Copyright ©2000 with permission from Springer

directions. The total current in this three-dimensional model has the following form:

$$I(t) = \frac{F}{V_m}\left\{\frac{1}{1+\exp(-\varphi)}\right\}\left[Y\left(\left(\frac{\Delta x_0\sqrt{D_e}+\Delta z_0\sqrt{D_{C^+}}}{2\sqrt{\pi t}}\right)+\sqrt{D_{C^+}D_e}\right)-D_{C^+}\sqrt{2tD_e}\right] \tag{5.54}$$

Compared to the expression of the current in the two-dimensional model [Eq. (5.53)], the three-dimensional model [Eq. (5.54)] has an additional subtractive term. This has been interpreted as an "edge effect" describing the overlapping of the cation diffusion (which now proceeds in x and y directions) near the corner and thereby its influence on the entire diffusion process. Equations (5.53) and (5.54) are a consequence of the diffusion of electrons and cations into a semi-infinite space. This diffusion is accompanied by the growth of the reaction zone, and it is not exclusively confined to a real semi-infinite diffusion space, but can also be expected to apply to the beginning of the conversion of a crystal of limited size. Since the crystal considered in the simulation has a very small size (some μm) and a time scale was assumed where total conversion of material may take place, an unhindered diffusion is expected only for the initial period of the reaction. Now, the most important question is: how do the diffusion barriers given with the crystal size and shape determine the advancement of the electrochemical conversion through the crystal? For a certain geometry (cube or cuboid) of the microcrystal, the ratio D_{e-}/D_C+determines the spatial course of the development of the reaction zone and thereby the resulting current-time functions. In Fig. 5.104, the isoconcentration lines along the (x, y) plane at half height are depicted for two differently shaped cuboids after a partial reduction of the particle. The progress of the reaction zone in such situation follows different geometries, causing tremendous changes in the chronoamperometric curve. This is also obvious from the different shapes of the cyclic voltammograms given in Fig. 5.105. The voltammograms are a result of a cyclic oxidation and reduction of differently

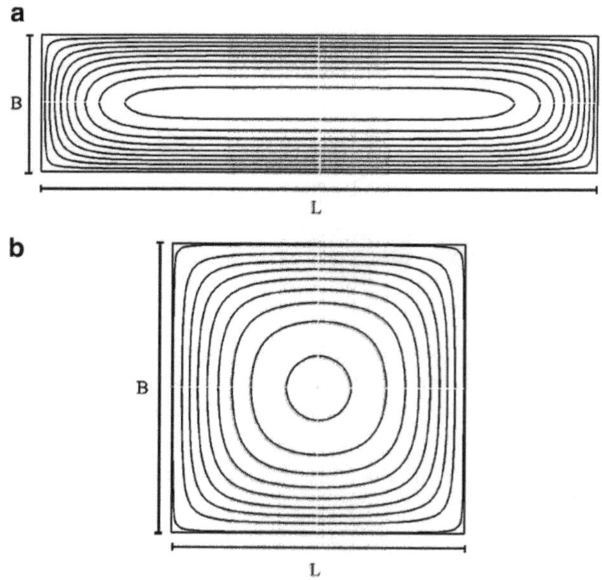

Fig. 5.104 Concentration profile inside partially (10 %) converted crystals: cross section through x–y plane at half height; (a) crystal of $L = H = 10$ μm, $B = 40$ μm and (b) $L = B = H = 10$ μm size [B142]. Copyright ©2000 with permission from Springer

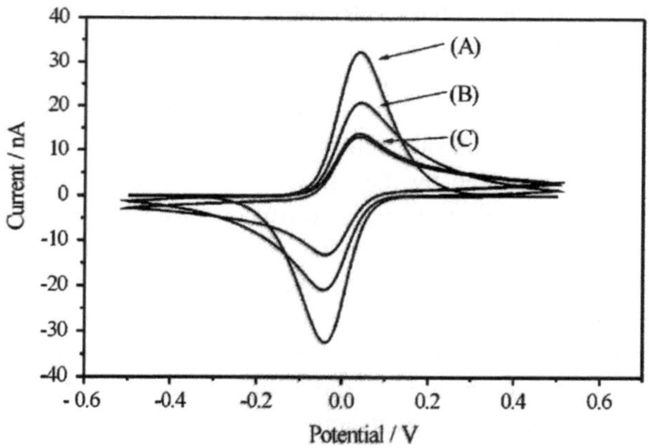

Fig. 5.105 Cyclic voltammograms of differently shaped cuboid particles of constant volume with a square base of different size. Concentration profile inside partially (10 %) converted crystals: cross section through x–y plane at half height; $D_{e-} = D_{C^+} = 10^{-8}$ cm^2 s^{-1}, $V_m = 153.8$ cm^3 mol^{-1}; (A) crystal of $L = B = 28$ μm, $H = 10$ μm, (B) $L = B = H = 20$ μm; (C) $L = B = 16$ μm $H = 31$ μm [B142]. Copyright ©2000 with permission from Springer

shaped cuboids of a uniform volume. Depending on the ratio of crystal height to length and width, different diffusion regimes dominate and determine the time for the total conversion of the solid. The geometry of the spatial development of the reaction zone additionally depends on the ratio of the diffusion coefficients of the electrons and cations. Due to the different values of diffusion coefficients of electrons and cations, the reaction at the three-phase boundary only initially determines the course of the reaction. During the initial three-phase reaction, the faster process leads to an exhaustive conversion along the corresponding interface either the crystal–electrode or the crystal–solution interface. The further course of the conversion is determined by the slowest diffusion process. In principle, the results of this study [B142] can be utilized to determine the diffusion coefficients from experimental data and to derive the geometric parameters of the attached crystals. This, however, would require experiments with single size and single-shaped crystals. In reality there will always be a size and shape distribution for which a theory still needs to be developed.

5.11.6 *Redox Mixed Phases and Miscibility Gaps*

In all previously considered reactions of type (XXXIII), it was assumed that the solid phase transforms in a series of completely miscible solid solutions from {ox} into the final form $\{C_n\text{red}\}$. In fact, in intercalation electrochemistry there are known many examples in which a continuous series of mixed crystals exist between the oxidized and reduced forms. Several hexacyanometalates are discussed as examples in Sect. 5.3. The phase transformation of {ox} to $\{C_n\text{red}\}$ via $\{\text{ox}_x(C_n\text{red})_{1-x}\}$ gives an opportunity to describe these processes on the basis of mixed phase thermodynamics. The model of *redox* mixed phases (*redox* solid solutions) allows treating also cases with miscibility gaps that frequently occur in solid systems. Considering only the thermodynamic conditions in the solid phases, Reaction (XXXIII) can be defined as follows:

$$c_{\text{ox}} + c_{\text{red}} = \rho;\ c_{\text{ox}} = c_{\text{red}} \exp(\varphi);\ \varphi = \frac{nF}{RT}\left(E - E_{\text{c}}^{\ominus'}\right) \tag{5.55}$$

In the case of partial miscibility of {ox} and $\{C_n\text{red}\}$, the formal concentrations c_{ox} and c_{red} are linked to the molar fractions of the corresponding compounds via the following relations:

$$c_{\text{ox}} = \frac{m_{\text{ox}}}{m_{\text{ox}} + m_{\text{red}}};\ c_{\text{red}} = \frac{m_{\text{red}}}{m_{\text{ox}} + m_{\text{red}}} \tag{5.56}$$

where m_{ox} and m_{red} are mole numbers of {ox} and {red}, respectively. If the mutual solubility is limited, the concentrations (or more precisely activities) of both components can be changed only within a restricted range. Consequently, the Nernst equilibrium [Eq. (5.55] will not be satisfied at all potentials. This will lead to an immiscibility polarization as shown in Fig. 5.106. If {ox} is gradually reduced to $\{C_n\text{red}\}$, the activity of $\{C_n\text{red}\}$ may increase up to a certain limiting value $Z_{\text{red/ox}} = m_{\text{red}}/(m_{\text{red}} + m_{\text{ox}})$, which depends on the maximal solubility of $\{C_n\text{red}\}$ in {ox}. In that case, the potential of the saturated mixed crystal is given by

Fig. 5.106 Schematic plot of the logarithm of the activity ratios of the oxidized to the reduced form in the solid state as a function of the applied potential [B79]. Copyright ©1997 with permission from Springer

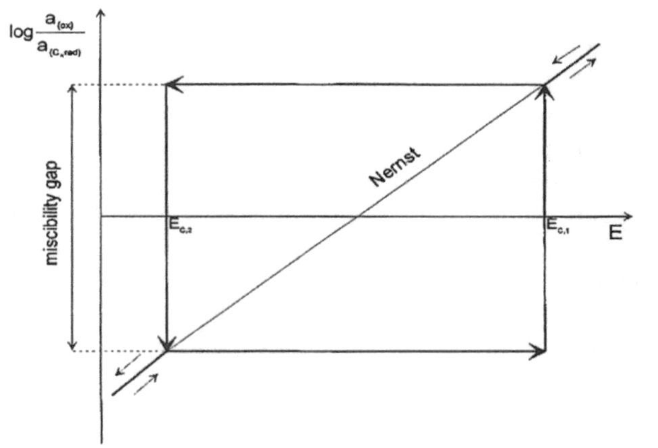

$$E_{\mathrm{C},1} = E_{\mathrm{c}}^{\ominus'} + 2.3\frac{RT}{nF}\log\left(\frac{1 - Z_{\mathrm{red/ox}}}{Z_{\mathrm{red/ox}}}\right) \tag{5.57}$$

where $Z_{\mathrm{red/ox}}$ is the maximal solubility of {$C_n$red} in {ox}. The second critical potential ($E_{\mathrm{C},2}$) depends on $Z_{\mathrm{ox/red}} = m_{\mathrm{ox}}/(m_{\mathrm{red}} + m_{\mathrm{ox}})$, that is, the maximal solubility of {ox} in {$C_n$red}:

$$E_{\mathrm{C},2} = E_{\mathrm{c}}^{\ominus'} + 2.3\frac{RT}{nF}\log\left(\frac{Z_{\mathrm{ox/red}}}{1 - Z_{\mathrm{ox/red}}}\right) \tag{5.58}$$

The ratio $a_{\{\mathrm{ox}\}}/a_{\{C n\mathrm{red}\}}$ in the crystal can be changed continuously for $E > E_{\mathrm{C},1}$ and $E < E_{\mathrm{C},2}$, but between $E_{\mathrm{C},1}$ and $E_{\mathrm{C},2}$, it will remain constant as shown in Fig. 5.106.

The miscibility gap can be also explained in terms of differences in the Gibbs energies of the two solid phases. Let us first assume that the reduction of {ox} to {C_nRed} forms mixed crystals of the type {$\mathrm{ox}_x(C_n\mathrm{red})_{1-x}$} until the maximum solubility of {C_nRed} in {ox} is reached. The mixed crystal with this maximum concentration of {C_nRed} may be written as {$\mathrm{ox}_{x'}(C_n\mathrm{red})_{1-x'}$}. Then there might exist a solubility gap between the composition {$\mathrm{ox}_{x'}(C_n\mathrm{red})_{1-x'}$} and {$\mathrm{ox}_{y'}(C_n\mathrm{red})_{1-y'}$} ($x'$ is much bigger than y'). The mixed crystal {$\mathrm{ox}_{y'}(C_n\mathrm{red})_{1-y'}$} can be further reduced until {C_nRed} is reached. The mixed phase {$\mathrm{ox}_{x'}(C_n\mathrm{red})_{1-x'}$} has a structure α, and the mixed phase {$\mathrm{ox}_{y'}(C_n\mathrm{red})_{1-y'}$} a structure β. α and β are so different that a miscibility gap arises. The difference in the Gibbs free energies $\Delta G_{\mathrm{A/B}}$ of the two phases α and β is equivalent to the difference in the formal potentials of both systems, and it can be split in two terms, i.e., $\Delta G_{\mathrm{A/B}} = \Delta G_{\mathrm{red/ox}} + \Delta G_{\mathrm{lattice}}$, where $\Delta G_{\mathrm{lattice}}$ is the difference in the free energies due to the different structures of the lattice of phases α and β. Generally, $\Delta G_{\mathrm{lattice}}$ may have very different values. If β is much more stable than α, this will shift the formal potential of the redox system in this phase so much in comparison to its value in phase α that within the miscibility gap the phase β will be formed at the expense of α until α is converted into β (line 1 in Fig. 5.107). An

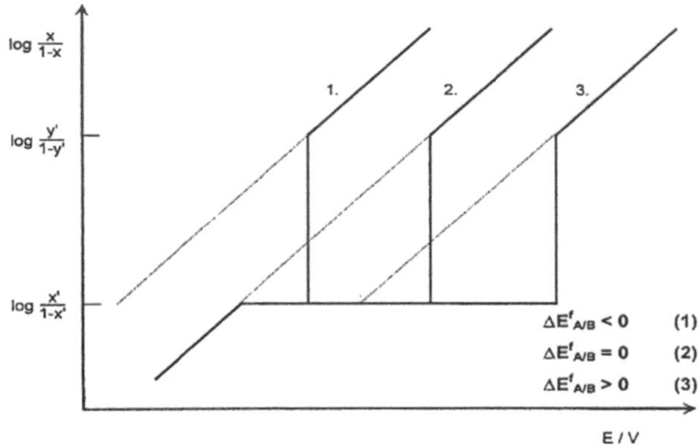

Fig. 5.107 Plot of log(x/$1-x$) for a solid solution system $\{ox_x(C_n red)_{1-x}\}$ versus potential [B158]. Copyright ©2000 with permission from Springer

interesting situation arises when $\Delta G_{lattice}$ is much smaller than $\Delta G_{red/ox}$. In such a case, we can say that $\Delta G_{A/B} = \Delta G_{red/ox}$, which would mean that the formal potentials of the redox systems in α and β are the same (line 2 in Fig. 5.107). This leads to the situation that, within the miscibility gap, there is no sufficient stabilization of β for phase segregation. In other words, within the miscibility gap, the potential corresponds to the composition neither of α nor of β. Thus, no net oxidation of α to form β can take place. In that case, the miscibility gap is simply an inert zone where no reaction can proceed. A third case is possible when the phase α is much more stable than β. It follows that the formal potential of ox/red in α is more negative than in β. This again will lead to a separation of the potential ranges in which reduction of ox can occur. After completing the reduction of {ox} to the composition $\{ox_{x'}(C_n red)_{1-x'}\}$, there will appear an inert zone until the formation of $\{ox_{y'}(C_n red)_{1-y'}\}$ may take place. This inert zone is bigger than the miscibility gap by the difference in formal potentials of ox/red in α and β (see line 3 in Fig. 5.107).

The mathematical modeling was considered identically as in [B87] using the differential Eq. (5.41) under the initial and boundary conditions given in Eq. (5.42). The numerical solution has been applied to cyclic staircase voltammetry and linear scan voltammetry [B79]. The influence of the volume of the substance and of the redox kinetics on the voltammetric features have been investigated, too. The influence of the limited miscibility of redox components of the mixed crystal on staircase cyclic voltammograms simulated for semi-infinite diffusion is depicted in Fig. 5.108. Curve A is the voltammogram for complete miscibility, while curve B shows the case of a miscibility gap assuming $Z_{red/ox} = Z_{ox/red} = 0.3$. It can be seen that the miscibility gap causes a splitting of the voltammogram into two peaks. The first peak appears when the surface of the crystal is charged to the first critical potential $E_{C,1}$. Though the potential of the electrode is continuously changed, the ratio of the oxidized to the reduced form remains constant and equal to the value of $E_{C,1}$, until the value of the second critical potential $E_{C,2}$ is reached. When the potential is equal to $E_{C,2}$, at this moment the composition of the crystal suddenly changes from 30 % $\{C_n red\}$ and 70 % {ox} to 70 % $\{C_n red\}$ and 30 % {ox}. This sudden reduction of 40 % of {ox} prompts a sharp peak to appear in the cathodic

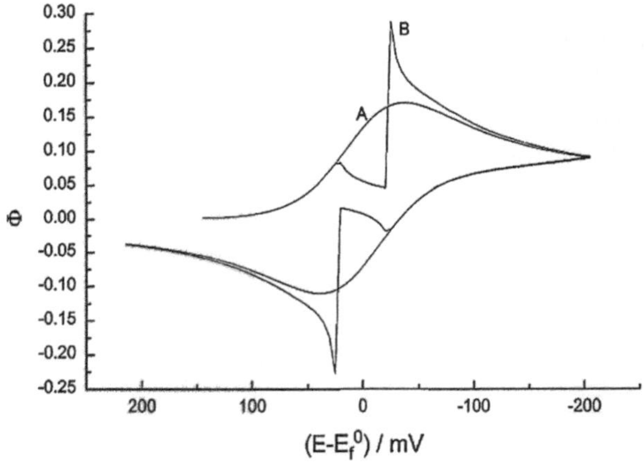

Fig. 5.108 Staircase cyclic voltammograms of ideally miscible (A) and partly immiscible (B) solid redox components. Miscibility limits: $Z_{red/ox} = Z_{ox/red} = 0.5$ (A) and 0.3 (B). Reversible redox reaction, $\Delta E = 5$ mV [B79]. Copyright ©1997 with permission from Springer

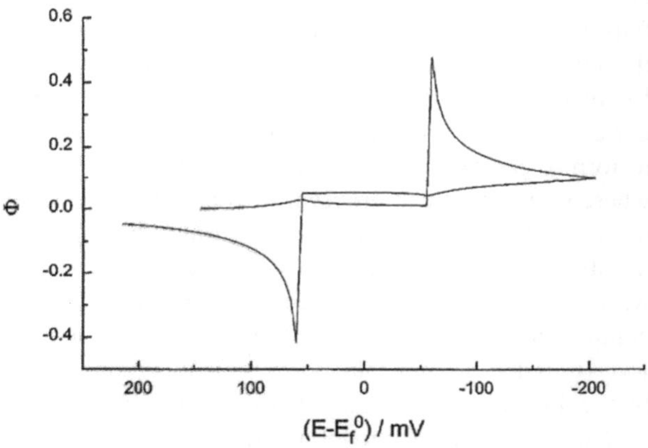

Fig. 5.109 Staircase cyclic voltammetry of a reversible solid-state redox reaction influenced by the limited miscibility of the components: $Z_{red/ox} = Z_{ox/red} = 0.1$ [B79]. Copyright ©1997 with permission from Springer

branch of the voltammogram. Under anodic polarization, a second sharp peak appears. The separation between these two peaks increases as the miscibility of the components decreases (see Fig. 5.109).

The influence of the finite crystal volume on the staircase voltammogram is shown in Fig. 5.110. The miscibility parameters are the same as those in Fig. 5.109, but the results in Fig. 5.110 are calculated for a very small hemispherical particle with the dimensionless radius $5r_0v^{0.5}(D\Delta E)^{-0.5} = 20$. The major difference between Figs. 5.109 and 5.110 is that such small particles can be electrolyzed completely within a single potential scan. Thus, the peaks are narrower, containing no diffusion "tails," and the current between cathodic and anodic peaks tends to zero.

An integral study which contains also results for the influence of the charge transfer kinetics can be found elsewhere [B79]. The outlined treatment of the effect of a miscibility gap on voltammograms is a purely thermodynamic treatment. However, when a phase transformation occurs due to a miscibility gap, a more or less pronounced nucleation process must be observed. Therefore, it is not surprising

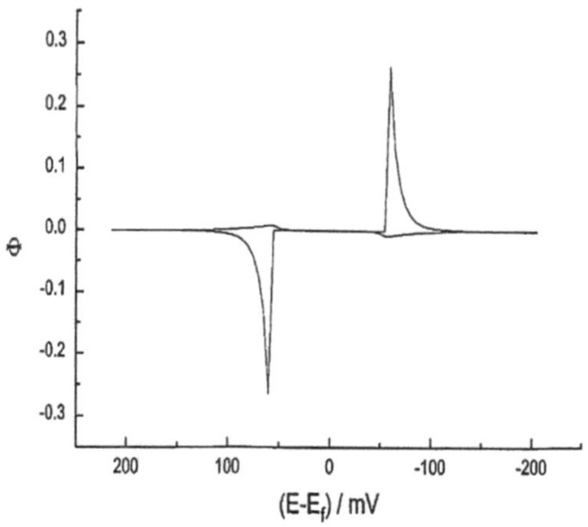

Fig. 5.110 Staircase cyclic voltammetry of hemispherical solid electroactive microparticle with a dimensionless radius $5r_0v^{0.5}(D\Delta E)^{-0.5} = 20$. Reversible redox reaction influenced by the limited immiscibility of the components $Z_{red/ox} = Z_{ox/red} = 0.1$ [B79]. Copyright ©1997 with permission from Springer

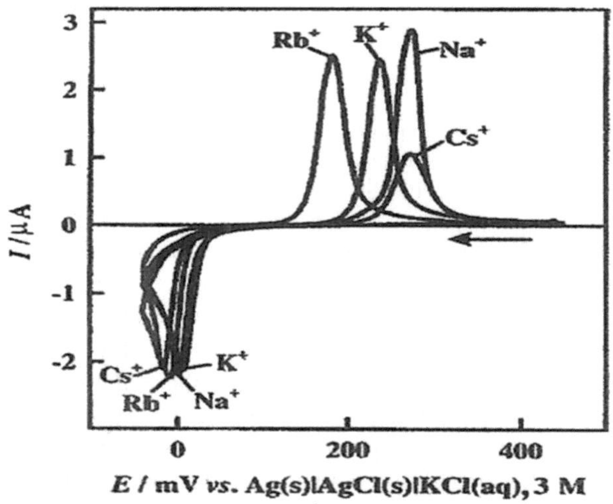

Fig. 5.111 Cyclic voltammograms of TCNQ microcrystals in 0.1 mol L^{-1} aqueous solutions of NaCl, KCl, $RbNO_3$, and CsCl. Scan rate is $v = 50$ mV s^{-1} [B101]. Copyright ©199 with permission from Royal Society of Chemistry

that voltammograms that resemble very much those simulated for systems with miscibility gaps have been observed experimentally for the reduction of TCNQ (7,7,8,8-tetracyanoquinodimethane) [B56, B101] under conditions of immobilized microparticles. Figure 5.111 shows cyclic voltammograms for the reduction of TCNQ to different alkali salts $Cat^{+}(TCNQ)^{-}$. The authors of these papers could show that the electrochemical transformation of TCNQ to the salts exhibits the typical features of nucleation–growth kinetics. The essential prerequisite for the occurrence of the nucleation phenomenon is a miscibility gap, and the authors have shown that the crystallographic structures of the educts and products are indeed very different [B101].

In a further paper [B158] on the theoretical effects of miscibility gaps, the focus was on the influence of the structure of the interface of the two immiscible solid

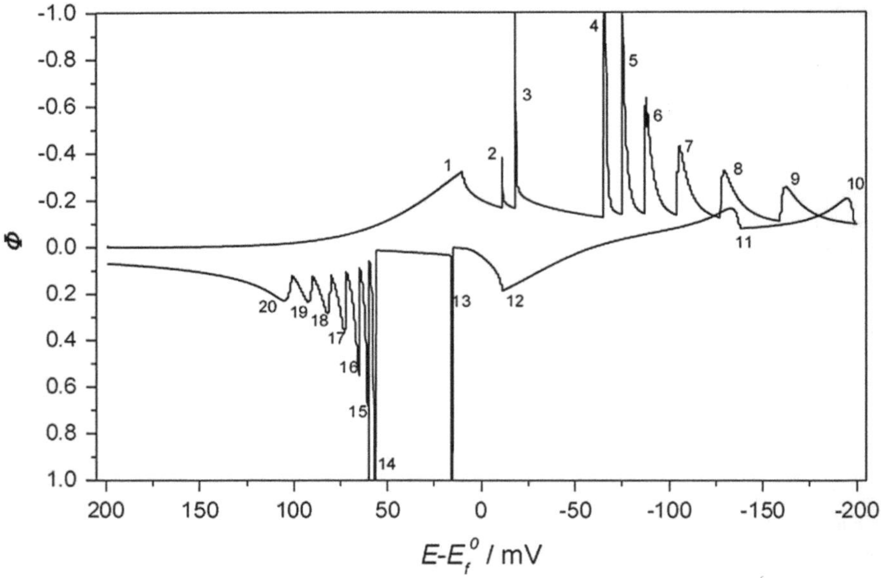

Fig. 5.112 Cyclic voltammograms of partly immiscible solid redox compounds. Reversible redox reaction. Miscibility limits: $Z_{\mathrm{Cred/ox}} = Z_{\mathrm{ox/Cred}} = 0.4$, $d = 0.4$, $\Delta E = 0.2$ mV. No transition zone was assumed [B158]. Copyright ©2000 with permission from Springer

phases. When an interfacial *interface* is assumed to exist at the place where the two structures transform, a smooth voltammetric response is expected. However, when a rather sharp interface exists, concentration barriers are built up that lead to very unusual voltammetric spikes as depicted in Fig. 5.112.

Shortly after the publication of these theoretical results [B158], Parkinson et al. [B169] published voltammograms of electrodes covered with the solid compound (4-dimethylamino-2-dihydroxyphenyl)squaraine that exhibit such curious spikes in addition to the typical feature of a miscibility gap (cf. Fig. 5.113).

5.11.7 Electrochemically Driven Formation of Bilayered Systems

In Sect. 5.5 the electrochemically initiated substitution of the high-spin iron ions of Prussian blue by foreign metal ions has been discussed from an experimental point of view (see Figs. 5.111, 5.112 and 5.113). This substitution reaction has been simulated on the basis of a theoretical model considering semi-infinite planar diffusion applied to a particle cylinder [B173]. In a solution containing K^+ as the sole cation, Prussian blue microcrystals adhering to a graphite electrode (see Fig. 5.114) can be both reduced and oxidized in cyclic voltammetry:

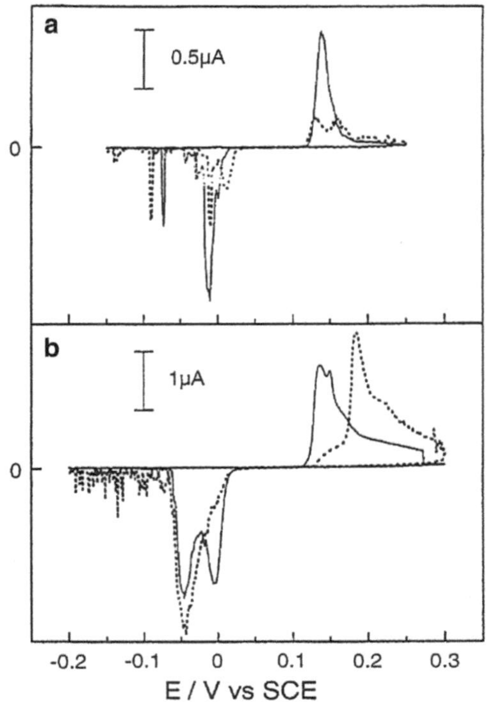

Fig. 5.113 Staircase cyclic voltammograms of 1-1 OHSQ/HOPG electrodes (dipping mode) in 0.1 mol L^{-1} aqueous solution of LiCl. In each panel, the first scan (*dotted line*) and the eleventh (**a**) or tenth (**b**) scan (*solid line*) for two different samples are shown. Scan rate $v = 10$ mV s^{-1} [B169]. Copyright ©2001 with permission from Elsevier

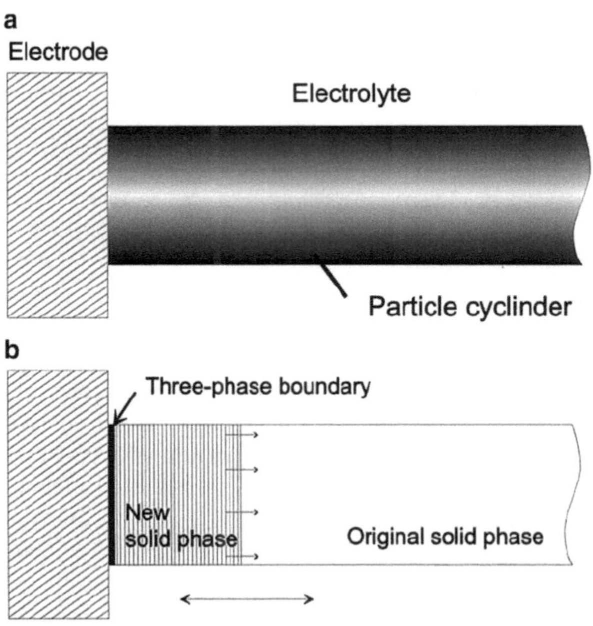

Fig. 5.114 Model of the semi-infinite planar diffusion applied to a particle cylinder attached to a solid electrode and in contact with an electrolyte solution (**a**). Initiated at the three-phase junction, the new solid phase advances into the bulk particle, following the diffusion of cations (**b**) [B173]. Copyright ©2001 with permission from Elsevier

$$\{KFe^{3+}{}_{N}Fe^{2+}{}_{C}(CN)_6\}_{solid} \rightleftarrows \{Fe^{3+}{}_{N}Fe^{3+}{}_{C}(CN)_6\}_{solid} + K^+ + e^- \qquad (XXXIV)$$

Prussian blue Berlin green

$$\{KFe^{3+}{}_{N}Fe^{2+}{}_{C}(CN)_6\}_{solid} + K^+ + e^- \rightleftarrows \{K_2Fe^{2+}{}_{N}Fe^{2+}{}_{C}(CN)_6\}_{solid} \qquad (XXXV)$$

Prussian blue Berlin white

($Fe^{3+}{}_{N}$ and $Fe^{2+}{}_{C}$ denote the N-coordinated high-spin and the C-coordinated low-spin iron ions, respectively). If these reactions are reversible, the following equations govern the concentration ratios:

$$\begin{aligned} \frac{c_0}{c_1} &= \exp\left\{\frac{F(E-E_1)}{RT}\right\}; \quad x=0, t\geq 0; \\ \frac{c_1}{c_2} &= \exp\left\{\frac{F(E-E_2)}{RT}\right\}; \quad x=0, t\geq 0 \end{aligned} \qquad (5.59)$$

where c_0, c_1, and c_2 are the concentrations of Berlin green, Prussian blue, and Berlin white, respectively. The sum of all three concentrations is equal to c^*. Both, the electron transfer and the mass transfer of the intercalated compounds by Reactions (XXXIV) and (XXXV) are given by the corresponding diffusion equations:

$$\frac{\partial c_i}{\partial t} = D\frac{\partial^2 c_i}{\partial x^2}; \quad x \geq 0, t \geq 0 \qquad (5.60)$$

Equation 5.60 has been solved by the standard finite difference method for all species in Reactions (XXXIV) and (XXXV).

When the electrochemical transformation of iron hexacyanoferrate into cadmium hexacyanoferrate takes place, the following reaction occurs:

$$\begin{aligned} &2\{Fe^{3+}{}_{N}Fe^{3+}{}_{C}(CN)_6\}_{solid} + 3Cd^{2+} \rightarrow \\ &Cd^{2+}\{Cd^{2+}{}_{N}Fe^{3+}{}_{C}(CN)_6\}_2 + 2\,Fe^{3+} \end{aligned} \qquad (XXXVI)$$

This reaction is characterized by k_f, the rate constant of the electrochemical transformation (or substitution). In order to quantify the progress of the reaction, a parameter ϖ is defined as $\varpi = \Phi^{ox}_{Cdhcf}/(\Phi^{ox}_{Cdhcf} + \Phi^{ox}_{Fehcf})$. This is the ratio of the oxidation currents of cadmium hexacyanoferrate to the sum of the oxidation currents of cadmium hexacyanoferrate and iron hexacyanoferrate. The dependences of the ϖ on the number of the potential cycles for several simulations assuming different values of k_f, and the same plot with the experimental values obtained under the same conditions as in the simulation, are given in Fig. 5.115. Assuming a dimensionless rate constant $k_f = 0.5$ of Reaction (XXXVI) provides the best fit of the experimental results. Simulated cyclic voltammograms based on this dimensionless rate constant are depicted in Fig. 5.116. Figure 5.117 depicts the dependence of the rate of the substitution reaction on the number of potential

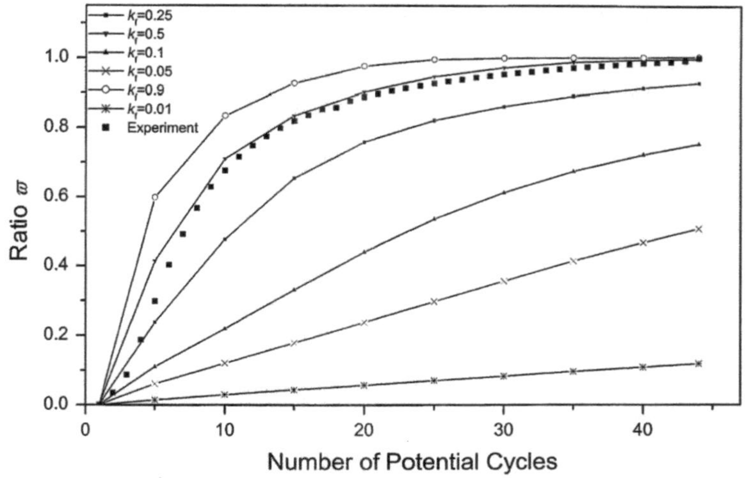

Fig. 5.115 Dependence of the ratio ϖ on the number of potential cycles for a couple of simulations assuming different values of the kinetic contribution k_f and for the real experiment [B173]. Copyright ©2001 with permission from Elsevier

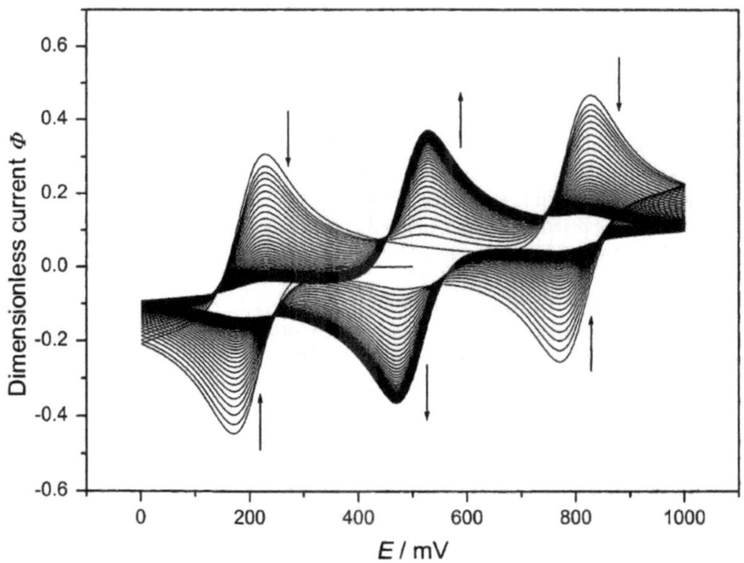

Fig. 5.116 Simulated multiple cyclic voltammograms for the solid-state conversion of Prussian blue (first and third redox couple) to cadmium hexacyanoferrate (second redox couple). $v = 100$ mV s^{-1}, $k_f = 0.25$, d$E = 2$ mV [B173]. Copyright ©2001 with permission from Elsevier

cycles. As can be seen in Fig. 5.117, after an induction period of 4–6 potential cycles, the rate of substitution is increasing to reach a maximum at around 15 cycles, corresponding to the increased amount of inserted ions, and then it is decreasing due to the decreasing concentration of starting material available for substitution.

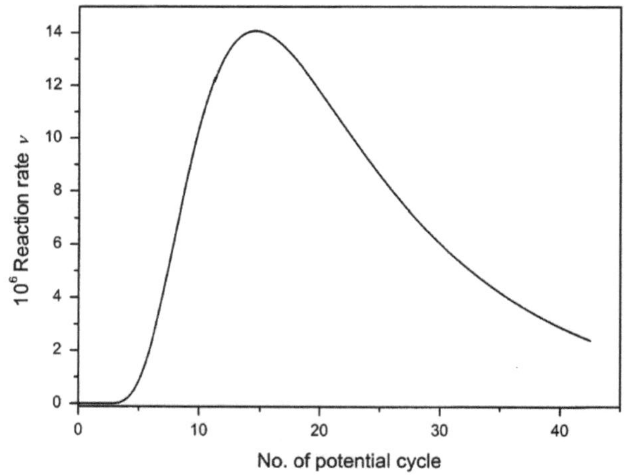

Fig. 5.117 Dependence of the reaction rate on the number of the potential cycle during the solid-state conversion of Prussian blue to cadmium hexacyanoferrate. Simulated data are the same as in Fig. 5.99 [B173]. Copyright ©2001 with permission from Elsevier

References

[B1] Scholz F, Nitschke L, Henrion G (1989) A new procedure for fast electrochemical analysis of solid materials. Naturwiss 76:71–72

[B2] Scholz F, Nitschke L, Henrion G, Damaschun F (1989) A new technique to study the electrochemistry of minerals. Naturwiss 76:167–168

[B3] Scholz F, Nitschke L, Henrion G (1989) Identification of solid materials with a new electrochemical technique – the abrasive stripping voltammetry. Fresenius Z Anal Chem 334:56–58

[B5] Scholz F, Nitschke L, Kemnitz E, Olesch T, Henrion G, Hass D, Bagchi RN, Herrmann R, Pruss N, Wilde W (1989) A simple and convenient solid state microanalytical technique for identification and characterization of the high temperature superconductor $YBa_2Cu_3O_{7-x}$. Fresenius Z Anal Chem 335:571–572

[B7] Scholz F, Nitschke L, Henrion G (1990) Abrasive stripping voltammetric analysis of tin-bismuth. Electroanalysis 2:85–87

[B8] Scholz F, Müller WD, Nitschke L, Rabi F, Livanova L, Fleischfresser C, Thierfelder C (1990) Fast and nondestructive identification of dental alloys by abrasive stripping voltammetry. Fresenius J Anal Chem 338:37–40

[B9] Scholz F, Lange B (1990) High-performance abrasive stripping voltammetry. Fresenius J Anal Chem 338:293–294

[B10] Björnbom P, Bursell M (1990) A method for studying microelectrodes by means of micromanipulators as applied to carbon agglomerates from oxygen reduction electrode catalyst. J Electrochem Soc 137:363–365

[B11] Scholz F, Lange B, Jaworski A, Pelzer J (1991) Analysis of powder mixtures with the help of abrasive stripping voltammetry and coulometry. Fresenius J Anal Chem 340:140–144

[B13] Bond AM, Scholz F (1991) Calculation of thermodynamic data from voltammetry of solid lead and mercury dithiocarbamate complexes mechanically attached to a graphite electrode. J Phys Chem 95:7460–7465

[B14] Bond AM, Scholz F (1991) Electrochemical, thermodynamic and mechanistic data derived from voltammetric studies on insoluble metallocenes, mercury halide and sulfide compounds, mixed silver halide crystals and other metal complexes following their mechanical transfer to a graphite electrode. Langmuir 7:3197–3204

[B17] Bond AM, Scholz F (1992) Field-based identification of minerals using a battery-operated electrochemical measuring system and mechanical transfer of the solid to a graphite electrode. J Geochem Explor 42:227–235
[B18] Scholz F, Rabi F, Müller WD (1992) The anodic dissolution of dental amalgams studied with abrasive stripping voltammetry. Electroanalysis 4:339–346
[B22] Komorsky-Lovrić Š, Lovrić M, Bond AM (1992) Comparison of the square-wave stripping voltammetry of lead and mercury following their electrochemical or abrasive deposition onto a paraffin impregnated graphite electrode. Anal Chim Acta 258:299–305
[B24] Lange B, Scholz F, Bautsch HJ, Damaschun F, Wappler G (1993) Thermodynamics of the xanthoconite-proustite and pyrostilpnite-pyrargyrite transitions as determined by abrasive stripping voltammetry. Phys Chem Miner 19:486–491
[B25] Lange B, Scholz F, Weiß A, Schwedt G, Behnert J, Raezke KP (1993) Abrasive stripping voltammetry – the electrochemical alternative for pigment analysis. Int Lab 23:23–26
[B28] Bond AM, Colton R, Daniels F, Fernando DR, Marken F, Nagaosa Y, Van Steveninck RFM, Walter JN (1993) Voltammetry, electron microscopy, and X-ray electron probe microanalysis at the electrode-aqueous electrolyte interface of solid microcrystalline *cis*- and *trans*-$Cr(CO)_2(dpe)_2$ and *trans*-$[Cr(CO)_2(dpe)_2]^+$ complexes (dpe = $Ph_2PCH_2CH_2PDh_2$) mechanically attached to carbon electrodes. J Am Chem Soc 115:9556–9562
[B29] Bond AM, Marken F (1994) Mechanistic aspects of the electron and ion transport processes across the electrode|solid solvent (electrolyte) interface of microcrystalline decamethylferrocene attached mechanically to a graphite electrode. J Electroanal Chem 372:125–135
[B31] Bond AM, Colton R, Marken F, Walter JN (1994) Mechanistic study of the voltammetry of nonconducting microcrystalline *cis*- and *trans*-$Cr(CO)_2(dpe)_2$ complexes (dpe = $Ph_2PCH_2CH_2PDh_2$) mechanically attached to a graphite electrode and immersed in different aqueous electrolyte media: identification by infrared spectroscopy of *cis*-[Cr $(CO)_2(dpe)_2]^+$ stabilized at the electrode-solid-solution interface. Organometallics 13:5122–5131
[B32] Zhang S, Meyer B, Moh G, Scholz F (1995) Development of analytical procedures based on abrasive stripping coulometry and voltammetry for solid state phase microanalysis of natural and synthetic tin-, arsenic-, and antimony bearing sulfosalts and sulfides of thallium, tin, lead, and silver. Electroanalysis 7:319–328
[B33] Dostal A, Schröder U, Scholz F (1995) Electrochemistry of chromium (II) hexacyanochromate(III)) and electrochemically induced isomerization of solid iron (II) hexacyanochromate(III)) mechanically imobilized on the surface of a graphite electrode. Inorg Chem 34:1711–1717
[B34] Scholz F, Schröder U, Meyer S, Brainina KZ, Zakharchuk NF, Sobolev NV, Kozmenko OA (1995) The electrochemical response of radiation defects of non-conducting materials. An electrochemical access to age determinations. J Electroanal Chem 385:139–142
[B35] Dostal A, Meyer B, Scholz F, Schröder U, Bond AM, Marken F, Shaw SJ (1995) Electrochemical study of microcrystalline solid Prussian blue particles mechanically attached to graphite and gold electrodes: electrochemically induced lattice reconstruction. J Phys Chem 99:2096–2103
[B36] Meyer B, Ziemer B, Scholz F (1995) In situ X-ray diffraction study of the electrochemical reduction of tetragonal lead oxide and orthorhombic Pb(OH)Cl mechanically immobilized on a graphite electrode. J Electroanal Chem 392:79–83
[B37] Grygar T, Šubrt J, Boháček J (1995) Electrochemical dissolution of goethite by abrasive stripping voltammetry. Coll Czech Chem Commun 60:950–959
[B40] Grygar T (1995) Kinetics of electrochemical reductive dissolution of iron(III) hydroxy-oxides. Coll Czech Chem Commun 60:1261–1273
[B41] Scholz F, Dostal A (1995) The formal potentials of the solid metal hexacyanometalates. Angew Chem 107:2876–2878 (Angew Chem Int Ed Engl 34(1995):2685–2687)

[B43] Komorsky–Lovrić S (1995) Voltammetry of microcrystals of cobalt and manganese phthalocyanines. J Electroanal Chem 397:211–215

[B45] Downard AJ, Bond AM, Hanton LR, Heath GA (1995) A comparison of solution, microcrystalline solid, and thin film phase voltammetry of $[Co(mtas)_2](X)_n$ (mtas = bis (2–(dimethylarsino)–phenyl)methylarsine; $X = BF_4^-$, n = 3; $X = ClO_4^-$, n = 2, 3; $X = BPh_4^-$, $n = 2$). Inorg Chem 34:6387–6395

[B46] Meyer B (1995) Entwicklung von festkörperelektrochemischen Methoden für die Mikroanalyse von Mineralen und synthetischen Festkörpern. Ph.D. thesis, Humboldt-Universität, Berlin

[B47] Ura H, Nishina T, Uchida I (1995) Electrochemical measurements of single particles of Pd and $LaNi_5$ with a microelectrode technique. J Electroanal Chem 396:169–173

[B49] Reddy SJ, Dostal A, Scholz F (1996) Solid state electrochemical studies of mixed nickel-iron hexacyanoferrates with the help of abrasive stripping voltammetry. J Electroanal Chem 403:209–212

[B50] Grygar T (1996) Electrochemical dissolution of iron(III) hydroxy-oxides: more information about the particles. Coll Czech Chem Commun 61:93–106

[B52] Grygar T (1996) The electrochemical dissolution of iron(III) and chromium(III) oxides and ferrites under conditions of abrasive stripping voltammetry. J Electroanal Chem 405:117–125

[B53] Shaw SJ, Marken F, Bond AM (1996) Simultaneous electrochemical and quartz crystal microbalance studies of nonconducting microcrystalline particles of *trans*-$Cr(CO)_2(dpe)_2$ and *trans*-$[Cr(CO)_2(dpe)_2]^+$ (dpe = $Ph_2PCH_2CH_2PDh_2$) attached to gold electrodes. J Electroanal Chem 404:227–235

[B54] Shaw SJ, Marken F, Bond AM (1996) Detection of new features associated with the oxidation of microcrystalline tetrathiafulvalene attached to gold electrodes by the simultaneous application of electrochemical and quartz crystal microbalance techniques. Electroanalysis 8:732–741

[B55] Dostal A, Hermes M, Scholz F (1996) The formation of bilayered nickel-iron, cadmium-iron and cadmium-silver hexacyanoferrates by an electrochemically driven insertion-substitution mechanism. J Electroanal Chem 415:133–141

[B56] Bond AM, Fletcher S, Marken F, Shaw SJ, Symons PG (1996) Electrochemical and X-ray diffraction study of the redox cycling of nanocrystals of 7,7,8,8-tetracyanoquinodimethane. Observation of a solid phase transformation controlled by nucleation and growth. J Chem Soc Faraday Trans 92(20):3925–2933

[B57] Meyer B, Zhang S, Scholz F (1996) The quantitative analysis of mixed crystals CuS_xSe_{1-x} with abrasive stripping voltammetry and a redetermination of the solubility product of CuSe and the standard potential of the Cu/CuSe electrode. Fresenius J Anal Chem 356:267–270

[B59] Reddy SJ, Hermes M, Scholz F (1996) The application of abrasive stripping voltammetry for a simple and rapid screening of pesticides. Electroanalysis 8:955–958

[B60] Lange B, Lovrić M, Scholz F (1996) The catalytic action of adsorbed thiocyanate ions and thiourea in the electron transfer from glassy carbon to solid copper(I) selenide and copper(I) sulfide particles. J Electroanal Chem 418:21–28

[B62] Meyer B, Scholz F (1997) Redetermination of the transformation enthalpies of the xanthoconite-proustite, pyrostilpnite-pyrargyrite and trechmannite-smithite phase transitions. Phys Chem Miner 24:50–52

[B63] Komorsky-Lovrić Š, Lovrić M, Scholz F (1997) Sulfide ion electrooxidation catalysed by cobalt phthalocyanine microcrystals. Mikrochim Acta 127:95–99

[B64] Lovrić M, Komorsky-Lovrić Š, Scholz F (1997) Staircase voltammetry with finite diffusion space. Electroanalysis 9:575–577

[B65] Bond AM, Colton R, Mahon PJ, Tan WT (1997) Tetrabutylammonium cation expulsion versus perchlorate electrolyte anion uptake in the electrochemical oxidation of

microcrystals of $[(C_4H_9)_4N][Cr(CO)_5I]$ mechanically attached to a gold electrode: a voltammetric and quartz crystal microbalance study. J Solid State Electrochem 1:53–61
[B66] Grygar T (1997) Dissolution of pure and substituted goethites controlled by the surface reaction under conditions of abrasive stripping voltammetry. J Solid State Electrochem 1:77–82
[B68] Komorsky-Lovrić Š (1997) Voltammetry of azobenzene microcrystal. J Solid State Electrochem 1:94–99
[B69] Lovrić M, Scholz F (1997) A model for the propagation of a redox reaction through microcrystals. J Solid State Electrochem 1:108–113
[B72] Bond AM, Marken F, Hill E, Compton RG, Hügel H (1997) The electrochemical reduction of indigo dissolved in organic solvents and as a solid mechanically attached to a basal plane pyrolytic graphite electrode immersed in aqueous electrolyte solution. J Chem Soc Perkin Trans 2:1735–1742
[B73] Fiedler DA, Besenhard JO, Fooken MH (1997) Rapid electrochemical characterization of battery electrode materials in the solid state. J Power Sources 69:157–160
[B75] Bond AM, Colton R, Marken F, Walter JN (1997) Voltammetric, specular reflectance infrared, and X-ray electron probe characterization of redox and isomerization processes associated with the $[Mn(CO)_2(\eta^3\text{–}P_2P')Br]^{+/0}$ ($P_2P' = \{Ph_2P(CH_2)_2\}PPh$), $[Mn(CO)_2(\eta^3\text{–}P_3P')Br]^{+/0}$, ($P_3P' = \{Ph_2PCH_2\}_3P$), and $[\{Mn(CO)(\eta^2\text{–dpe})Br\}_2(\mu\text{–dpe})]^{2+/0}$ (dpe = $Ph_2P(CH_2)_2PPh_2$) solid state systems. Organometallics 16:5006–5014
[B77] Bond AM, Colton R (1997) Electrochemical studies of metal carbonyl compounds. Coord Chem Rev 166:161–180
[B78] Uchida I, Fujiyoshi H, Waki S (1997) Microvoltammetric studies on single particles of battery active materials. J Power Sources 68:139–144
[B79] Scholz F, Lovrić M, Stojek Z (1997) The role of redox mixed phases $\{ox_x(C_n red)_{1-x}\}$ in solid state electrochemical reactions and the effect of miscibility gaps in voltammetry. J Solid State Electrochem 1:134–142
[B80] Perdicakis M, Grosselin N, Bessière J (1997) Voltammetric behaviour of single particles of silver chalcogenides in the micrometric range. Comparison with measurements at modified carbon electrodes. Electrochim Acta 42:3351–3358
[B81] Grygar T (1998) Phenomenological kinetics of irreversible electrochemical dissolution of metal-oxide microparticles. J Solid State Electrochem 2:127–136
[B84] Komorsky-Lovrić Š, Scholz F (1998) Stripping chronopotentiometry of immobilized microparticles. J Electroanal Chem 445:81–87
[B86] Fiedler DA (1998) Rapid evaluation of the rechargeability of γ-MnO_2 in alkaline media by abrasive stripping voltammetry. J Solid State Electrochem 2:315–320
[B87] Lovrić M, Hermes M, Scholz F (1998) The effect of electrolyte concentration in the solution on the voltammetric response of insertion electrodes. J Solid State Electrochem 2:401–404
[B88] Oldham KB (1998) Voltammetry at a three-phase junction. J Solid State Electrochem 2:367–377
[B90] Bond AM, Colton R, Humphrey DG, Mahon PJ, Snook GA, Tedesco V, Walter JN (1998) Systematic studies of 17-electron rhenium(II) carbonyl phosphine complexes. Organometallics 17:2977–2985
[B92] Grygar T, Bezdička P (1998) Electrochemical dissolution of CrIII and CrIV oxides. J Solid State Electrochem 3:31–38
[B95] Kahlert H, Retter U, Lohse H, Siegler K, Scholz F (1998) On the determination of the diffusion coefficients of electrons and potassium ions in copper(II) hexacyanoferrate(II) composite electrodes. J Phys Chem B 102:8757–8765
[B98] Nalini B, Sriman Narayanan S (1998) Ruthenium(III) diphenyldithiocarbamate as mediator for the electrocatalytic oxidation of sulfhydryl compounds at graphite electrode. Bull Electrochem 14:241–245

[B99] Ravi Shankaran D, Sriman Narayanan S (1998) Studies on copper hexacyanoferrate modified electrode and its application for the determination of cysteine. Bull Electrochem 14:267–270

[B100] Nalini B, Sriman Narayanan S (1998) Electrocatalytic oxidation of ascorbic acid at a graphite electrode modified with a new Ruthenium(III) dithiocarbamate complex. Bull Electrochem 14:267–270

[B101] Bond AM, Fletcher S, Symons PG (1998) The relationship between the electrochemistry and the crystallography of microcrystals – the case of TCNQ (7,7,8,8-tetracyanoquinodimethane). Analyst 123:1891–1904

[B102] Bond M, Deacon GB, Howitt J, MacFarlane DR, Spiccia L, Wolfbauer G (1998) Voltammetric determination of the reversible redox potential for the oxidation of the highly active polypyridyl ruthenium photovoltaic sensitizer *cis*–$Ru^{(II)}(dcbpy)_2(NCS)_2$. J Electrochem Soc 146:648–656

[B103] Dostal A (1998) Festkörperreaktionen an metallhexacyanometallat-modifizierten Elektroden. Ph.D. thesis, Humboldt–Universität, Berlin

[B104] Grygar T (1998) Elektrochemické rozpouštění práškových oxidů kovů. Ph.D. thesis, Řež

[B105] Albering JH, Grygar T (1999) A note regarding the rate-determining step of reduction of solid manganate(V). J Solid State Electrochem 3:117–120

[B106] Bond M, Fiedler DA, Lamprecht A, Tedesco V (1999) Electron-, anion-, and proton-transfer processes associated with the redox chemistry of $Fe(\eta^5\text{-}C_5Ph_5)(\eta^6\text{-}C_6H_5)C_5Ph_4)$ and its protonated form $[Fe(\eta^5\text{-}C_5Ph_5)(\eta^6\text{-}_6H_5)C_5Ph_4H)]BF_4$ at microcrystal-electrode-solvent (electrolyte) interfaces. Organometallics 18:642–649

[B107] Lovrić M, Scholz F (1999) A model for the coupled transport of ions and electrons in redox conductive microcrystals. J Solid State Electrochem 3:172–175

[B108] Zezula I, Galova M (1999) Application of abrasive stripping voltammetry in corrosion science. I. Determination of the corrosion potential of metals. J Solid State Electrochem 3:231–233

[B109] Komorsky–Lovrić S, Galic I, Penovski R (1999) Voltammetric determination of cocaine microparticles. Electroanalysis 11:120–123

[B111] Zakharchuk NF, Naumov N, Stösser R, Schröder U, Scholz F, Mehner H (1999) Solid state electrochemistry, X-ray powder diffraction, magnetic susceptibility, electron spin resonance, Mössbauer and diffuse reflectance spectroscopy of mixed iron(III)-cadmium (II) hexacyanoferrates. J Solid State Electrochem 3(5):264–276

[B113] Narayanan SS, Scholz F (1999) A comparative study of the electrocatalytic activities of some metal hexacyanoferrates for the oxidation of hydrazine. Electroanalysis 11:465–469

[B114] Zhuang Q, Scholz F, Pragst F (1999) The voltammetric behaviour of solid 2,2-diphenyl-1-picrylhydrazyl (DPPH) microparticles. Electrochem Commun 1:406–4103

[B115] Shankaran DR, Narayanan SS (1999) Characterization and application of an electrode modified by mechanically immobilized copper hexacyanoferrate. Fresenius J Anal Chem 364:686–689

[B117] Grygar T (1999) Electrochemical reactions of $La(Ni,Cr)O_3$ and $La(Ni,Fe)O_3$ in acid aqueous solution. J Solid State Electrochem 3:412–416

[B118] Grygar T, Bezdička P, Caspary EG (1999) Electrochemical dissolution of immobilized alpha–$(Fe_xCr_{1-x})_2O_3$ microparticles. J Electrochem Soc 146:3234–3238

[B119] Suárez MF, Marken F, Compton RG, Bond AM, Miao WJ, Raston CL (1999) Evidence for nucleation-growth, redistribution, and dissolution mechanisms during the course of redox cycling experiments on the C-60/NBu_4C_{60} solid-state redox system – voltammetric, SEM, and in-situ AFM studies. J Phys Chem 103:5637–5644

[B120] Shankaran DR, Narayanan SS (1999) Chemically-modified sensor for amperometric determination of sulfur-dioxide. Sens Actuators B Chem 55:191–194

[B124] Suárez MF, Bond AM, Compton RG (1999) Significance of redistribution reactions detected by in situ atomic force microscopy during early stages of fast scan rate redox cycling experiments at a solid 7,7,8,8-tetracyanoquinodimethane-glassy carbon electrode-aqueous (electrolyte) interface. J Solid State Electrochem 4:24–33

[B125] Bezdička P, Grygar T, Klapste B, Vondrak J (1999) MnO_x/C composites as electrode materials. I. Synthesis, XRD and cyclic voltammetric investigation. Electrochim Acta 45:913–920

[B126] Komorsky-Lovrić Š, Mirčeski V, Scholz F (1999) Voltammetry of organic microparticles. Mikrochim Acta 132:67–77

[B128] Eklund JC, Bond AM (1999) Photocatalytic reactions at microcrystalline *fac*-Mn $(CO)_3(\eta^2\text{-}Ph_2PCH_2PPh_2)Cl$-electrode-aqueous (electrolyte) interfaces. J Am Chem Soc 121:8306–8312

[B129] Bond AM, Lamprecht A, Tedesco V, Marken F (1999) Novel features associated with the electrochemically driven bis(eta(5)-pentaphenyl-cyclopentadienyl)iron(II)-iron(III) redox transformation at an electrode-microcrystal-solvent (electrolyte) interface. Inorg Chim Acta 291:21–31

[B130] Shankaran DR, Narayanan SS (1999) Evaluation of mechanically immobilised nickel hexacyanoferrate electrode as an amperometric sensor for thiosulphate determination. Fresenius J Anal Chem 365:663

[B133] Scholz F, Schädel S, Schultz A, Schauer F (2000) Chronopotentiometric study of laccase catalysed oxidation of quinhydrone microcrystals immobilised on a gold electrode surface and of the oxidation of a phenol derivatised graphite electrode surface. J Electroanal Chem 480:241–248

[B134] Ramaraj R, Kabbe C, Scholz F (2000) Electrochemistry of microparticles of tris(2,2-′-bipyridine)ruthenium(II) hexafluorophosphate. Electrochem Commun 2:190–194

[B135] Zhuang QK, Scholz F (2000) Electrochemical driven introduction of copper ions into the ring of 5,10,15,20-tetraphenyl-21H, 23H-porphyrin mechanically attached as solid microparticles to a graphite electrode. J Porphyr Phthalocyanines 4:202–208

[B137] Schwudke D, Stößer R, Scholz F (2000) Solid-state electrochemical, X-ray and spectroscopic characterization of substitutional solid solutions of iron-copper hexacyanoferrates. Electrochem Commun 2:301–306

[B138] Jiang J, Kucernak A (2000) The electrochemistry of platinum phthalocyanine microcrystals. I. Electrochemical behaviour in acetonitrile electrolytes. Electrochim Acta 45:2227–2239

[B139] Bond AM, Marken F, Williams CT, Beattie DA, Keyes TE, Foster RJ, Vos JG (2000) Unusually fast electron and anion transport processes observed in the oxidation of "electrochemically open" microcrystalline $[\{M(bipy)_2\}\{M'(bipy)_2\}(\text{-}L)](PF_6)_2$ complexes (M, M′ = Ru, Os; bipy = 2,2′–bipyridyl; L = 1,4-dihydroxy-2,5-bis(pyrazol-1-yl)benzene. J Phys Chem 104:1977–1983

[B141] Bakardjieva S, Bezdička P, Grygar T, Vorm P (2000) Reductive dissolution of microparticulate manganese oxides. J Solid State Electrochem 4:306–313

[B142] Schröder U, Oldham KB, Myland JC, Mahon PJ, Scholz F (2000) Modelling of solid-state voltammetry of immobilized microcrystals assuming an initiation of the electrochemical reaction at a three-phase junction. J Solid State Electrochem 4:314–324

[B143] Doménech-Carbó A, Domenéch-Carbó MT, Gimeno-Adelantado JV, Moya-Moreno M, Bosch-Reig F (2000) Voltammetric identification of lead(II) and (IV) in mediaeval glazes in abrasion-modified carbon paste and polymer film electrodes. Application to the study of alterations in archaeological ceramics. Electroanalysis 12:120–127

[B146] Doménech-Carbó A, Doménech-Carbó MT, Moya-Moreno M, Gimeno-Adelantado JV, Bosch-Reig F (2000) Identification of inorganic pigments from paintings and polychromed sculptures immobilized into polymer electrodes by stripping differential pulse voltammetry. Anal Chim Acta 407:275–289

[B147] Kucernak AR, Chowdhury PB, Wilde CP, Kelsall GH, Zhu YY, Williams DE (2000) Scanning electrochemical microscopy of a fuel-cell electrocatalyst deposit onto highly oriented pyrolytic graphite. Electrochim Acta 45:4483–4491

[B148] Zakharchuk N, Meyer S, Lange B, Scholz F (2000) A comparative study of lead oxyde modified graphite paste electrodes and solid graphite electrodes with mechanically immobilised lead oxides. Croat Chem Acta 73:667–704

[B154] Bond AM, Miao W, Raston CL (2000) Identification of processes that occur after reduction and dissolution of C_{60} adhered to gold, glassy carbon, and platinum electrodes placed in acetonitrile (electrolyte) solution. J Phys Chem B 104:2320–2329

[B155] Foster RJ, Keyes TE, Bond AM (2000) Protonation effects on the structure and homogeneous charge transport dynamics of solid state osmium bis(bipyridyl)tetrazine chloride films. J Phys Chem B 104:6389–6396

[B156] Bond AM, Miao W, Raston CL, Sandoval CA (2000) Electrochemical, EPR, and magnetic studies on microcrystals of the $[C_{60}\subset(\text{p-benzyl-calix[5]arene})_2]\cdot 8$toluene and its one-electron-reduced encapsulation complex. J Phys Chem B 104:8129–8137

[B158] Lovrić M, Hermes M, Scholz F (2000) Solid state electrochemical reactions in systems with miscibility gaps. J Solid State Electrochem 4:394–401

[B159] Zhuang QK (2000) The mechanistic studies of the electrochemical in situ association of copper ions with 5,10,15,20-tetraphynyl-21H,23H-porphyrin. Microchem J 65:333–340

[B160] Shankaran DR, Narayanan SS (2000) Amperometric determination of sulphur dioxide at a nickel hexacyanoferrate modified electrode. Res J Chem Environ 6:13–16

[B163] Bárcena Soto M (2000) Thermodynamische Untersuchungen zur Insertionselektrochemie von Metallhexacyanoferraten. Ph.D. thesis, Humboldt-Universität, Berlin

[B167] Jiang J, Kucernak A (2001) The electrochemistry of platinum phthalocynine microcrystals. II. A microelectrode observation of nucleation-growth controlled solid-solid phase transformation in non-aqueous solvent. Electrochim Acta 46:1223–1231

[B169] Takeda N, Stawasz ME, Parkinson BA (2001) Electrochemical oxidation and ex-situ STM observation of bis(4-dimethylamino-2-dihydroxyphenyl)squarine dye layers on HOPG electrodes. J Electroanal Chem 498:19–33

[B170] Bond M, Miao W, Raston CL, Ness TJ, Barnes MJ, Atwood JL (2001) Electrochemical and structural studies on microcrystals of the $(C_{60})_x$(CTV) inclusion complexes (x = 1, 1.5; CTV = cyclotriveratrylene). J Phys Chem B 105:1687–1695

[B172] Bond AM, Feldberg SW, Miao W, Oldham KB, Raston CL (2001) Modelling of solid-state, dissolution and solution-phase reactions at adhered solid-electrode-solvent (electrolyte) interfaces: electrochemistry of microcrystals of C_{60} adhered to an electrode in contact with dichloromethane (Bu_4NclO_4). J Electroanal Chem 501:22–32

[B173] Hermes M, Lovrić M, Hartl M, Retter U, Scholz F (2001) On the electrochemically driven formation of bilayered systems of solid Prussian-blue-type metal hexacyanoferrates: a model for Prussian blue cadmium hexacyanoferrate supported by finite difference simulations. J Electroanal Chem 501:193–204

[B175] Doménech-Carbó MT, Casas-Catalán MJ, Doménech-Carbó A, Mateo-Castro R, Gimeno-Adelantado JV, Bosch-Reig F (2001) Analytical study of canvas painting collection from the *Basilica de la Virgen de los Desaparados* using SEM/EDX, FT–IR, GC and electrochemical techniques. Fresenius J Anal Chem 369:571–575

[B176] Doménech-Carbó A, Doménech-Carbó MT, Gimeno-Adelantado V, Bosch-Reig F, Saurí-Peris MC, Casas-Catálan MJ (2001) Electrochemical analysis of the alterations in copper pigments using charge transfer coefficient/peak potential diagrams. Application to microsamples of baroque wall paintings attached to polymer film electrodes. Fresenius J Anal Chem 369:576–581

[B177] Cepriá G, Abadias O, Pérez-Arantegui J, Castillo JR (2001) Electrochemical behavior of silver-copper alloys in voltammetry of microparticles: a simple method for screening purposes. Electroanalysis 13:477–483

[B178] Zhuang QK, Scholz F (2001) In-situ electrochemical synthesis of the metalloporphyrins with help of the voltammetry of immobilized microparticles. Chem J Chin Univ (Gaodeng Xuexiao Huaxue Xuebao) 22:171–174

[B179] Keyes TE, Foster RJ, Bond AM, Miao W (2001) Electron self-exchange in the solid-state: cocrystals of hydroquinone and bipyridyl triazole. J Am Chem Soc 123:2877–2884

[B181] Grygar T, Bakardjieva S, Bezdička P, Vorm P (2001) Electrochemical dissolution of mixed oxides of Mn and Fe: the relationship between phase composition and reactivity. Ceramics-Silikaty 45:55–61

[B184] Hasse U, Scholz F (2001) In situ atomic force microscopy of the reduction of lead oxide nanocrystals immobilized on an electrode surface. Electrochem Commun 3:429–434

[B185] Brown RJC, Kucernak AR (2001) The electrochemistry of platinum phthalocyanine microcrystals. III. Electrochemical behaviour in aqueous electrolytes. Electrochim Acta 46:2573–2582

[B187] Scholz F, Schwudke D, Stößer R, Boháček J (2001) The interaction of Prussian blue and dissolved hexacyanoferrate ions with goethite (α-FeOOH) studied to assess the chemical stability and physical mobility of Prussian blue in soils. Ecotoxicol Environ Saf Environ Res 49:245–254

[B189] Doménech-Carbó A, Doménech-Carbó MT, Osete-Cortina L (2001) Identification of manganese(IV) centers in archaeological glass using microsample coatings attached to polymer film electrodes. Electroanalysis 13:927–935

[B190] Jiang J, Kucernak AR (2001) The electrochemistry of platinum phthalocyanine microcrystals. IV. Temperature dependence of the electrochemical behaviour in non-aqueous solution. Electrochim Acta 46:3445–3456

[B191] Grygar T, Bezdička P, Piszora P, Wolska E (2001) Electrochemical reactivity of Li-Mn-O and Li-Fe-Mn-O spinels. J Solid State Electrochem 5:487–494

[B192] Galova M, Orinakova R, Grygar T, Lux L, Hezelova M (2001) Relation between the dissolution reactivity of powder substances and efficiency of their electroplating. Particulate Sci Technol 19:85–94

[B193] Cepriá G, Aranda C, Pérez-Arantegui J, Lacueva F, Castillo JR (2001) Voltammetry of immobilized microparticles: a powerful analytical technique to study the physical and chemical composition of brass. J Electroanal Chem 513:52–58

[B194] Grygar T, Bezdička P, Vorm P, Jordanova N, Krtil P (2001) Spinel solid solutions in the Li-Fe-Mn-O system. J Solid State Chem 161:152–160

[B195] Doménech-Carbó A, Doménech-Carbó MT, Gimeno-Adelantado JV, Bosch-Reig F, Saurí-Peris MC, Sánchez-Ramos S (2001) Electrochemistry of iron oxide pigments (earths) from pictorial microsamples attached to graphite-polyester composite electrodes. Analyst 126:1764–1772

[B196] Wooster TJ, Bond AM, Honeychurch MJ (2001) Resistance transitions detected by analysis of the voltammetry of tetrathiafulvalene microparticles adhered to electrode surfaces under conditions of dynamic resistance compensation. Electrochem Commun 3:746–752

[B198] van Oorschot IHM, Grygar T, Dekkers MJ (2001) Detection of low concentrations of fine-grained iron oxides by voltammetry of microparticles. Earth Planet Sci Lett 193:631–642

[B199] Shankaran DR, Narayanan SS (2001) Mechanically immobilized copper hexacyanoferrate modified electrode for electrocatalysis and amperometric determination of glutathione. Bull Korean Chem Soc 22:816–820

[B201] Shankaran DR, Narayanan SS (2001) Amperometric sensor for thiols based on mechanically immobilised nickel hexacyanoferrate modified electrode. Bull Electrochem 17:277–280

[B202] Shankaran DR, Narayanan SS (2001) Amperometric sensor for hydrazine determination based on mechanically immobilized nickel hexacyanoferrate modified electrode. Russ J Electrochem 37:1149–1153

[B204] Grygar T, Marken F, Schröder U, Scholz F (2002) Electrochemical analysis of solids. Coll Czech Chem Commun 67:163–208

[B208] Jiang J, Kucernak A (2002) Nanostructured platinum as an electrocatalyst for the electrooxidation of formic acid. J Electroanal Chem 520:64–70

[B209] Bárcena Soto M, Scholz F (2002) The thermodynamics of the insertion electrochemistry of solid metal hexacyanometallates. J Electroanal Chem 521:183–189

[B211] Inzelt G (2002) Cyclic voltammetry of solid diphenylamine crystals immobilized on an electrode surface and in the presence of an aqueous solution. J Solid State Electrochem 6:265–271

[B212] Shankaran DR, Narayanan SS (2002) Amperometric sensor for glutathione based on mechanically immobilised cobalt hexacyanoferrate modified electrode. Bull Chem Soc Jpn 75:501–505

[B217] Doménech-Carbó A, Doménech-Carbó MT, Osete-Cortina L, Gimeno-Adelantado JV, Bosch-Reig F, Mateo-Castro R (2002) Electrochemical identification of metal ions in archaeological ceramic glazes by stripping voltammetry at graphite/polyester composite electrodes. Talanta 56:161–174

[B218] Fehér K, Inzelt G (2002) Electrochemical quartz crystal microbalance study of formation and redox transformation of poly(diphenylamine). Electrochim Acta 47:3551–3559

[B222] Grygar T, Dedecek J, Hradil D (2002) Analysis of low concentrations of ferric oxides in clays by diffuse reflectance spectroscopy and voltammetry. Geologica Carpathica 53:1–7

[B223] Long JW, Ayers KE, Rolison DR (2002) Electrochemical characterization of high-surface-area catalysts and other nanoscale electroactive materials at sticky-carbon electrodes. J Electroanal Chem 522:58–65

[B225] Doménech-Carbó A, Pérez-Ramírez J, Ribera A, Kapteijn F, Mul G, Moulijn JA (2002) Characterization of iron species in ex-framework FeZSM-5 by electrochemical methods. Catal Lett 78:303–312

[B226] Doménech-Carbó A, Pérez-Ramírez J, Ribera A, Mul G, Kapteijn F, Arends IWCE (2002) Electrochemical characterization of iron sites in ex-framework FeZSM-5. J Electroanal Chem 519:72–84

[B243] Jinag J, Kucernak A (2002) Electrooxidation of small organic molecules on mesoporous precious metal catalysts. I: CO and methanol on platinum. J Electroanal Chem 533:153–165

[B246] Pérez-Ramírez J, Mul G, Kapteijn F, Moulijn JA, Overweg AR, Doménech-Carbó A, Ribera A, Arends IWCE (2002) Physicochemical characterization of isomorphously substituted FeZSM-5 during activation. J Catal 207:113–126

[B247] Widmann A, Kahlert H, Petrovic-Prelevic I, Wulff H, Yakhmi JV, Bagkar N, Scholz F (2002) Structure, insertion electrochemistry and magnetic properties of a new type of substitutional solid solutions of copper, nickel and iron hexacyano-ferrates/hexacyanocobaltates. Inorg Chem 42:5706–5715

[B251] Keane L, Hogan C, Forster RJ (2002) Dynamics of charge transport through osmium tris dimethoxy bipyridyl solid deposits. Langmuir 18:4826–4833

[B252] Walsh DA, Keyes TE, Forster RJ (2002) Redox switching in solid deposits: triazole bridged osmium dimers. J Electroanal Chem 538(539):75–85

[B253] Shankaran DR, Narayanan SS (2002) Cobalt hexacyanoferrate-modified electrode for amperometric assay of hydrazine. Russ J Electrochem 38:987–991

[B255] Wooster TJ, Bond AM, Honeychurch MJ (2003) An analogy of an ion-selective electrode sensor based on the voltammetry of microcrystals of tetracyanoquinodimethane or tetrathiafulvalene adhered to an electrode surface. Anal Chem 75:586–592

[B260] Marken F, Cromie S, McKee V (2003) Electrochemically driven reversible solid state metal exchange processes in polynuclear copper complexes. J Solid State Electrochem 7:141–146

[B261] Cepriá G, Bolea E, Laborda F, Castillo JR (2003) Quick, easy, and inexpensive way to detect small metallic particles in suspension using voltammetry of immobilized microparticles. Anal Lett 36:921–929
[B263] Tan WT, Lim EB, Bond AM (2003) Voltammetric studies on microcrystalline C_{60} adhered to an electrode surface by solvent casting and mechanical transfer methods. J Solid State Electrochem 7:134–140
[B265] Doménech-Carbó A, Dménech-Carbó MT, Saurí-Peris MC, Gimeno-Adelantado JV, Bosch-Reig F (2003) Electrochemical identification of anthraquinone-based dyes in solid microsamples by square wave voltammetry using graphite/polyester composite electrodes. Anal Bioanal Chem 375:1169–1175
[B267] Tan WT, Ng GK, Bond AM (2000) Electrochemical oxidation of microcrystalline tetrathiafulvalene (TTF) at an electrode-solid-aqueous (KBr) interface. Malays J Chem 2:34–42
[B272] Inzelt G (2003) Formation and redox behaviour of polycarbazole prepared by electropolymerisation of solid carbazole crystals immobilized on an electrode surface. J Solid State Electrochem 7:503–510
[B276] Hasse U, Nießen J, Scholz F (2003) Atomic force microscopy of the electrochemical reductive dissolution of sub-micrometer sized crystals of goethite immobilized on a gold electrode. J Electroanal Chem 556:13–22
[B294] Zhang J, Bond AM (2004) Voltammetric studies with adhered microparticles and the detection of a dependence of organometallic $cis^+ \rightarrow trans^+$ first-order isomerization rate constants on the identity of the ionic liquid. J Phys Chem B 108:7363–7372
[B295] Snook G, Bond AM, Fletcher S (2003) The catalysis of solid state intercalation processes by organic solvents. J Electroanal Chem 554–555:157–165
[B296] Neufeld AK, Madsen I, Bond AM, Hogan CF (2003) Phase, morphology, and particle size changes associated with the solid-solid electrochemical interconversion of TCNQ and semiconducting CuTCNQ (TCNQ = tetracyanoquinodimethane). Chem Mater 15:3573–3585
[B297] Guo S, Zhang J, Elton DM, Bond AM (2004) Fourier transform large-amplitude alternating current cyclic voltammetry of surface-bound azurin. Anal Chem 76:166–177
[B299] Wooster TJ, Bond AM (2003) Ion selectivity obtained under voltammetric conditions when a TCNQ chemically modified electrode is presented with aqueous solutions containing tetraalkylammonium cations. Analyst 128:1386–1390
[B301] Doménech-Carbó A, Alvaro M, Ferrer B, García H (2003) Electrochemistry of mesoporous organosilica of MCM-41 type containing 4,4′-bipyridinium units: voltammetric response and electrocatalytic effect on 1,4-dihydrobenzoquinone oxidation. J Phys Chem B 107:12781–12788
[B302] Grygar T, Kuckova S, Hradil D, Hradilova J (2003) Electrochemical analysis of natural solid organic dyes and pigments. J Solid State Electrochem 7:706–713
[B303] Fay N, Dempsey E, Kennedy A, McCormac T (2003) Synthesis and electrochemical characterisation of $[Ru(bpy)_3]_3[P_2W_{18}O_{62}]$. J Electroanal Chem 556:63–74
[B304] Doménech-Carbó A, Doménech-Carbó MT, Osete-Cortina L, Gimeno-Adelantado JV, Sánchez-Ramos S, Bosch-Reig F (2003) Quantitation of metal ions in archaeological glass by abrasive stripping square-wave voltammetry using graphite/polyester composite electrodes. Electroanalysis 15:1465–1475
[B305] Wildgoose GG, Pandurangappa M, Lawrence NS, Jiang L, Jones TGJ, Compton RG (2003) Anthraquinone-derivatised carbon powder: reagentless voltammetric pH electrodes. Talanta 60:887–893
[B306] Zhang J, Bond AM, Belcher J, Wallace KJ, Steed JW (2003) Electrochemical studies on the modular podand 1,3,5-tris(3-((ferrocenylmethyl)amino)pyridiniumyl)-2,4,6-triethylbenzene hexafluorophosphate in conventional solvents and ionic liquids. J Phys Chem B 107:5777–57

[B307] Galova M, Grygar T, Pikna L, Lux L (2003) Electrochemical study on the reactivity of Fe powders. Chem Anal 48:293–303

[B308] Zhang J, Bond AM (2003) Conditions required to achieve the apparent equivalence of adhered solid- and solution-phase voltammetry for ferrocene and other redox-active solids in ionic liquids. Anal Chem 75:2694–2702

[B309] Pandurangappa M, Lawrence NS, Compton RG (2002) Homogeneous chemical derivatisation of carbon particles: a novel method for functionalising carbon surfaces. Analyst 127:1568–1571

[B310] Pandurangappa M, Lawrence NS, Jiang L, Jones TGJ, Compton RG (2003) Physical adsorption of *N,N′*-diphenyl-*p*-phenylenediamine onto carbon particles: application to the detection of sulfide. Analyst 128:473–479

[B311] Grygar T, Bezdička P, Hradil D, Pikna L (2003) Electrochemical analysis of metal oxides. Solid State Chem Solid State Phenomena 90(91):45–49

[B312] Cepriá G, Usón A, Pérez-Arantegui J, Castillo JR (2003) Identification of iron(III) oxides and hydroxyl-oxides by voltammetry of immobilised microparticle. Anal Chim Acta 477:157–168

[B313] Uzun D, Ozser ME, Yuney K, Icil H, Demuth M (2003) Synthesis and photophysical properties of *N,N′*-bis(4-cyanophenyl)-3,4,9,10-perylene-bis(dicarboximide) and *N,N′*-bis(4-cyanophe-nyl)-1,4,5,8-naphthalenediimide. J Photochem Photobiol A Chem 156:45–54

[B314] Alonso-Sedano AB, Tascón-García L, Vázquez-Barbado D, Sánchez Batanero P (2004) Electrochemical study of copper and iron compounds in the solid state by using voltammetry of immobilized microparticles: application to copper ferrite characterization. J Electroanal Chem 566:433–441

[B315] Wildgoose GG, Leventis HC, Streeter I, Lawrence NS, Wilkins SJ, Jiang L, Jones TGJ, Compton RG (2004) Abrasively immobilised multiwalled carbon nanotube agglomerates: a novel electrode material approach for the analytical sensing of pH. ChemPhysChem 5:669–677

[B316] Rees NV, Wadhawan JD, Klymenko OV, Coles BA, Compton RG (2004) An electrochemical study of the oxidation of 1,3,5-tris[4-[(3-methylphenyl)phenylamino]phenyl] benzene. J Electroanal Chem 563:191–202

[B317] Doménech-Carbó A, Torres FJ, Alarcón J (2004) Electrochemical characterization of cobalt cordierites attached to paraffin-impregnated graphite electrodes. J Solid State Electrochem 8:127–137

[B318] Doménech-Carbó A, Sánchez-Ramos S, Yusá-Marco DJ, Moya-Moreno M, Gimeno-Adelantado J, Bosch-Reig F (2004) Electrochemical determination of boron in minerals and ceramic materials. Anal Chim Acta 501:103–111

[B319] Doménech-Carbó A, Doménech-Carbó MT (2006) Chronoamperometric study of proton transfer/electron transfer in solid state electrochemistry of organic dyes. J Solid State Electrochem 10:949–958

[B320] Doménech-Carbó A, Doménech-Carbó MT (2008) In situ AFM study of proton-assisted electrochemical oxidation/reduction of microparticles of organic dyes. Electrochem Commun 10:1238–1241

[B321] Doménech-Carbó A, Doménech-Carbó MT, Calisti M, Maiolo V (2010) Identification of naphtoquinonic and anthraquinonic dyes via sequential potential steps applied to the voltammetry of microparticles methodology. J Solid State Electrochem 14:465–477

[B322] Doménech-Carbó A, Doménech-Carbó MT, Calisti M, Maiolo V (2010) Sequential identification of organic dyes using the voltammetry of microparticles approach. Talanta 81:404–414

[B323] Doménech-Carbó A, Martini M, de Carvalho LM, Doménech-Carbó MT, Silva M (2013) Screening of pharmacologic adulterant classes in herbal formulations using voltammetry of microparticles. J Pharm Biomed Anal 74:194–204

[B324] Doménech-Carbó A, Martini M, de Carvalho LM, Viana C, Doménech-Carbó MT, Silva M (2013) Standard additions-dilution method for absolute quantification in voltammetry of microparticles. Application for determining psychoactive 1,4-benzodiazepine and antidepressants drugs as adulterants in phytotherapeutic formulations. J Pharm Biomed Anal 80:159–163

[B331] Doménech-Carbó A, Doménech-Carbó MT, Peiró MA (2011) Dating archaeological lead artifacts from measurement of the corrosion content using the voltammetry of microparticles. Anal Chem 83:5639–5644

[B332] Doménech-Carbó A, Doménech-Carbó MT, Peiró MA, Martínez I, Barrio J (2012) Application of the voltammetry of microparticles for dating archaeological lead using polarization curves and electrochemical impedance spectroscopy. J Solid State Electrochem 16:2349–2356

[B333] Doménech-Carbó A, Doménech-Carbó MT, Peiró MA (2011) 'One-touch' voltammetry of microparticles for the identification of corrosion products in archaeological lead. Electroanalysis 23:1391–1400

[B334] Pavlov D, Monakhov B (1987) Effect of Sb on the electrochemical properties of $Pb/PbSO_4/H_2SO_4$ and $Pb/PbO/PbSO_4/H_2SO_4$ electrodes. J Electroanal Chem 218:135–153

[B335] Pavlov D, Monakhov B, Salmi K, Sundholm G (1991) Ring–disk electrode studies of soluble intermediates formed during the polarization of Pb in H_2SO_4. Electrochim Acta 36:953–963

[B336] Cai W-B, Wan Y-Q, Liu H-T, Zhou W-F (1995) A study of the reduction process of anodic PbO_2 film on Pb in sulfuric acid solution. J Electroanal Chem 387:95–100

[B337] Pavlov D (1981) Semiconductor mechanism of the processes during electrochemical oxidation of PbO to PbO_2. J Electroanal Chem 118:167–185

[B338] Bartlett PN, Dunford T, Ghanem MA (2002) Templated electrochemical deposition of nanostructured macroporous PbO_2. J Mater Chem 12:3130–3135

[B339] Pavlov D, Monakhov B (1989) Mechanism of action of Sn on the passivation phenomena in the lead-acid battery positive plate (Sn-free effect). J Electrochem Soc 136:27–33

[B340] Doménech-Carbó A, Doménech-Carbó MT, Peiró M, Osete-Cortina L (2011) Authentication of archaeological lead artifacts using voltammetry of microparticles: the case of the *Tossal de Sant Miquel* Iberian plate. Archaeometry 53:1193–1211

[B341] Doménech-Carbó A, Doménech-Carbó MT, Pasíes T, Bouzas MC (2012) Modeling corrosion of archaeological silver-copper coins using the voltammetry of immobilized particles. Electroanalysis 24:1945–1955

[B342] Rodgers JD, Jedral W, Bunce NJ (1999) Electrochemical oxidation of chlorinated phenols. Environ Sci Technol 33:1453–1457

[B343] Panizza M, Cerisola G (2003) Influence of anode material on the electrochemical oxidation of 2-naphthol: Part 1. Cyclic voltammetry and potential step experiments. Electrochim Acta 48:3491–3497

[B344] Pavlov D, Kirchev A, Monahov B (2005) Mechanism of the oxygen cycle reactions proceeding at the negative plates of VRLA batteries. J Power Sources 144:521–527

[B345] Wu YM, Li WS, Long XM, Wu FH, Chen HY, Yan JH, Zhang CR (2005) Effect of bismuth on hydrogen evolution reaction on lead in sulfuric acid solution. J Power Sources 144:338–345

[B346] Chen P, McCreery RL (1996) Control of electron transfer kinetics at glassy carbon electrodes by specific surface modification. Anal Chem 68:3958–3965

[B347] Barisci JN, Wallace GG, Baughman RH (2000) Electrochemical studies of single-wall carbon nanotubes in aqueous solutions. J Electroanal Chem 488:92–98

[B348] Doménech-Carbó A (2012) Electrochemical analysis: voltammetry of microparticles. In: Dillmann P, Adriaens A, Angelini E, Watkinson D (eds) Green book on corrosion of cultural heritage artifacts, Chap 9. European Federation of Corrosion, Maney, Leeds, pp 165–189

[B349] Doménech-Carbó A (2012) Electrochemical techniques. In: Edwards HGM, Vandenabeele P (eds) Analytical archaeometry, selected topics, Chap 7. The Royal Society of Chemistry, London, pp 210–241
[B350] Doménech-Carbó A, Doménech-Carbó MT, Lee Y, Osete-Cortina L (2012) Potential application of voltammetry of microparticles for dating porcine blood-based binding media used in Taiwanese architectural polychromies. Chem Asian J 7:2268–2273
[B351] Doménech-Carbó A, Doménech-Carbó MT, Costa V (2009) Electrochemical methods for archaeometry, conservation and restoration. In: Scholz F (ed) Monographs in electrochemistry. Springer, Berlin
[B352] Doménech-Carbó A, Doménech-Carbó MT, Edwards HGM (2008) Quantitation from Tafel analysis in solid-state voltammetry. Application to the study of cobalt and copper pigments in severely damaged frescoes. Anal Chem 80:2704–2716
[B353] Doménech-Carbó A, Doménech-Carbó MT, Martínez-Lázaro L (2010) Layer-by-layer identification of copper alteration products in metallic works of art using the voltammetry of microparticles. Anal Chim Acta 680:1–9
[B354] Doménech-Carbó A, Doménech-Carbó MT, Peiró-Ronda MA (2011) 'One-touch' voltammetry of microparticles for the identification of corrosion products in archaeological lead. Electroanalysis 23:1391–1400
[B355] Doménech-Carbó A, Lastras M, Rodríguez F, Osete-Cortina L (2013) Mapping of corrosion products of highly altered archeological iron using voltammetry of microparticles. Microchem J 106:41–50
[B357] Doménech-Carbó A, Moya-Moreno M, Doménech-Carbó MT (2004) Standard addition method applied to solid–state striping voltammetry. Determination of zirconium in minerals and ceramic materials. Anal Bioanal Chem 380:146–156
[B358] Doménech-Carbó A, Doménech-Carbó MT, Gimeno-Adelantado JV, Bosch-Reig F (2006) H-point standard addition method applied to solid state stripping voltammetry. Application to the determination of lead and tin in archaeological glazes. Anal Bioanal Chem 385:1552–1561
[B359] Bosch-Reig F, Doménech-Carbó A, Doménech-Carbó MT, Gimeno-Adelantado JV (2007) H-Point standard addition method applied to voltammetry of microparticles. Quantitation of dyes in pictorial samples. Electroanalysis 19:1575–1584
[B360] Doménech-Carbó A, Koshevoy IO, Montoya N, Pakkanen TA (2010) Electrochemically assisted anion insertion in Au(I)-Cu(I) heterotrimetallic clusters bearing ferrocenyl groups: application to the flouride/chloride discrimination in aqueous media. Electrochem Commun 12:206–209
[B361] Doménech-Carbó A, Koshevoy IO, Montoya N, Pakkanen TA (2010) Electrochemical anion sensing using electrodes chemically modified with Au(I)-Cu(I) heterotrimetallic alkynyl cluster complexes containing ferrocenyl groups. Anal Bioanal Chem 397:2013–2022
[B362] Barbante GJ, Hogan CF, Hughes AB (2009) Solid state spectroelectrochemistry of microparticles of ruthenium diimine complexes immobilised on optically transparent electrodes. J Solid State Electrochem 13:599–608
[B363] Brown RJC, Kucernak AR (2005) The photoelectrochemistry of platinum phthalocyanine films in aqueous media. J Solid State Electrochem 9:459–468
[B364] Komorsky-Lovrić Š, Quentel F, L'Her M, Scholz F (2008) Voltammetry of microparticles of lutetium bisphthalocyanine. J Solid State Electrochem 12:165–169
[B365] Brown RJC, Brett DJL, Kucernak ARJ (2009) An electrochemical quartz crystal microbalance study of platinum phthalocyanine thin films. J Electroanal Chem 633:339–346
[B366] Doménech-Carbó A, Maciuk A, Figadère B, Poupon E, Cebrián-Torrejón G (2013) Solid state electrochemical assay of heme-binding molecules for screening of drugs with antimalarial potential. Anal Chem 85:4014–4021

[B367] Doménech-Carbó A, García H, Doménech-Carbó MT, Llabrés-i-Xamena F (2006) Electrochemical nanometric patterning of MOF particles: anisotropic metal electrodeposition in Cu/MOF. Electrochem Commun 8:1830–1834
[B368] Doménech-Carbó A, García H, Doménech-Carbó MT, Llabrés-i-Xamena F (2007) Electrochemistry of metal-organic frameworks: a description from the voltammetry of microparticles approach. J Phys Chem C 111:13701–13711
[B369] Babu KF, Kulandainathan MA, Katsounaros I, Burrows AD, Raithby PR, Marken F (2010) Electrocatalytic activity of Basolite™ F300 metal-organic-framework structures. Electrochem Commun 12:632–636
[B370] Cummings CI, Roweth AH, Ching AKZ, Jenkings ATA, Mitchels JM, Shariki S, Liew SY, Thielemans W, Walsh DA, Marken F (2010) Facile cation electro-insertion into layer-by-layer assembled iron phytate films. Electrochem Commun 12:1722–1726
[B371] Halls JE, Ahn SD, Jiang D, Keenan LL, Burrows AD, Marken F (2012) Proton uptake vs. redox driven release from metal-organic-frameworks: alizarin red S reactivity in UMCM–1. J Electroanal Chem 689:168–175
[B372] Halls JE, Hernan-Gomez A, Burrows AD, Marken F (2012) Metal-organic frameworks post-synthetically modified with ferrocenyl groups: framework effects on redox processes and surface conduction. Dalton Trans 41:1475–1480
[B373] Komorsky-Lovrić Š, Vukašinović N, Penovski R (2003) Voltammetric determination of microparticles of some local anesthetics and antithusics immobilized on the graphite electrode. Electroanalysis 15:544–547
[B374] Komorsky–Lovrić Š, Nigović B (2004) Identification of 5–aminosalicylic acid, ciprofloxacin and azithromycin by abrasive stripping voltammetry. J Pharm Biomed Anal 36:81–89
[B375] Komorsky-Lovrić Š, Nigović B (2006) Electrochemical characterization of simvastatin by abrasive stripping and square-wave voltammetry. J Electroanal Chem 593:125–130
[B376] Komorsky-Lovrić Š, Novak I (2009) Estimation of antioxidative properties of tea leaves by abrasive stripping electrochemistry using paraffin-impregnated graphite electrode. Coll Czech Chem Commun 74:1467–1475
[B377] Komorsky-Lovrić Š, Novak I (2011) Abrasive stripping square–wave voltammetry of blackberry, raspberry, strawberry, pomegranate, and sweet and blue Potatoes. J Food Sci 76:C916–C920
[B378] Corduneanu O, Janeiro P, Oliveira-Brett AM (2006) On the electrochemical oxidation of resveratrol. Electroanalysis 18:757–762
[B379] Oliveira Brett AM, da Silva LA, Fujii H, Mataka S, Thienmann T (2003) Detection of the damage caused to DNA by a thiophene-*S*-oxide using an electrochemical DNA-biosensor. J Electroanal Chem 549:91–99
[B380] de Carvalho LM, Hilgemann M, Spengler C, do Nascimento PC, Bohrer D (2010) Voltametria de micropartículas imobilizadas: fundamentos e aplicações analíticas (Voltammetry of immobilized microparticles: fundamentals and analytical applications). Química Nova 33:1765–1772
[B381] de Carvalho LM, Martini M, Moreira APL, de Lima APS, Correia D, Falcao T, Garcia SC, de Bairros AV, do Nascimento PC, Bohrer D (2011) Presence of synthetic pharmaceuticals as adulterants in slimming phytotherapeutic formulations and their analytical determination. For Sci Int 204:6–12
[B382] Cavalheiro ETG, Brett CMA, Oliveira-Brett AM, Fatibello-Filho O (2012) Bioelectroanalysis of pharmaceutical compounds. Bioanal Rev 4:31–53
[B383] Marselli B, Garcia-Gomez J, Michaud P-A, Rodrigo MA, Comninellis C (2003) Electrogeneration of hydroxyl radicals on boron-doped diamond electrodes. J Electrochem Soc 150:D79–D83
[B384] Enache TA, Chiorcea-Paquim A-M, Fatibello-Filho O, Oliveira-Brett AM (2009) Hydroxyl radicals electrochemically generated in situ on a boron–doped diamond electrode. Electrochem Commun 11:1342–1345

[B385] Kapałka A, Foti G, Comninellis C (2009) The importance of electrode material in environmental electrochemistry: formation and reactivity of free hydroxyl radicals on boron–doped diamond electrodes. Electrochim Acta 54:2018–2023
[B386] Hilgemann M, Scholz F, Kahlert H, de Carvalho LM, Barcellos da Rosa M, Lindequist U, Wurster M, do Nascimento PC, Bohrer D (2010) Electrochemical assay to quantify the hydroxyl radical scavenging activity of medicinal plant extracts. Electroanalysis 22:406–412
[B387] Gulaboski R, Lovrić M, Mirčeski V, Bogeski I, Hoth M (2008) Protein-film voltammetry: a theoretical study of the temperature effect using square-wave voltammetry. Biophys Chem 137:49–55
[B388] Bogeski I, Mirčeski V, Hoth M (2008) Probing the redox activity of T-lymphocytes deposited at electrode surfaces with voltammetric methods. Clin Chem Lab Med 46:197–203
[B389] Li G, Liu X, Du L, Wang E (2008) Direct electrochemistry of horseradish peroxidase immobilized in calcium carbonate microsphere doped with phospholipids. Electroanalysis 20:1421–1426
[B390] Hermes M, Scholz F, Walther R, Schild L, Wolke C, Lendeckel U (2011) Electrochemical signals of mitochondria: a new probe of their membrane properties. Angew Chem Int Ed 50:6872–6875
[B391] Doménech-Carbó A, Doménech-Carbó MT, Saurí-Peris MC (2005) Electrochemical identification of flavonoid dyes in work of art samples by abrasive voltammetry at paraffin-impregnated graphite electrodes. Talanta 66:769–782
[B392] Janeiro P, Oliveira Brett AM (2005) Solid state electrochemical oxidation mechanisms of morin in aqueous media. Electroanalysis 17:733–738
[B393] Doménech-Carbó A, Doménech-Carbó MT, Saurí-Peris MC (2005) Identification of curcuma and safflower dyes by voltammetry of microparticles using paraffin-impregnated graphite electrodes. Microchim Acta 152:75–84
[B394] Doménech-Carbó A, Doménech-Carbó MT, Vázquez de Agredos-Pascual ML (2007) Electrochemical monitoring Maya blue preparation from Maya's ancient procedures. J Solid State Electrochem 11:1335–1346
[B395] Hu X-W, He J-B, You Y-H, Zhang Y-M, Zhang S-L (2011) Spectroelectrochemistry of solid indirubin and its sulfonated form. Electrochim Acta 56:1219–1226
[B396] Ngameni E, Tonle IK, Nanseu CP, Wandji R (2000) Voltammetry study of 2-hydroxy-3-isopropenyl-1,4-naphthoquinone using a carbon paste electrode. Electroanalysis 12:847–852
[B397] He J-B, Ma G-H, Chen J-C, Yao Y, Wang Y (2010) Voltammetry and spectroelectrochemistry of solid indigo dispersed in carbon paste. Electrochim Acta 55:4845–4850
[B398] Gómez-Romero P, Sanchez C (2005) Hybrid materials. Functional properties. From Maya blue to 21st century materials. New J Chem 29:57–58
[B399] Doménech-Carbó A (2010) Electrochemistry of porous materials. Taylor & Francis, Boca Raton, FL
[B400] Doménech-Carbó A, Doménech-Carbó MT, Sánchez del Río M, Vázquez de Agredos-Pascual ML, Lima E (2009) Maya blue as a nanostructured polyfunctional hybrid organic-inorganic material: the need to change paradigms. New J Chem 33:2371–2379
[B401] Doménech-Carbó A, Doménech-Carbó MT, Vázquez de Agredos–Pascual ML (2006) Dehydroindigo: a new piece into the Maya blue puzzle from the voltammetry of microparticles approach. J Phys Chem B 110:6027–6039
[B402] Doménech–Carbó A, Doménech–Carbó MT, Vázquez de Agredos–Pascual ML (2007) Indigo/dehydroindigo/palygorskite complex in Maya blue: an electrochemical approach. J Phys Chem C 111:4585–4595
[B403] Doménech-Carbó A, Doménech-Carbó MT, Sánchez del Río M, Goberna S, Lima E (2009) Evidence of topological indigo/dehydroindigo isomers in Maya blue-like complexes prepared from palygorskite and sepiolite. J Phys Chem C 113:12118–12131

[B404] Doménech-Carbó A, Doménech-Carbó MT, Vázquez de Agredos-Pascual ML (2007) Chemometric study of Maya blue from the voltammetry of microparticles approach. Anal Chem 79:2812–2821
[B405] Doménech-Carbó A, Doménech-Carbó MT, Vázquez de Agredos-Pascual ML (2009) Correlation between spectral, SEM/EDX and electrochemical properties of Maya blue. A chemometric study. Archaeometry 51:1015–1034
[B406] Doménech-Carbó A, Doménech-Carbó MT, Vázquez de Agredos-Pascual ML (2011) From Maya blue to 'Maya yellow': a connection between ancient nanostructured materials from the voltammetry of microparticles. Angew Chem Int Ed 50:5741–5744
[B407] Doménech-Carbó A, Doménech-Carbó MT, Vidal-Lorenzo C, Vázquez de Agredos-Pascual ML (2012) Insights into the Maya blue technology: greenish pellets from the ancient city of La Blanca. Angew Chem Int Ed 51:700–703
[B408] Doménech-Carbó A, Doménech-Carbó MT, Sánchez del Río M, Vázquez de Agredos-Pascual ML (2009) Comparative study of different indigo-clay Maya blue-like systems using the voltammetry of microparticles approach. J Solid State Electrochem 13:869–878
[B409] Doménech-Carbó A, Martini M, De Carvalho LM, Doménech–Carbó MT (2012) Square wave voltammetric determination of the redox state of a reversibly oxidized/reduced depolarizer in solution and in solid state. J Electroanal Chem 684:13–19
[B410] Doménech-Carbó A, Doménech-Carbó MT, Osete-Cortina L, Montoya N (2012) Application of solid-state electrochemistry techniques to polyfunctional organic-inorganic hybrid materials: the Maya blue problem. Micropor Mesopor Mater 166:123–130
[B411] Doménech-Carbó A, Montoya N, Alarcón J (2012) Study of solid state kinetics using voltammetry of immobilized particles. Application to tetragonal to monoclinic transition in nanoparticulate zirconia and praseodymia-doped zirconia. Electrochim Acta 67:24–32
[B412] Walcarius A (2003) Implication of zeolite chemistry in electrochemical science and applications of zeolite-modified electrodes. In: Auerbach, SM, Carrado KA, Dutta PK (eds) Handbook of zeolite science and technology, Chap 14. Marcel Dekker, New York, pp 721–783
[B413] Walcarius A (2006) Zeolite-modified paraffin-impregnated graphite electrode. J Solid State Electrochem 10:469–478
[B414] Rolison DR, Nowak RJ, Welsh TA, Murray CG (1991) Analytical implications of zeolites in overlayers at electrodes. Talanta 38:27–35
[B415] Walcarius A (1996) Zeolite-modified electrodes: analytical applications and prospects. Electroanalysis 8:971–986
[B416] Švancara I, Vytřas K, Kalcher K, Walcarius A, Wang J (2009) Carbon paste electrodes in facts, numbers, and notes: a review on the occasion of the 50-years jubilee of carbon paste in electrochemistry and electroanalysis. Electroanalysis 21:7–28
[B417] Bessel DA, Rolison DR (1997) Topological redox isomers: surface chemistry of zeolite-encapsulated Co(salen) and [Fe(bpy)3]2+ complexes. J Phys Chem B 101:1148–1157
[B418] Doménech-Carbó A, Formentín P, García H, Sabater MJ (2002) On the existence of different zeolite-associated topological redox isomers. Electrochemistry of Y zeolite-associated Mn(salen)N_3 complex. J Phys Chem B 106:574–582
[B419] Doménech-Carbó A, Alvaro M, Ferrer B, García H (2003) Electrochemistry of mesoporous organosilica of MCM-41 type containing 4,4′-bipyridinium units: voltammetric response and electrocatalytic effect on 1,4-dihydrobenzoquinone oxidation. J Phys Chem B 107:12781–12788
[B426] Doménech-Carbó A, Cebrián-Torrejón G, De Miguel L, Tordera V, Rodrigues D, Assad-Kahn S, Fournet A, Figadère B, Vázquez R, Poupon E (2014) dsDNA, ssDNA, G-quadruplex DNA, and nucleosomal DNA electrochemical screening using canthin-6-one alkaloid–modified electrodes. Electrochim Acta 115:546–552. doi:10.1016/j.electacta.2013.11.025

[B427] Allanore A, Levalaine H, Valentin G, Birat JP, Delcroix P, Lapicque F (2010) Observation and modeling of the reduction of hematite particles to metal in alkaline solution by electrolysis. Electrochim Acta 55:4007–4013

[B428] Aeschbacher M, Sander M, Schwarzenbach RP (2010) Novel electrochemical approach to assess the redox properties of humic substances. Environ Sci Technol 44:87–93

[B429] Aeschbacher M, Vergari D, Schwarzenbach RP, Sander M (2011) Electrochemical analysis of proton and electron transfer equilibria of the reducible moities in humic acids. Environ Sci Technol 45:8385–8394

[B430] Gorski CA, Aeschbacher M, Soltermann D, Voeglin A, Baeyens B, Fernández MM, Hofstetter TB, Sander M (2012) Redox properties of structural Fe in clay minerals. 1. Electrochemical quantification of electron-donating and -accepting capacities of smectites. Environ Sci Technol 46:9360–9368

[B431] Gorski CA, Khüpfel L, Voeglin A, Sander M, Hofstetter TB (2012) Redox properties of structural Fe in clay minerals. 2. Electrochemical and spectroscopic characterization of electron transfer irreversibility in ferruginous smectite SWa-1. Environ Sci Technol 46:9369–9377

[B432] Doménech-Carbó A, Doménech-Carbó MT, Valle-Algarra FM, Domine ME, Osete-Cortina L (2013) On the dehydroindigo contribution to Maya blue. J Mater Sci 48:7171–7183

[B433] Doménech-Carbó A, Valle-Algarra FM, Doménech-Carbó MT, Domine ME, Osete-Cortina L, Gimeno-Adelantado JV (2013) Redox tuning and species distribution in Maya blue–type materials: a reassessment. App Mater Sci Interfaces 5:8134–8145

[B444] Doménech-Carbó A, Labuda J, Scholz F (2013) Electroanalytical chemistry for the analysis of solids: characterization and classification (Technical Report). Pure Appl Chem 85:609–631

[B445] Gulppi MA, Vejar N, Tamayo L, Azocar MI, Vera C, Silva C, Zagal JH, Scholz F, Páez MA (2014) Electrochem Commun 41:8–11

[B446] Roznyatovskaya NV, Vassiliev SYu, Yusipovich AI, Tsirlina GA, Roznyatovskii VV (2005) Aqueous electrochemistry of binuclear copper complex with Robson-type ligand: dissolved versus surface-immobilized reactant. J Solid State Electrochem 9:581–589

[B451] Doménech-Carbó A, Doménech-Carbó MT, Capelo S, Pasíes T, Martínez-Lázaro I (2014) Dating archaeological copper/bronze artifacts using the voltammetry of microparticles. Angew Chem Int Ed 53:262–9266

[B452] Doménech-Carbó A, Martini M, Valle-Algarra FM (2014) Determination of the depth profile distribution of guest species in microporous materials using the voltammetry of immobilized particles methodology: application to lapachol attachment to palygorskite and kaolinite. PhysChemChemPhys 16:19024–19034

Chapter 6
Immobilized Droplets

In principle, the voltammetry of immobilized droplets is in many respects very similar to the voltammetry of immobilized microparticles. However, the fluidity of droplets in contrast to the rigidity of solids leads to specific features which have to be discussed before the details of electrochemical studies can be presented.

Although the adhesion of a liquid droplet to a solid electrode surface in the presence of a third liquid phase, usually the electrolyte solution, is, from a physical point of view, not fundamentally different to that of solid particles, the magnitude of forces keeping the droplet adhering is different. In addition, the fluidity of the droplet allows a fast establishment of the equilibrium of forces, which is not the case with solid particles.

Figure 6.1 depicts the situation when a three-phase electrode is composed of a droplet of liquid, phase II, immobilized on a solid electrode surface (phase I) and immersed in an electrolyte solution (phase III). The relation between the interfacial tensions between the three phases and the contact angle is described by the Young[1] equation:

$$\cos \Theta = \frac{\sigma_{\text{I-III}} - \sigma_{\text{I-II}}}{\sigma_{\text{II-III}}} \tag{6.1}$$

From this equation follows that the shape of the droplet is a function of the interfacial tensions between the phases I and II, II and III, and I and III. When the contact angle is very small or even going to zero, the liquid II would form a film on the solid, which is clearly not desirable in the experiments that we are discussing, because the three-phase junction would be lost. On the other side, a too large contact angle would destabilize the droplet or, for θ approaching 180°, even detach the droplet immediately. The consequences of electrode potentials for the adhesion

[1] Young Th (1805) Philos Trans 65–87

F. Scholz et al., *Electrochemistry of Immobilized Particles and Droplets*,
DOI 10.1007/978-3-319-10843-8_6

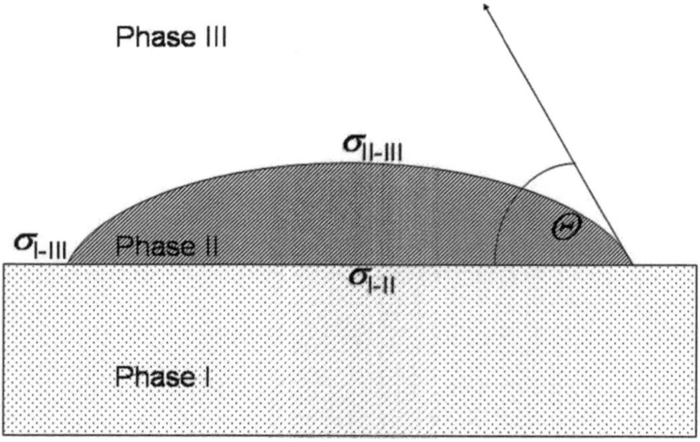

Fig. 6.1 Three-phase electrode with an immobilized droplet (phase II) attached to a solid electrode (phase I) and in an electrolyte solution (phase III). θ is the contact angle and σ is the interfacial tension between the phases indicated in the subscripts

of oil droplets[2] and of liposomes[3] suspended in an electrolyte solution have been demonstrated with mercury electrodes.

In electrochemistry, another complication arises from the well-known relation between the electrode potential and the interfacial tension:

$$\sigma = -\frac{1}{2} C_d E^2 + \text{const.} \tag{6.2}$$

where C_d is the differential double layer capacity and E the electrode potential. This means that changing the electrode potential will alter in any case the interfacial tensions $\sigma_{\text{I–II}}$ and $\sigma_{\text{I–III}}$. Consequently, as the interface between the two liquid phases II and III is in series with interface I/II, the interfacial tension $\sigma_{\text{II–III}}$ changes too, provided the droplet has a sufficient ionic conductivity. If the conductivity of the droplet is high in all its volume elements, the entire interface II/III will be affected by the electrode potential, whereas in case that only the edge of the droplet acquired ionic conductivity (by natural partition of the electrolyte from phase III or by an electrochemical reaction), there will even be an anisotropy of the interfacial tension between II and III from the edge to the apex of the droplet (see Sect. 6.2). Further complications may arise from the fact that the electrochemical reaction will change the composition of the droplet phase and possibly also of the electrolyte phase near the three-phase junction and near the droplet surface. A change of composition will always affect the interfacial tension, and this parameter may even become strongly anisotropic over the interface. This could deform the

[2] Ivošević N, Žutić V (1998) Langmuir 14:231–234

[3] Hellberg D, Scholz F, Schauer F, Weitschies W (2002) Electrochem Commun 4:305–309

interface, and it may induce the so-called Marangoni[4] streaming. The latter is an effect of an induced flowing of a liquid along an interface from places with low interfacial tension to places with a higher interfacial tension. This way a convective mass transport may act at a three-phase electrode. The feedback effects between electrode potential, interfacial tension, mass transport, and electrode reaction are certainly very complicated. Fortunately, the changes of interfacial tensions caused by the electrode potential are in many cases practically negligible, esp. when lipophilic droplets are attached to rather lipophilic surfaces like that of graphite electrodes or even better for the paraffin-impregnated electrode (PIGE). In these cases, the value of the interfacial tension between the droplet and the underlying base material is such that there is hardly any shape change detectable upon potential variations. However, when an organic liquid droplet is attached to a metal electrode, platinum, or gold perhaps, it is impossible to keep the droplet stable in voltammetric experiments. The alteration of the interfacial tensions by potential scanning leads to a rapid movement of the droplet on the surface and to spreading or detachment, i.e., to irreproducible conditions at the electrode surface. The situation is slightly better when an aqueous droplet is attached to a platinum electrode since the latter is rather hydrophilic. Ulmeanu et al. [B180] reported experiments with a very small aqueous electrolyte droplet attached to a platinum microelectrode in a Teflon holder.

The mechanisms of electrode reactions observed at three-phase electrodes with immobilized droplets resemble those observed with immobilized particles. Table 6.1 gives an overview of some frequently observed cases. There are certainly more possible cases, and especially catalytic reactions at such three-phase electrodes are not considered here. Examples for these reactions are discussed in the following chapters. The first case listed in Table 6.1 is when an oxidizable nonionic compound X is dissolved in a polar organic solvent. The oxidation to X^+ may be accompanied by the transfer of anions B^- from the aqueous to the organic phase (case 1a) if the free energy for that transfer is smaller than that for the transfer of X^+ to the aqueous phase (case 1b). When a nonpolar solvent is used, the transfer of X^+ to the aqueous phase is the most probable reaction (case 2). When a reducible nonionic compound X is dissolved in a polar organic solvent, the reaction to X^- may be followed by the transfer of cations A^+ to the organic phase (case 3a) or by the transfer of X^- to the aqueous phase (case 3b). Again, the free energies of the ion transfer of X^- to the aqueous phase and of A^+ to the organic phase decide what will happen. The case of a reducible compound in a nonpolar organic solvent is not listed. Of course, the transfer of X^- to the aqueous phase would be the most probable reaction path. Case 4 in Table 6.1 concerns oxidizable organic liquids. The oxidation can result as in case 1a to the transfer of anions B^- from the aqueous to the organic phase; however, now the organic liquid is converted to an ionic liquid (in principle, also a precipitation of the salt $[O^+B^-]$ in the organic liquid may occur). Alternatively, the cations O^+ can be transferred to the aqueous phase, where they

[4] Marangoni CGM (1871) Ann Phys Chem (Poggendorf) 143:337–354

Table 6.1 Overview of possible reaction mechanisms at three-phase electrodes with immobilized droplets

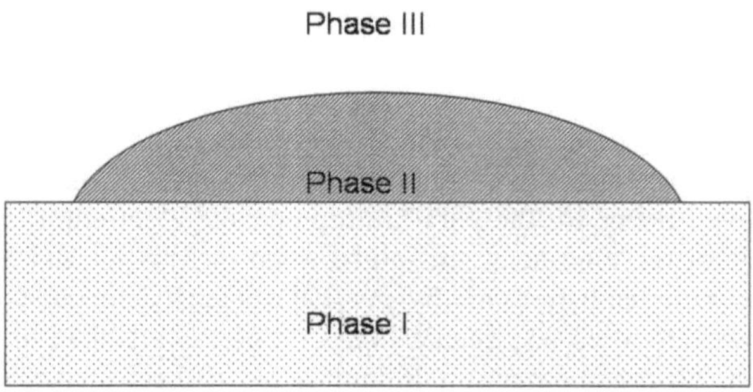

	Phase II	Phase III	Mechanism	References
1	Polar organic solvent O with dissolved nonionic oxidizable compound X	Aqueous electrolyte A^+B^-	1a) $X_o + B_{aq}^- \leftrightarrows X_o^+ + B_o^- + e^-$ 1b) $X_o \leftrightarrows X_{aq}^+ + e^-$	[B132]
2	Nonpolar organic solvent O with dissolved nonionic oxidizable compound X	Aqueous electrolyte A^+B^-	$X_o \leftrightarrows X_{aq}^+ + e^-$	[B132]
3	Polar organic solvent O with dissolved nonionic reducible compound X	Aqueous electrolyte A^+B^-	3a) $X_o + e^- + A_{aq}^+ \leftrightarrows X_o^- + A_o^+$ 3b) $X_o + e^- \leftrightarrows X_o^-$	[B271]
4	Organic oxidizable liquid O	Aqueous electrolyte A^+B^-	4a) $O_o + B_{aq}^- \leftrightarrows O_o^+ + B_o^- + e^-$ ($O_o^+ + B_o^-$ form an ionic liquid) 4b) $O_o \leftrightarrows O_{aq}^+ + e^-$ (follow-up precipitation of $[O^+B^-]$ is possible)	[B166]
5	Organic reducible liquid O	Aqueous electrolyte A^+B^-	$O_o + e^- + H_{aq}^+ \leftrightarrows OH_o$ (the organic liquid remains nonionic)	[B273]

remain dissolved or they may form a precipitate $[O^+B^-]$ with the anions. Again the free energies of ion transfer determine the pathway. Case 5 in Table 6.1 describes the reduction of an organic liquid O to the organic liquid OH that results from a transfer of electrons from phase I to phase II and a proton transfer from phase III to phase II. Also this case has been verified experimentally to exist.

The very simplified Table 6.1 shows already the diversity of possible reaction mechanisms, but it also suggests that the underlying principles are rather general. It is hoped that this table gives an orientation for further investigations. Detailed explanations are discussed on the following pages.

6.1 The Electrochemistry of Redox Liquids

The first report that involved the voltammetry of immobilized redox liquids was a study of the *n*-hexyldithiocarbamate complex of lead(II) [B13]. It was found that the voltammetry of the immobilized droplets of this compound was the same as that of its microcrystals. Initiated by the idea of studying electrochemical processes in biphasic media such as emulsions, Marken et al. has undertaken the first intentional investigations involving immobilized microdroplets of redox-active liquids [B74]. The immobilization of droplets was meant to provide a controlled environment to separately elucidate processes involving the direct and simultaneous contact of immiscible liquids to an electrode surface and to an aqueous solution. Since then a wealth of information regarding electron transfer and ion transfer processes as well as chemical reactions of the deposited redox liquids has been gathered. As it is presented in the latest review by Banks et al. [B285], the *N*,*N*,*N*′, *N*′-tetraalkylphenylenediamines, TRPDs (R = alkyl), are the by far the most investigated redox liquids (see Fig. 6.2 and Table 6.2).

These compounds combine an extremely low miscibility with water, a suitable oxidation potential, the formation of colored radical species, and the possibility to tailor their physical and chemical properties by changing the length and the nature of the alkyl chains. In contrast to the solid, water-soluble tetramethylphenylenediamine, TMPD, the long-chain compounds are liquids that remain extremely insoluble in water even after oxidation and the formation of the radical cationic form. The long alkyl chains not only ensure the low solubility and the highly lipophilic character of the compound, they also hinder regular packing of the molecules and thus prevent crystallization. That all makes the TRPDs very well suited as model compounds for the study of the voltammetry of immobilized redox liquids.

The tetra-*N*-alkylated phenylenediamine backbone can be reversibly oxidized in two one-electron transfer steps, the first, which involves the formation of the radical cation, being by far the most intensively studied reaction:

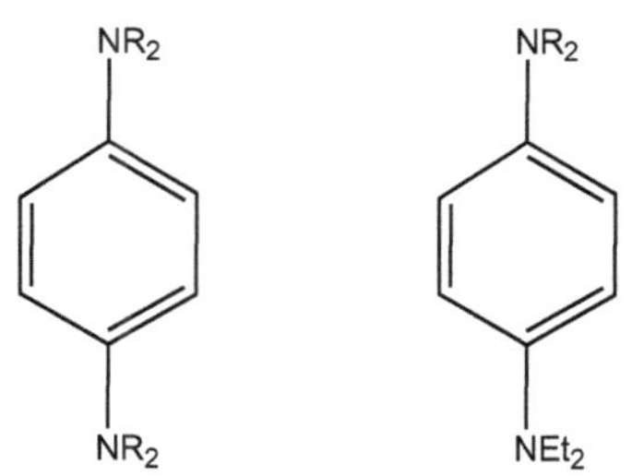

R = *n*-butyl (TBPD)
n-hexyl (THPD)
n-heptyl (THepPD)
n-nonyl (TNPD)

R = *n*-butyl (DEDBPD)
n-hexyl (DEDHPD)
n-heptyl (DEDHePD)

Fig. 6.2 Structure of the *N*, *N*,*N*′,*N*′-tetraalkyl-*para*-phenylenediamines (*p*-TRPD) and the *N*,*N*-diethyl-*N*′,*N*′-dialkyl-*para*-phenylenediamines (*p*-DEDRPD)

Table 6.2 Redox liquids that have been studied voltammetrically as microdroplets immobilized on basal plane pyrolytic carbon electrodes (Reproduced from [B285])

Redox liquid	References
N,N,N′,N′-Tetrahexyl-*para*-phenylenediamine (*p*-THPD)	[B74, B97, B149, B150, B166, B168, B171, B182, B230, B254, B284, B287, B288]
N,N,N′,N′-Tetrahexyl-*meta*-phenylenediamine (*m*-THPD)	[B166]
N,N,N′,N′-Tetrabutyl-*para*-phenylenediamine (*p*-TBPD)	[B230, B254, B284]
N,N,N′,N′-Tetraheptyl-*para*-phenylenediamine (*p*-THePD)	[B230, B254, B284]
N,N,N′,N′-Tetraoctyl-*para*-phenylenediamine (*p*-TOPD)	[B213, B214, B254]
N,N,N′,N′-Tetranonyl-*para*-phenylenediamine	[B230, B284]
N,N-Diethyl-*N′,N′*-dibutyl-*para*-phenylenediamine (*p*-DEDBPD)	[B284]
N,N-Diethyl-*N′,N′*-dihexyl-*para*-phenylenediamine (p-DEDHPD)	[B284]
N,N-Diethyl-*N′,N′*-diheptyl-*para*-phenylenediamine (p-DEDHePD)	[B284]
N,N,N′-Trihexyl-*para*-phenylenediamine (*p*-TriHPD)	[B171]
N,N,N′,N′-Tetrakis(6-methoxyhexyl)-*para*-phenylenediamine (p-TMHPD)	[B166]
N^1-[4-(Dihexylamino)phenyl]-N^1,N^4,N^4-trihexyl-1,4-phenylenediamine (DPTPD)	[B289]
3-Methylthiophene[a]	[B197]
n-Butylferrocene	[B241]
4-Nitrophenyl nonyl ether	[B286]
Vitamin K1	[B273, B287]

[a]Studied as single millimetric drop immobilized on a PIGE

$$\mathrm{TRPD} \leftrightarrows \mathrm{TRPD}^{\bullet+} + e^- \qquad \text{(I)}$$

$$\mathrm{TRPD}^{\bullet+} \leftrightarrows \mathrm{TRPD}^{2+} + e^- \qquad \text{(II)}$$

The electrochemical oxidation of TRPD in the biphasic system of the immobilized microdroplets has been studied from very different viewpoints including the electrochemically and chemically facilitated ion transfer reactions, reactive chemistry, photochemistry, and catalysis. In the following sections, these different aspects will be appreciated, demonstrating the potential of the voltammetry of immobilized microdroplets as a simple and yet powerful electrochemical tool.

6.1.1 Electrochemically Driven Ion Transfer Reactions

N,N,N′,N′-tetrahexyl-*para*-phenylenediamine, THPD, was the first compound of the phenylenediamine family to be studied in detail [B74, B97]. The authors have compared the electrochemistry of THPD, dissolved in acetonitrile, and the electrochemistry of immobilized microdroplets of the compound in contact to an aqueous electrolyte solution. They have found that the oxidation of the immobilized droplets was essentially connected with the uptake of anions from the surrounding electrolyte solution in order to maintain electroneutrality within the droplets:

$$\mathrm{THPD_{(o)}} + \mathrm{X^-_{(aq)}} \leftrightarrows \mathrm{THPD^+_{(o)}} + \mathrm{X^-_{(o)}} + \mathrm{e^-} \qquad \text{(III)}$$

The authors also observed that the mid-peak potential of this reaction and the shape of the voltammetric signals (see Fig. 6.3) depend on the nature of the electrolyte anion, whereas the nature of the cations did not contribute significantly. As a consequence, the authors suggested utilizing this effect, which they attributed to specific interactions between the phenylenediamine cation and the inserted anion, for the detection of anions in aqueous solutions [B74].

Initiated by these studies, a series of papers have appeared dedicated to the systematic investigation of the voltammetry of differently substituted phenylenediamines. Thus, the anion insertion and chemical reactions in microdroplets of *para*-tetrakis(6-methoxyhexyl)phenylenediamine and *para*- and *meta*-tetrahexylphenylenediamine (Fig. 6.4) have been compared [B166]. In this study, the effects

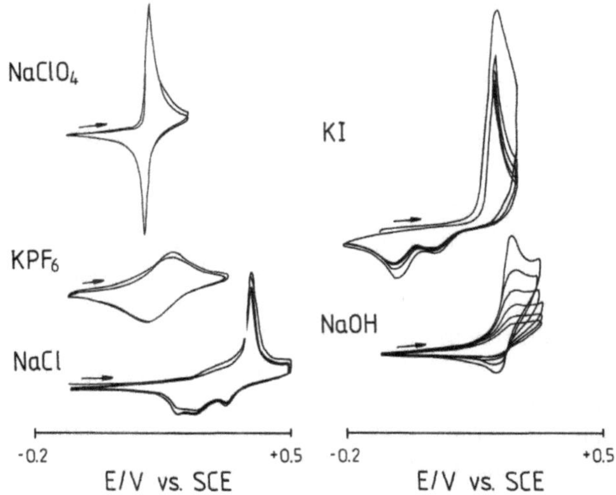

Fig. 6.3 Multicycle voltammograms for the oxidation of 1.24 μg THPD deposited on a 4.9-mm diameter basal plane pyrolytic carbon electrode and immersed in aqueous solutions of different electrolyte salts (electrolyte concentration 0.1 mol L^{-1}, 6 cycles, 22 °C, scan rate 10 mV s^{-1}) [B74]. Copyright ©1997 with permission from Elsevier

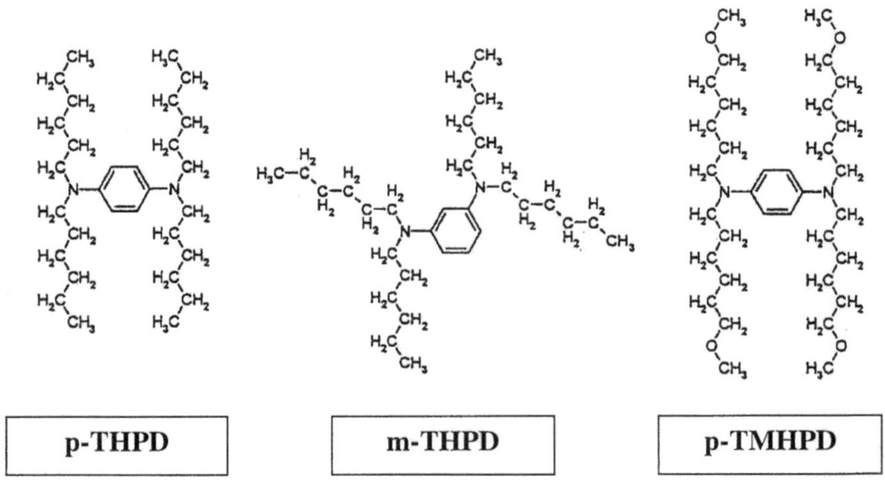

Fig. 6.4 Structures of *para-N, N, N′,N′* tetrahexylphenylenediamine (*p*-THPD), *meta-N, N, N′,N′* tetrahexylphenylenediamine (*m*-THPD) and *para*-tetrakis(6-methoxyhexyl) phenylenediamine (*p*-TMHPD) [B166]

of the altered molecular structures of the liquid precursors were investigated. The authors observed dramatic changes in the voltammetric behavior of the liquids, which they interpreted in terms of changes in the interaction between the TRPD radical cation and the transferred anion within the organic liquid as a function of the chain length of the phenylenediamine and the position of the amine groups at the benzene unit. Due to the high density of the redox centers within the redox liquids, very much in contrast to the electrochemistry of compounds dissolved in droplets of inert solvents, such specific and nonspecific interactions between the constituents in the liquids are very likely. The very narrow peak shape that was found for the oxidation of THPD microdroplets [B74] supports this assumption. Very interesting voltammetric features have been reported for the insertion of halides into the redox liquids (Fig. 6.5). Upon oxidation in the presence of bromide and iodide ions, THPD showed an additional reduction signal connected with a large peak-to-peak separation. The occurrence of this reduction signal was found to be fully chemically reversible, and in the case of bromide, it disappeared at elevated temperature (40 °C). To interpret this behavior, the authors have presented a number of possible explanations that comprise (i) possible crystallization processes upon oxidation and (ii) the co-insertion of water into the organic liquid.

The search for the relationship between the nature of the electrolyte ions transferred across the phenylenediamine–water interface and the characteristics of the voltammetric signals has always been a key issue in the published studies. Generally, the authors have found correlations between the measured formal potential, the lipophilicity of the inserted ions, and the lipophilicity of the redox liquid (see Table 6.3). Thus, increasing the number of carbon units in the alkyl chains of the TRPD, for a given anion, leads to a shift of the midpoint potentials

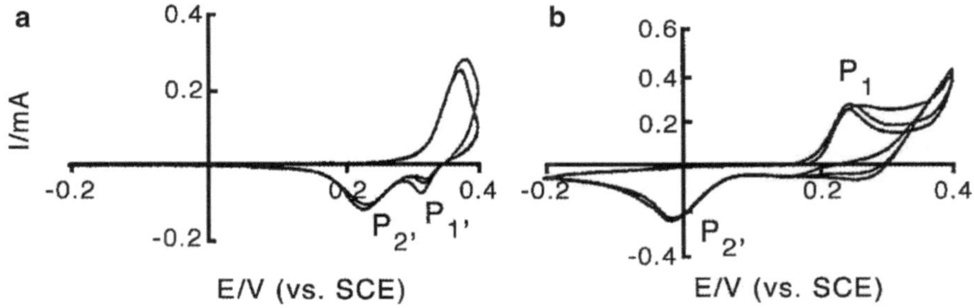

Fig. 6.5 Cyclic voltammograms (first three scans) recorded at 100 mV s^{-1} for the oxidation of 5.3×10^{-9} mol *p*-THPD deposited by solvent evaporation onto a 4.9-mm diameter basal plane pyrolytic carbon electrode immersed in (**a**) 0.1 mol L^{-1} NaBr and (**b**) 0.1 mol L^{-1} KI [B166]. Copyright ©2001 with permission from Springer

Table 6.3 E_{mid}/mV data[a] obtained from cyclic voltammograms[b] for the oxidation of DEDRPD and TRPD oils[c] [B284]

	DEDBPD	TBPD[d]	DEDHPD	DEDHePD	THPD[d]	THePD[d]	TNPD[d]
PF_6^-	–30[i]	–20[i]	3[i]	18[i]	26[i]	41[i]	64[i]
ClO_4^-	45[ii]	63[i]	86[i]	91[i]	109[i]	126[i]	146[i]
SCN^-	105[ii‡]	129[ii]	141[i]	153[i]	178[i]	188[i]	201[i]
Br^-	125[ii]	185[ii]	259[ii‡]	269[*]	296[i]	318[i]	324[i]
NO_3^-	117[ii‡]	180[ii‡]	245[*‡]	253[*]	298[i]	313[i]	327[i]
OCN^-	117[ii‡]	188[ii]	265[ii‡]	290[ii]	365[ii]	381[ii]	385[ii]
Cl^-	126[ii]	193[ii]	266[ii‡]	310[i]	371[ii]	383[ii]	383[i]
SO_4^{2-}	135[ii]	203[ii]	281[ii]	318[ii]	444[†]	467[†]	481[ii]
F^-	113[ii]	198[ii]	274[ii]	330[ii]	465[†]	489[†]	513[ii]
IO_3^-	134[ii]	238[ii]	268[ii]	336[ii]	470[ii]	481[ii]	501[ii]

[a] All data are reported in mV vs SCE; error =5 mV
[b] Oil (5 nmol) was deposited as microdroplets on a 4.9-mm diameter basal plane pyrolytic carbon electrode and immersed into 0.1 mol L^{-1} aqueous electrolyte
[c] The superscript i or ii identifies the particular oil/anion pair as behaving as either case ([i]) or case ([ii]). The data were obtained using a san rate of 100 mV s^{-1}
[d] Data taken from [B230 and B254]. Case ([i]): lipophilic anions, case ([ii]): lipophobic anions
[i] Indicates that the oxidation follows Reaction I
[ii] Indicates that the oxidation follows Reaction III
[*] Indicates that the voltammetric behavior is case ([i]) at fast scan rates but case ([ii]) at slow scan rates
[†] Indicates this case is effectively type ([ii]) behavior, only with very fast kinetics
[‡] Indicates those systems that give rise to split waves at slow scan rates

toward more positive values, viz., DEDBPD < TBPD < DEDHPD < DEDHePD < THPD < THePD < TOPD < TNPD [B284]. Also, increasing the hydrophilicity of the supporting electrolyte anion leads to more positive midpoint potentials. Inspired by a new approach to access thermodynamic quantities of electrochemically driven ion transfer reactions at nitrobenzene droplets ([B132], see Sect. 6.2), the voltammetry of immobilized redox liquids has been compared with experimental data achieved when droplets of TRPD dissolved in nitrobenzene were used instead of the pure redox liquid.

Although the basic methodology of both approaches, the voltammetry of compounds dissolved in inert solvents and the voltammetry of unsupported redox liquids, is practically identical and it can be expected that their basic features are comparable, there are fundamental differences: performing voltammetric measurements at droplets of inert solvents that contain an electroactive species generally means that the concentration of the electroactive species within the droplet is significantly lower than the concentration of the solvent. Due to the low concentration, interactions between the molecules of the active compound are very often insignificant. Oxidation or reduction processes of the dissolved species, connected with an ion transfer between the droplet and the adjacent aqueous solution, generally do not completely change the physical and chemical phase characteristics of the organic droplet. An exception is the change of the ionic conductivity due to the formation or consumption of ionic species and their transfer into or out of the organic phase. However, increasing the concentration of the redox-active species at the expense of the solvent gradually raises the impact of the reaction at the droplet phase. The end of this line is marked by the voltammetry of redox liquids or, if the target compound is solid, by the voltammetry of immobilized microparticles. Here, chemical or electrochemical reactions will not only cause changes on the molecular level, but they will also alter the properties of the entire organic phase. Thus, an entirely new phase evolves, in some cases even accompanied by a crystallization or precipitation [B213]. Additionally to the phase transition, the high density of redox centers increases the impact of intermolecular interactions on the electrochemical behavior [B74].

In a comparative study, the electrochemically driven ion transfer processes across liquid–liquid interfaces of droplets of *para-N*, *N*, *N′*,*N′* tetraalkylphenylenediamines and of the same compounds diluted in droplets of nitrobenzene have been elucidated [B254]. As expected, the voltammetric responses of the two redox steps of the diluted compound are electrochemically fully reversible: peak shape and virtually zero peak separation showing ideal thin-film behavior (Fig. 6.6b). The voltammograms of the undiluted compound deviate from that behavior. Besides the narrow peak shape of the first oxidation process, additional voltammetric features are found for the $THPD^{+}/THPD^{2+}$ redox couple. At slow scan rates, the reduction of $THPD^{2+}$ in the pure redox liquid is characterized by a large overpotential (peak P_2, Fig. 6.6a). At elevated temperature, however, and at high scan rates (above 100 mV s^{-1}), this peak vanishes and is replaced by a new signal, P_1. This signal can be assigned to the reversible reaction

$$\left[THPD^{+}ClO_4^{-}\right]_{(o)} + ClO_4^{-}(aq) \leftrightarrows \left[THPD^{2+}(ClO_4^{-})_2\right]_{(o)} + e^{-} \qquad \text{(IV)}$$

as it has been observed for the $THPD^{+}$ oxidation in nitrobenzene droplets (Fig. 6.6b). The authors attributed the occurrence of the two chemically reversible

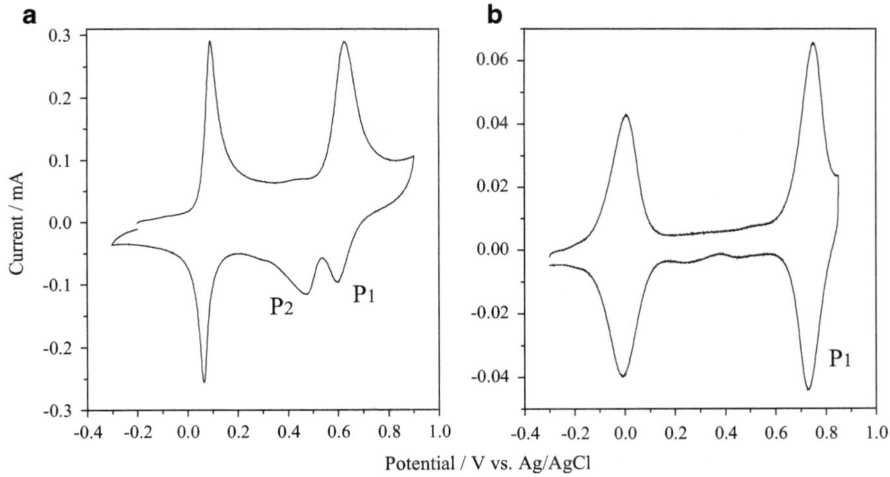

Fig. 6.6 (**a**) Cyclic voltammogram for the oxidation of 3.5 nmol THPD deposited on a 4.9-mm diameter basal plane pyrolytic graphite electrode and immersed in a 0.1 mol L^{-1} $NaClO_4$ electrolyte solution, $\theta = 42$ °C; (**b**) cyclic voltammogram for the oxidation of THPD dissolved in microdroplets of nitrobenzene ($c_{THPD} = 25$ mmol L^{-1}), deposited at a basal plane pyrolytic graphite electrode, and immersed in a 0.1 mol L^{-1} $NaClO_4$ electrolyte solution; scan rate = 100 mV s^{-1} [B254]. Copyright ©2002 with permission from American Chemical Society

reduction processes P_1 and P_2 to a phase transition of the $[THPD^{2+}(ClO_4^-)_2]$ ionic liquid phase toward a higher packing density caused by geometrical configuration changes of the THPD scaffold upon oxidation. Naturally, such a reorientation would not be expected to take place in the diluted system.

Apart from these differences, however, the thermodynamics of the voltammetric reactions of the immobilized tetraalkylphenylenediamines has been found to be very similar to that of the compounds dissolved in nitrobenzene. The authors have shown that for lipophilic anions (section (i) in Fig. 6.7), the formal potentials of Reaction (III) is strongly correlated to the Gibbs free energy of the ion transfer across the water–nitrobenzene interface. So, despite of the differences in the appearance of the voltammetric signals, the redox processes are predominately ruled by the Gibbs free energy of the transfer of the electrolyte anions across the water–redox liquid interface, $\Delta\phi^{(o)\ominus}_{(aq),\,X^-}$ (see Eq. 6.3).

$$E = E^{\ominus} + \Delta\phi^{(o)\ominus}_{(aq),\,X^-} - \frac{RT}{F}\ln a_{X^-_{(aq)}} + \frac{RT}{F}\ln\frac{a_{TRPD^{\bullet+}_{(o)}}a_{X^-_{(o)}}}{a_{TRPD_{(o)}}} \tag{6.3}$$

The extremely high hydrophilicity (high values of the Gibbs energy of transfer) of small anions like F^- or highly charged ions like sulfate makes the transfer of these ions energetically unfavorable. Instead, upon oxidation, the transfer of the

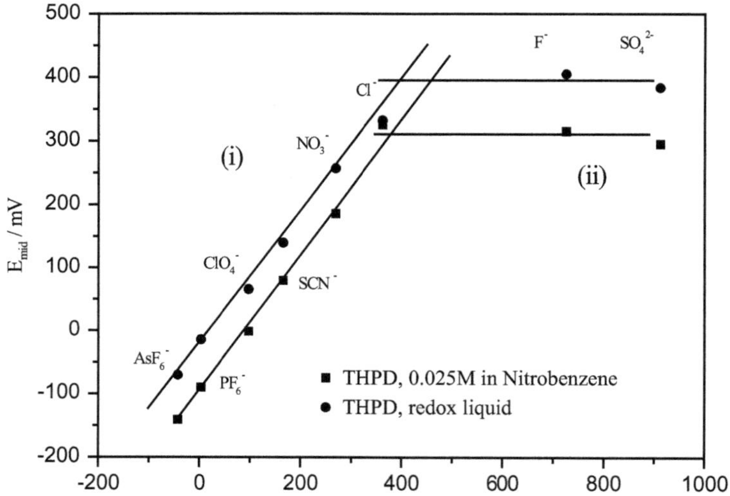

Fig. 6.7 Plot of the formal potentials of the first oxidation step (Reaction III) for THPD dissolved in nitrobenzene and for the pure redox liquid measured in the presence of different electrolyte anions versus the standard membrane potentials of the electrolyte anions at the water–nitrobenzene interface; pH = 9; oxidation of THPD in section (i) follows Reaction (III); in section (ii), it follows Reaction (V) [B254]. Copyright ©2002 with permission from American Chemical Society

oxidation product, the radical cation $TRPD^{+}_{(aq)}$, takes place from the organic phase into the aqueous solution [Reaction (V); see also section (ii) in Fig. 6.3]:

$$TRPD_{(o)} \leftrightarrows TRPD^{\bullet+}_{(aq)} + e^{-} \quad \text{(V)}$$

$$E = E^{\ominus} - \Delta\phi^{(o)\ominus}_{(aq),\ TRPD^{\bullet+}} + \frac{RT}{F}\ln\frac{a_{TRPD^{\bullet+}_{(aq)}}}{a_{TRPD_{(o)}}} \quad (6.4)$$

According to Eq. 6.4, the redox potential does not anymore depend on the Gibbs free energy of transfer of the electrolyte anions, $\Delta\phi^{(o)\ominus}_{(aq),\ X^-}$, but on that of the TRPD radical cation, $\Delta\phi^{(o)\ominus}_{(aq),\ TRPD^{\bullet+}}$. This process is chemically irreversible. It leads to the dissolution of the redox liquid.

In order to electrochemically drive the transfer of strongly hydrophilic anions from the aqueous into the organic phase by means of the voltammetry of immobilized microdroplets, the redox liquid is required to be extremely hydrophobic even in its cationic form. Marken et al. have been the first to report the transfer of sulfate into droplets of *para*-tetrakis(6-methoxyhexyl)phenylenediamine (TMHPD), which, however, was accompanied by dimerization and oligomerization reactions of the formed $TMHPD^{\bullet+}$ radical cation [B166]. In a later study, the same authors have reported the reversible insertion of sulfate, chromate, and dichromate into microdroplets of *para-N,N,N′,N′*-tetraoctylphenylenediamine, TOPD [B214], following Reaction (VI):

$$2\text{TOPD}_{(\text{o})} + \text{X}^{2-}_{(\text{aq})} \leftrightarrows 2\text{TOPD}^{+}_{(\text{o})} + \text{X}^{2-}_{(\text{o})} + 2\text{e}^{-} \qquad \text{(VI)}$$

The authors determined the mid-peak potential of the TOPD oxidation in the presence of a 0.1 mol L^{-1} Na_2SO_4 solution with $E_{\text{mid}} = 0.428$ V (vs. Ag/AgCl). However, in a similar study, Wadhawan et al. could not confirm these results [B230]. They systematically examined the electrochemically induced ion insertion and expulsion processes at microdroplets of TRPDs with R = *n*-butyl, *n*-hexyl, *n*-heptyl, and *n*-nonyl. In contrast to the *n*-octyl-substituted compound examined by Marken, they interpreted the voltammetric response, recorded for the oxidation of the long-chain TRPDs in the presence of sulfate, by means of Reaction (V), the expulsion of the radical cation from the organic phase into the aqueous solution. This assumption seems very reasonable since due to the high value of the Gibbs energy of transfer between water and nonpolar solvents ($\Delta G^{(\text{o})\ominus}_{(\text{aq}),\ \text{SO}_4^{2-}} = 88\ \text{kJ mol}^{-1}$, for the water to acetonitrile transfer[5]), the oxidation of the TRPDs, connected with the uptake of sulfate, would have to be expected to proceed at considerably higher potential values than those found for TOPD.

Recently, Marken et al. have also claimed the insertion of extremely hydrophilic anions such as phosphate and arsenate into microdroplets of *para-N*, *N*, *N′*,*N′* tetraoctylphenylenediamines [B213]. Upon oxidation, the redox liquid transformed into a crystalline compound, forming needle structures on the electrode surface (Fig. 6.8). It can be supposed that again a transfer of TOPD cations to the aqueous phase takes place with a follow-up precipitation of the salts at the electrode surface.

Fig. 6.8 Field emission gun scanning electron microscopic image of a deposit of 6.5 μg TOPD on a 1-cm^2 basal plane pyrolytic graphite electrode after 30-s immersion in aqueous 2.0 mol L^{-1} phosphate buffer at pH 2.3 with a potential of 0.5 V versus SCE applied [B213]. Copyright ©2002 with permission from Elsevier

[5] Marcus Y (1997) Ion properties. Marcel Dekker, New York

6.1.2 Voltammetric Study of Chemically Facilitated Ion Transfer Processes Across the Liquid–Liquid Interface

Due to the generally very low solubility of inorganic salts in the studied redox liquids, the extent of the partitioning of ionic species from water into the organic phase can be considered to be extremely low. The partitioning can, however, be facilitated by chemical processes like acid–base reactions of the organic compound.

Theoretically, due to their basicity, it should be possible to protonate both amine groups of the phenylenediamine backbone of a TRPD. However, the delocalization of charge via the conjugated π-system of the phenylene ring leads to a decreased electron density, hence to a considerably decreased basicity of the second amine group after the protonation of the first amine group. That means that under mild pH conditions, i.e., pH ≥ 0, generally only one amine group can be protonated.

Being performed in the biphasic oil–water system, the protonation reaction requires the uptake of protons from the surrounding aqueous into the organic phase. In order to maintain charge neutrality, the transfer is accompanied by the transfer of an equivalent number of anions (Reaction VII).

$$\mathrm{TRPD_{(o)}} + \mathrm{H^+_{(aq)}} + \mathrm{ClO^-_{4\,(aq)}} \leftrightarrows \mathrm{TRPDH^+_{(o)}} + \mathrm{ClO^-_{4\,(o)}} \qquad \text{(VII)}$$

As expected, the protonation has a tremendous effect on the voltammetry of the redox liquid [B97]. The voltammetric redox system I_{ox}/I_{red} (see Fig. 6.9), recorded

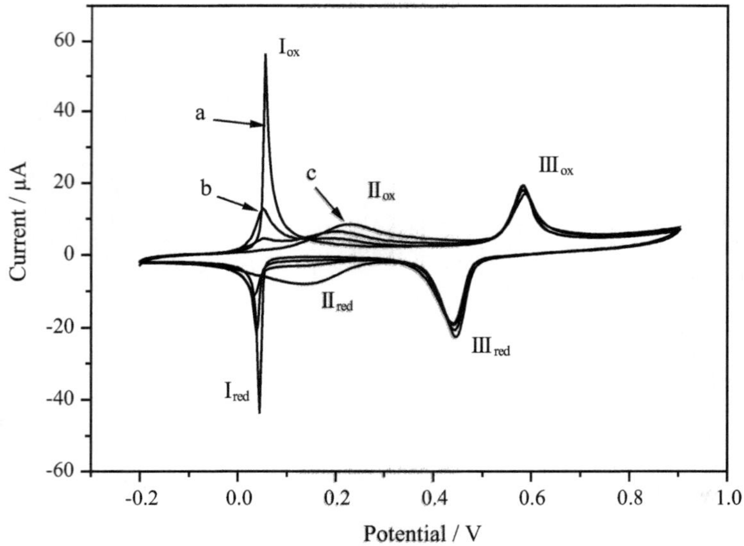

Fig. 6.9 Cyclic voltammograms for the oxidation of 3.5 nmol THPD deposited on a 4.9-mm diameter basal plane pyrolytic graphite electrode and immersed in a 1 mol L^{-1} $NaClO_4$ electrolyte solution; **a** pH = 8, **b** pH = 5.9, **c** pH = 3.7; scan rate = 10 mV s^{-1} [B254]. Copyright ©2002 with permission from American Chemical Society

at neutral pH and based upon Reaction (III), in Sect. 6.1.1, disappears with increasing proton concentration for a new voltammetric system, II_{ox}/II_{red}. The oxidation of the immobilized redox liquid, which under neutral and basic pH proceeds under anion uptake, at low pH takes place under expulsion of protons from the redox liquid into the aqueous phase:

$$\left[\mathrm{TRPDH^{+}ClO_4^{-}}\right]_{(o)} \leftrightarrows \left[\mathrm{TRPD^{+}ClO_4^{-}}\right]_{(o)} + \mathrm{H^{+}_{(aq)}} + \mathrm{e^{-}} \qquad \text{(VIII)}$$

The second oxidation step (signals III_{ox}/III_{red}, Fig. 6.9) remains unaffected, showing that only one amine group is involved in the protonation. As depicted in Fig. 6.10, the new voltammetric signal shows different voltammetric behavior.

A first attempt to determine the acidity constant pK_{A2} for THPD from voltammetric data has been made by Schröder and coworkers [B171]. The authors have derived pK_{A2} from the extent of ionic (protonated) liquid formation, as a function of the pH of the electrolyte solution, and determined from the analysis of the voltammetric responses process I_{ox}/I_{red} and II_{ox}/II_{red} (Fig. 6.9). This approach, however, is oversimplified since it does not include the Gibbs free energy of the transfer of the accompanying anions. This quantity needs to be taken into account in order to appropriately regard the influence of the electrolyte anion co-inserted into the redox liquid in the process of protonation.

As an important side effect, the chemically facilitated ion transfer into the redox liquid considerably changes the phase properties of the redox liquid. Originally being ionically nonconducting, the liquid now turns into an ionic liquid with the consequence that electrochemical reactions are not anymore initially confined to

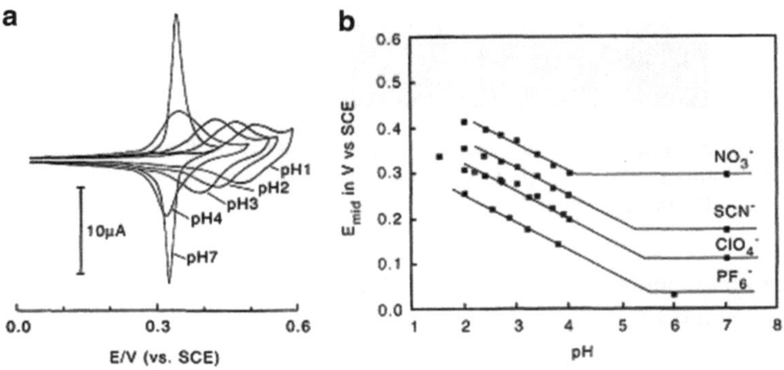

Fig. 6.10 (**a**) Cyclic voltammograms for the oxidation and re-reduction of 0.5 μg THPD deposited onto a 4.9-mm diameter basal plane pyrolytic graphite electrode and immersed in aqueous 0.1 mol L^{-1} KNO_3 solution. The pH was adjusted by addition of 0.1 mol L^{-1} HNO_3. Scan rate 10 mV s^{-1}. (**b**) Plot of the mid-peak potential, $E_{mid} = ½(E_{p,ox} + E_{p,red})$, for the oxidation and re-reduction of 0.5 μg THPD, immersed in aqueous 0.1 mol L^{-1} KNO_3 (pH adjusted with HNO_3), 0.1 mol L^{-1} KSCN (pH adjusted with HNO_3), 0.1 mol L^{-1} $NaClO_4$ (pH adjusted with $HClO_4$), 0.1 mol L^{-1} KPF_6 (pH adjusted with HNO_3) [B97]. Copyright ©1998 with permission from Wiley-VCH

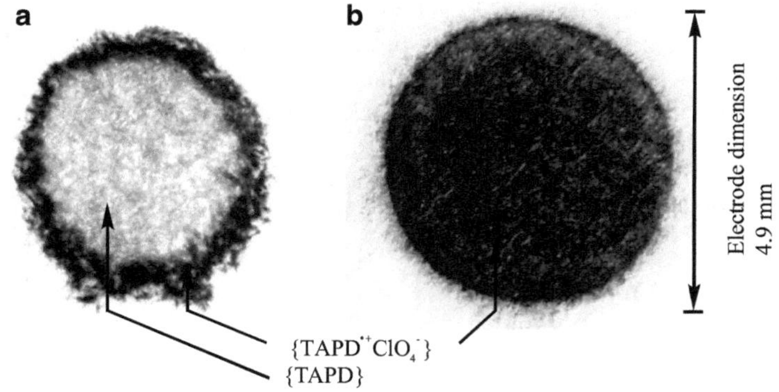

Fig. 6.11 Photographic image of a stamp imprint of the interface layer of TriHPD at a graphite electrode. The electrode was modified with a continuous layer of the redox liquid and electrolyzed at a potential of +500 mV for 60 s in order to form the colored radical cation. (**a**) the neutral redox liquid and (**b**) the protonated, ionic liquid [B171]. Copyright ©2001 with permission from American Chemical Society

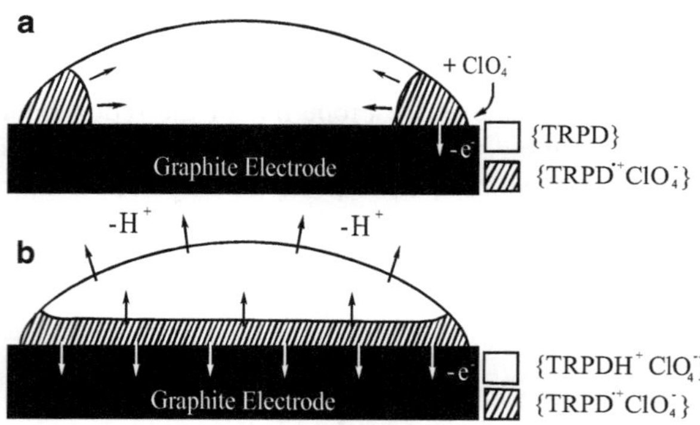

Fig. 6.12 Schematic drawing of the mechanism of the oxidation of deposits of TRPDs of the unprotonated form (**a**), and of the protonated form (**b**) [B171]. Copyright ©2001 with permission from American Chemical Society

the three-phase junction but can start at the entire droplet–electrode interface (Figs. 6.11 and 6.12).

6.1.3 Photochemical and Chemical Reactions at Immobilized Microdroplets of Redox Liquids

The known ability of $TMPD^{\bullet +}$, Würster's blue, to mediate the homogeneous solution oxidation of biological redox compound, like ascorbic acid, NADH, or

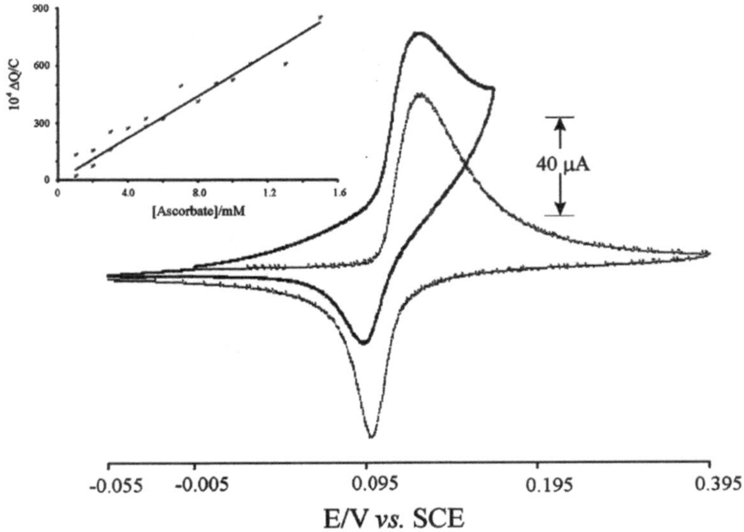

Fig. 6.13 Cyclic voltammograms (scan rate 10 mV s^{-1}) illustrating the electrocatalytic oxidation of ascorbate; 5.4 nmol of THPD deposited on the surface of a 4.9-mm diameter bppg electrode immersed into aqueous 0.1 mol L^{-1} sodium perchlorate solution, in the presence (*solid line*) and absence (*dotted line*) of 50 mmol L^{-1} ascorbic acid at pH 7. The insert illustrates the charge difference inferred from cyclic voltammogram (scan rate 5 mV s^{-1}) variation with ascorbic acid concentration for the oxidation of 5.4 nmol of THPD; 0.1 mol L^{-1} potassium hexafluorophosphate containing 0.1 mol L^{-1} phosphate buffer solution at pH 7 [B284]. Copyright ©2003 with permission from American Chemical Society

ferrocytochromes, inspired researchers in the lab of R. Compton to study the heterogeneous catalysis of immobilized microdroplets of long-chain TRPDs. Thus, Wadhawan et al. [B284] reported on the electrocatalytic oxidation of L-ascorbate at the electrode–oil–electrolyte triple-phase junction of THPD microdroplets. They demonstrate that the presence of ascorbate ions in 0.1 mol L^{-1} perchlorate or hexafluorophosphate electrolyte solutions has a dramatic effect on the voltammetry (Fig. 6.13). During the oxidation scan, more charge is passed compared to the case when ascorbic acid is not present. Second, on reverse scan, the magnitude of the back peak is considerably diminished in the presence of ascorbate ions. The authors suggest that the oxidation of ascorbate takes place at the liquid–liquid interface (see Fig. 6.14). A transfer of ascorbate into the oil droplet seems unlikely since the ascorbate ion is more hydrophilic than perchlorate and hexafluorophosphate. The voltammetry that is established in the presence of ascorbate ions is characterized by the competition of the catalytic oxidation of ascorbate [AH^-, Reaction (IXa)] and the electrochemical reduction of $THPD^+$.

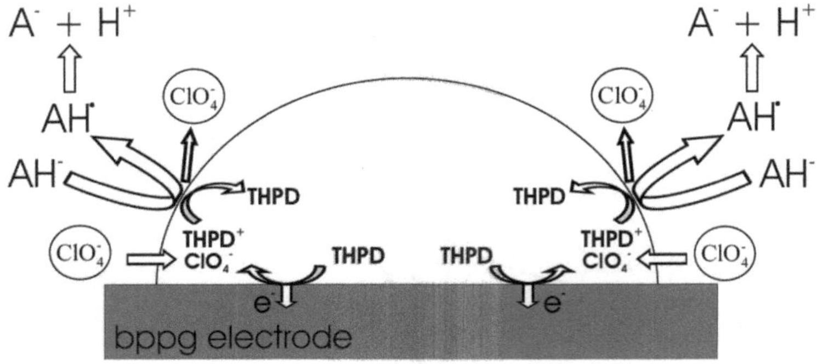

Fig. 6.14 Schematic diagram illustrating the electrocatalysis of L-ascorbate oxidation by $THPD^{+}ClO_4^{-}$. For simplicity, only one microdroplet is considered. Reproduced from [B284]. Copyright ©2003 with permission from American Chemical Society

$$\left[THPD^{\bullet+}ClO_4^{-}\right]_{(o)} + AH^{-}_{(aq)} \xrightarrow{k_{cat}} THPD_{(o)} + AH^{\bullet}_{(aq)} + ClO^{-}_{4(aq)} \qquad (IXa)$$

$$\left[THPD^{\bullet+}ClO_4^{-}\right]_{(o)} + e^{-} \leftrightarrows THPD_{(o)} + ClO^{-}_{4(aq)} \qquad (IXb)$$

From their observations the authors have derived a kinetic model describing the electrocatalytic oxidation of ascorbate at immobilized droplets of TRPD. The model is based on the assumptions that (i) $TRPD^{+}$ is electrochemically generated at the three-phase junction electrode–redox liquid–electrolyte; however, for simplicity reasons, the electrochemical oxidation is considered as in a "thin-film" regime; (ii) the catalytic oxidation of ascorbate proceeds at the droplet–electrolyte interface, and no transfer of ascorbate into the droplet takes place; and (iii) the microdroplet-modified electrode consists of a monodispersed set of N femtoliter, each of radius R_O. The increase in the charge passed in the presence of ascorbate (bulk concentration in the aqueous phase $[AH^{-}]_{bulk}$), ΔQ_{obs}, is then expressed by Eq. 6.5:

$$\Delta Q_{obs} = \frac{6d}{R_0} k_{cat} F \tau N_0 [AH^{-}]_{bulk} \qquad (6.5)$$

where d represents the spherical thickness of a $THPD^{\bullet+}$ ion, and τ is the time, in which the electron transfer across the oil–water interface can take place. It is defined by the scan rate of the voltammetric experiment, v, and the difference between the starting and finishing potentials of the sweep, ΔE, with $\tau = \Delta E v^{-1}$; N_0 is the number of moles TRPD immobilized at the electrode surface.

Based on these assumptions, the rate constants of the electrocatalytic oxidation of ascorbate by different TRPDs are given in Table 6.4. The table shows that the reactivity generally increases with decreasing length of the substituent, viz., in the order DEDBPD > DEDHPD > TBPD = THPD > DEDHePD = THePD > TNPD.

Table 6.4 Reaction-on-a-drop analysis (Eq. 6.5) of the reaction between DEDRPD and TRPD radical cations and L-ascorbate[a] (Reproduced from [B284])

	Gradient of plot/C M^{-1}	$6dk_{cat}/R_0$/M^{-1} s^{-1}	k_{cat}[b]/M^{-1} s^{-1}
DEDBPD[c]	0.2373 ± 0.012	10.9 ± 0.5	7,237 ± 366
DEDHPD[d]	0.1298 ± 0.012	5.8 ± 0.5	3,876 ± 358
TBPD[e]	0.0798 ± 0.012	2.3 ± 0.3	1,257 ± 230
THPD[f]	0.0553 ± 0.012	3.2 ± 0.7	2,139 ± 464
DEDHePD[g]	0.0289 ± 0.012	1.3 ± 0.5	866 ± 360
THePD[h]	0.0245 ± 0.012	1.8 ± 0.9	1,184 ± 580
TNPD[i]	0.0232 ± 0.012	3.0 ± 1.6	2,005 ± 1,039

[a]Data obtained from cyclic voltammograms recorded at a scan rate of 5 mV s^{-1}, in 0.1 mol L^{-1} phosphate buffer solution containing 0.1 mol L^{-1} potassium hexafluorophosphate, using a modified 4.9-mm diameter bppg electrode
[b]Assuming $d = 5$ Å and $R_0 = 2\,\mu m$
[c]Electrode modified with 5.7 nmol of DEDBPD, $\tau = 40$ s
[d]Electrode modified with 5.8 nmol of DEDHPD, $\tau = 40$ s
[e]Electrode modified with 6.0 nmol of TBPD, $\tau = 60$ s
[f]Electrode modified with 6.0 nmol of THPD, $\tau = 30$ s
[g]Electrode modified with 5.8 nmol of DEDHePD, $\tau = 40$ s
[h]Electrode modified with 6.0 nmol of THePD, $\tau = 24$ s
[i]Electrode modified with 5.7 nmol of TNPD, $\tau = 14$ s

The authors attribute this order to the greater number of cations per unit surface area for lower homologous.

For electrodes modified with tetrahexylphenylenediamines, Wadhawan et al. have reported photoelectrochemically induced catalytic processes [B284]. The authors made use of the property of THPD to undergo facile photoexcitation coupled to redox conversion when exposed to light of ca. 300-nm wavelength. This behavior, which was already known for the tetramethyl-substituted phenylenediamine, TMPD[6], has been investigated by Wadhawan with respect to (i) its occurrence at THPD microdroplet-modified electrodes, (ii) the role of the three-phase junction, and the effects introduced by anion insertion processes associated with the redox processes [B182]. As Fig. 6.15 demonstrates, both significant photoreduction and photooxidation currents are observed. From the dependence of the magnitude of the photocurrent on the coverage of the electrode surface, the authors have concluded that, as depicted in Fig. 6.16, the photoelectrochemical processes are confined to the triple-phase interface THPD–electrode–electrolyte.

In a later communication [B284], the authors have demonstrated that this photoactivity can be utilized for the photochemically induced catalytic reduction

[6]Richtol HH, Fitzgerald EA Jr, Wuelfing P Jr (1971) J Phys Chem. 75: 2737–2741; Avdievich NI, Jeevariajan AS, Forbes MDE (1996) J Phys Chem 100: 5334–5342

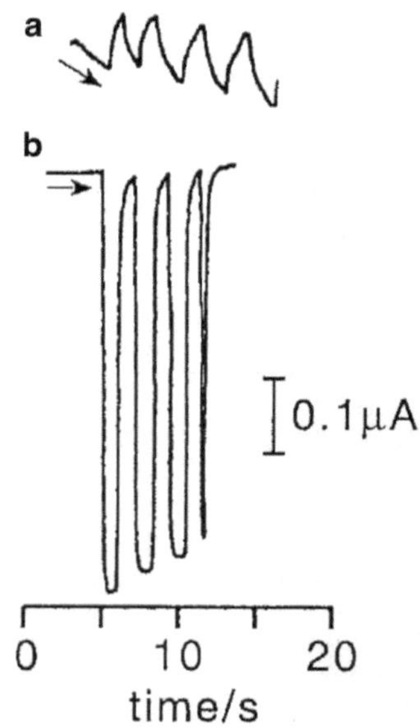

Fig. 6.15 Photocurrents observed for the oxidation at 0.4 V versus SCE (a) and for the photoreduction at 0.0 V versus SCE (b) of 80 nmol THPD (35 μg) deposited onto a 3 × 3-mm basal plane pyrolytic graphite electrode immersed in 0.1 mol L^{-1} $NaClO_4$ (10 mW cm^{-2}, 280 nm, solution flow rate 0.02 cm^3 s^{-1}). The arrows indicate the initial part of the trace with the light switched off [B182]. Copyright ©2001 with permission from Springer

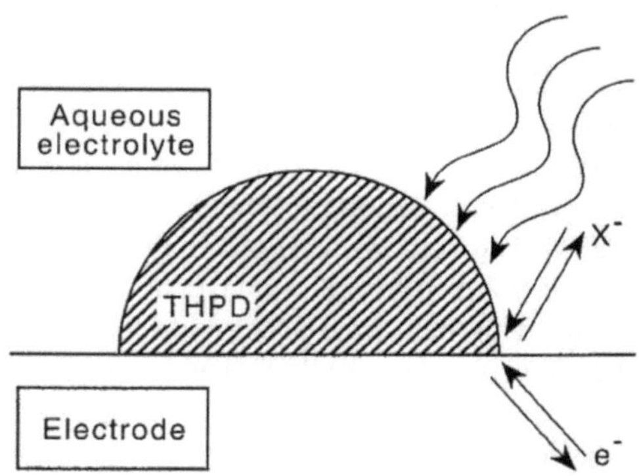

Fig. 6.16 Schematic drawing of the THPD–electrode–electrolyte triple interface reaction zone [B182]. Copyright ©2001 with permission from Springer

of the lipophilic anesthetic halothane (2-bromo-2chloro-1,1,1-trifluoroethane), Fig. 6.17. The mechanism of this process is proposed as follows:

$$\mathrm{THPD_{(o)}} + h\nu \leftrightarrows \mathrm{THPD^{*}_{(o)}} \tag{Xa}$$

$$\mathrm{THPD^{*}_{(o)}} + \mathrm{CF_3CHClBr_{(aq)}} \longrightarrow \mathrm{THPD^{\bullet +}_{(o)}} + \mathrm{CF_3CHClBr^{\bullet -}_{(aq)}} \tag{Xb}$$

$$\mathrm{CF_3CHClBr^{\bullet -}_{(aq)}} \longrightarrow \mathrm{CF_3CHCl^{\bullet}_{(aq)}} + \mathrm{Br^{-}_{(aq)}} \tag{Xc}$$

Noteworthy, under similar conditions, photoinduced electron transfer from excited chlorophyll a to vitamin K_1 has been observed [B287], which bears possible potential in the development of solar cells based on photosynthesis.

Recently, the same group interrogated the electrochemistry of vitamin K_1, deposited as an array of microdroplets onto the surface of a pyrolytic carbon

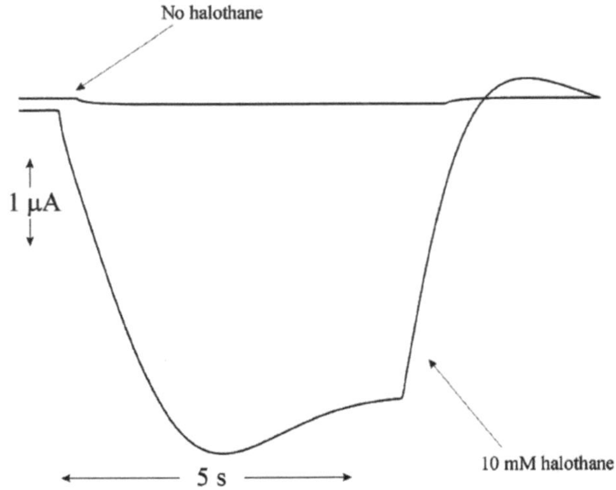

Fig. 6.17 Phototransients (2 mW cm^{-2} illumination at 320 nm) observed for 5.4 nmol of THPD deposited on the 3 × 3 mm^2 basal plane pyrolytic carbon electrode potentiostated at 0.0 V versus SCE, immersed 0.1 mol L^{-1} NaClO4 containing 10 mmol L^{-1} halothane (solution flow rate 0.003 cm^{3}s^{-1}) [B284]. Copyright ©2003 with permission from American Chemical Society

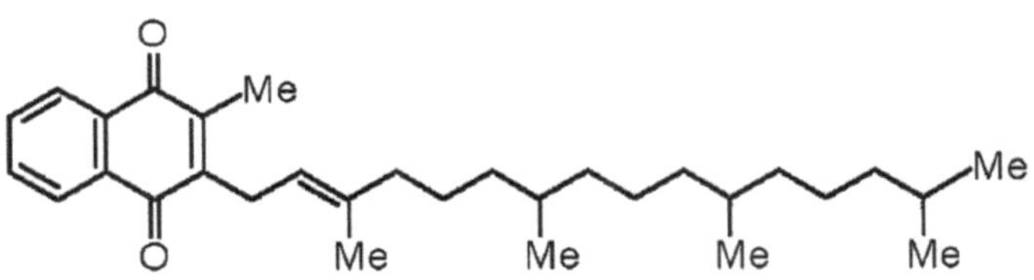

Fig. 6.18 Structure of vitamin K_1, phylloquinone [B273]

electrode. The results of this study [B273] are highly interesting from several viewpoints. The electrochemical reduction of vitamin K_1 at low pH (Reaction XI) is accompanied by a proton uptake and the protonation of the reduced phylloquinone (Fig. 6.18).

$$VK1_{(o)} + 2H^+_{(aq)} + 2e^- \leftrightarrows VK1H_{2(o)} \tag{XI}$$

That means that very much in contrast to most other examples of electrochemical reactions of immobilized microdroplets or particles, both the precursor liquid and the product are ionically nonconducting. Hence, it must be assumed that there cannot be a spatial proceeding of the reaction zone into the droplets. The electron and ion transfer remains confined to the three-phase junction. Most likely based on diffusion and convection processes, however, a conversion of the femtoliter droplets is achieved. When in the presence of acidic solutions alkali-metal cations are added to the electrolyte, the authors find a profound change in the voltammetric

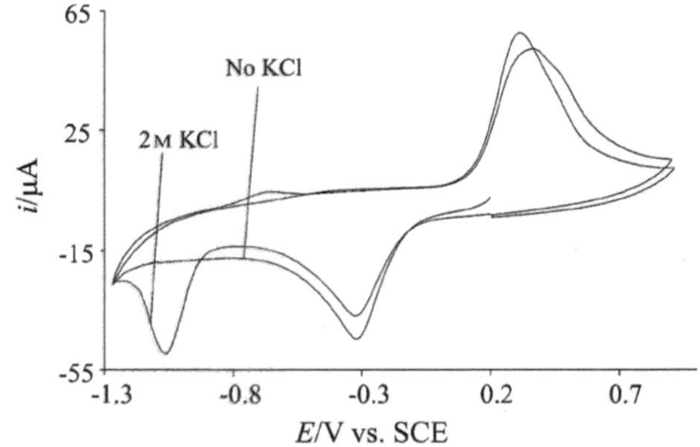

Fig. 6.19 Cyclic voltammograms (scan rate 20 mV s^{-1}) for the reduction of 5.6 nmol vitamin K_1 immobilized on a 4.9-mm diameter basal plane pyrolytic carbon electrode immersed into both 0.1 mol L^{-1} aqueous HCl and 0.1 mol L^{-1} HCl containing 2.0 mol L^{-1} KCl [B273]. Copyright ©2003 with permission from Wiley-VCH

response. The voltammogram (see Fig. 6.19) now shows an additional peak in the reductive scan. The appearance of two well-defined reductive signals was also connected with extra charge in the reductive scan, which can be explained only on the basis of an additional Faradaic process. After a thorough analysis of the occurring phenomena, Wain and coworkers conclude the following mechanism: initially, vitamin K_1 is reduced in a two-electron two-proton wave to yield quinole as in process III. The second Faradaic process is likely due to further reduction within the redox liquid. The authors suggest that, due to the strong naphthoquinole-potassium ion pair bonding and the formation of the strong H–H bond process XII occurs, leading to hydrogen evolution:

$$VK1H_{2(o)} + 2K^+_{(aq)} + 2e^- \leftrightarrows \left[(VK1)^{2-}(K^+)_2\right]_{(o)} + H_{2(o)} \qquad \text{(XII)}$$

In the presence of strongly alkaline solutions, the electrochemical reduction again takes place in a two-electron two-cation reaction that is sensitive to the nature of the transferred cation (Reaction XIII).

$$VK1_{(o)} + 2M^+_{(aq)} + 2e^- \leftrightarrows \left[(VK1)^{2-}(M^+)_2\right]_{(o)} \qquad \text{(XIII)}$$

Apart from a study of the redox switching of 4-nitrophenyl nonyl ether by Wain et al. [B286], this is the first time that alkali-metal uptake has been demonstrated in such environments.

A different kind of "reactive electrochemistry" has been presented by Gergely and Inzelt [B197]. In their communication, the authors report on the electropolymerization that takes place when droplets of 3-methylthiophene are oxidized in the presence of aqueous solutions of $LiClO_4$.

As Fig. 6.20 shows, during the first redox cycles, only the oxidation of the starting material can be seen. During repetitive voltammetric cycling, additional redox systems in the potential range between 0.3 and 0.8 V appear and start

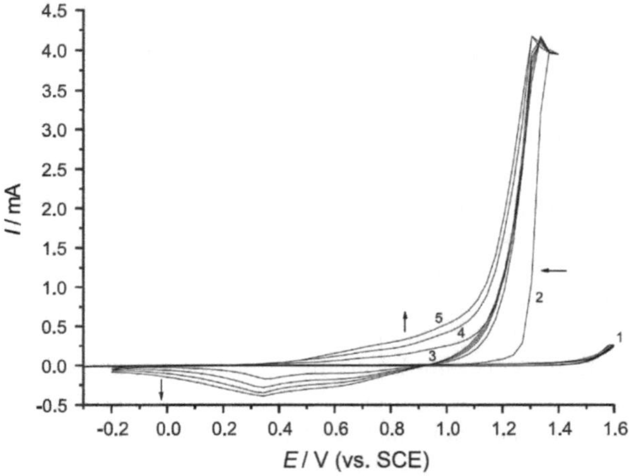

Fig. 6.20 Consecutive cyclic voltammetric curves obtained for 3-methylthiophene droplets attached to a paraffin-impregnated graphite electrode, PIGE, in the presence of an aqueous solution containing 0.5 mol L^{-1} $LiClO_4$. Cycles 1–5 (curves 2–5, curve 1 shows the background current), scan rate 100 mV s^{-1} [B197]. Copyright ©2001 with permission from Elsevier

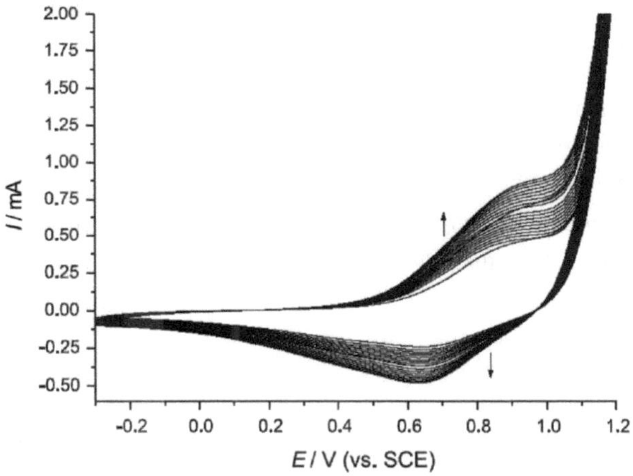

Fig. 6.21 Consecutive cyclic voltammetric curves obtained for 3-methylthiophene droplets attached to a paraffin-impregnated graphite electrode, PIGE, in the presence of an aqueous solution containing 0.5 mol L^{-1} $LiClO_4$. Cycles 15–34, scan rate 100 mV s^{-1} [B197]. Copyright ©2001 with permission from Elsevier

growing. This growing process goes on with further cycling (Fig. 6.21). It shows the formation of poly(3-methylthiophene) at the electrode surface. The evaluation of the growth of the polymer film by integration of the electric charge of the redox process caused by polymethylthiophene, however, showed that only about 1 % of the monomer was electropolymerized up to the 35th cycle.

Reactive chemistry in deposited redox liquid microdroplets can be interesting also from the analytical point of view. Thus, a reaction mechanism has been proposed that can be utilized for the accumulation and detection of sulfide. The approach makes use of the reaction of the oxidized N^1-[4-(dihexylamino)phenyl]-N^1,N^4,N^4-trihexyl-1,4-phenylenediamine (DPTPD) with HS^- leading to the formation of a methylene blue derivative [B289].

6.2 The Electrochemistry of Compounds Dissolved in Droplets

The electrochemistry of electroactive compounds dissolved in water-immiscible solvents and attached to the surface of solid electrodes in the form of droplets is an interesting topic. This chapter will be focused on one feature of such electrodes, namely, the *ion transfer* between the adjacent liquid phases, which always accompanies the electron transfer between the solid electrode and the electroactive compound. Figure 6.22 schematically depicts the situation at the electrode when a droplet of a solution of an electroactive compound is attached to an electrode surface. Very similar to the situation of immobilized solid particles that are undergoing insertion electrochemical reactions (see, e.g., the electrochemistry of metal hexacyanoferrates in previous chapters) and even more similar to the situation of immobilized electroactive droplets (Sect. 6.1), one can observe a reaction scenario that could be called "insertion electrochemistry with droplets." This approach offers a unique access to the standard Gibbs energy of ion transfer across a liquid–liquid interface [B132].

The standard Gibbs energy of ion transfer between two solvents is an important thermodynamic quantity since its value portrays the difference between the free energies of solvation of an ion in two solvents α and β. The standard Gibbs energy of ion transfer from phase α to phase β ($-\Delta G_{\alpha,\, i}^{\beta\, \ominus}$) is related to the standard potential of ion transfer ($\Delta\phi_{\alpha,\, i}^{\beta\, \ominus}$) and to the standard partition coefficient (P_i) through the following equations:

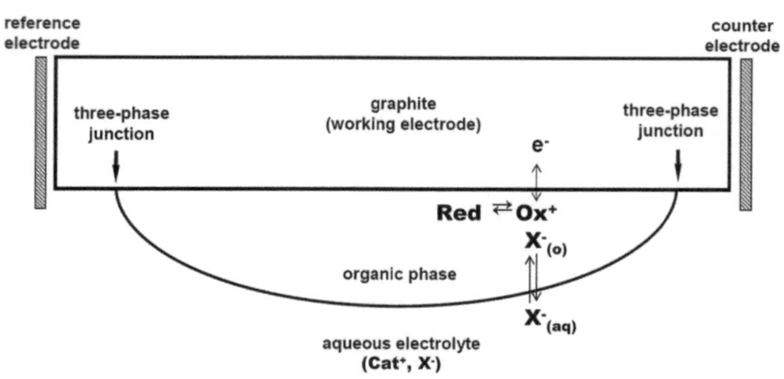

Fig. 6.22 Three-phase electrode consisting of a droplet of an organic solvent containing an electrochemically oxidizable compound Red. The droplet adheres to a solid electrode (graphite) and it is immersed in an aqueous electrolyte solution. The oxidation of Red at the graphite–organic liquid interface is accompanied by a transfer of X^- from aqueous to the organic phase [B280]. Copyright ©2004 with permission from Elsevier

$$\Delta\phi_{\alpha,\,i}^{\beta\,\ominus} = \frac{-\Delta G_{\alpha,\,i}^{\beta\,\ominus}}{z_i F} \tag{6.6}$$

$$P_i = \exp\left(\frac{-\Delta G_{\alpha,\,i}^{\beta\,\ominus}}{RT}\right) \tag{6.7}$$

The logarithm of the partition coefficient (log$P$) is a measure of the lipophilicity of compounds. It helps in understanding a variety of biological phenomena, such as passive transfer through membranes, enzyme-receptor interactions, drug activities, etc., and it is a major parameter in quantitative structure activities relationships (QSAR) and quantitative structure properties relationships (QSPR)[7]. While log$P$ of neutral compounds is relatively easily accessible by different partition techniques, such as shake-flask methods, partition chromatography, high-performance liquid chromatography, extraction (see footnote 7), etc., log$P$ of single ions was accessible until recently only by four-electrode voltammetry at the interface of two immiscible electrolyte solutions (ITIES)[8] . The inevitable presence of electrolytes in the two immiscible solvents, as well as the nonpolarizability of some important water–organic solvent interfaces[9], severely restricts the use of that technique.

In this chapter, the applicability of the three-phase electrode approach for determining standard Gibbs energies of transfer of ions will be discussed. To achieve an ion transfer across the organic droplet–aqueous solution interface, a small amount of a neutral electroactive compound is dissolved in an organic and water-immiscible solvent that is devoid of any electrolyte. A droplet (e.g., 1 μL) of that solution is attached to the surface of a suitable electrode. Graphite electrodes (e.g., PIGE) are superior compared to metal electrodes because the stability of the attachment of a nonaqueous droplet is very good. The electrode with the attached droplet is immersed in an aqueous electrolyte solution of the salt Cat^+A^-. In this solution, the reference and counter electrodes are placed as in a conventional voltammetric cell. A prerequisite for the proper operation of the experiment is the existence of the *three-phase junction line* where the three-phase working electrode, organic solution, and aqueous solution boarder to each other. Since no electrolyte is added to the organic solvent, the initial conductivity in the region near the edge of the droplet results only from free partition of the salt present in the aqueous solution. The applied potential between the working and the reference electrode can initially act in a restricted area only, i.e., at the three-phase junction where all three phases meet. Figure 6.23 depicts the potential drops at such a three-phase electrode.

---

[7] Testa B, van de Waterbeemd H, Folkers G, Gay R (2001) Pharmacokinetic optimization in drug research. Wiley-VCH, Weinheim

[8] Girault HHJ, Schiffrin DJ (1989) Electrochemistry of liquid-liquid interfaces. In: Bard, AJ (ed) Electroanalytical chemistry, a series of advances, vol 15. Dekker, New York, pp 1–141

[9] Marcus Y (1997) Ion properties. Dekker, New York

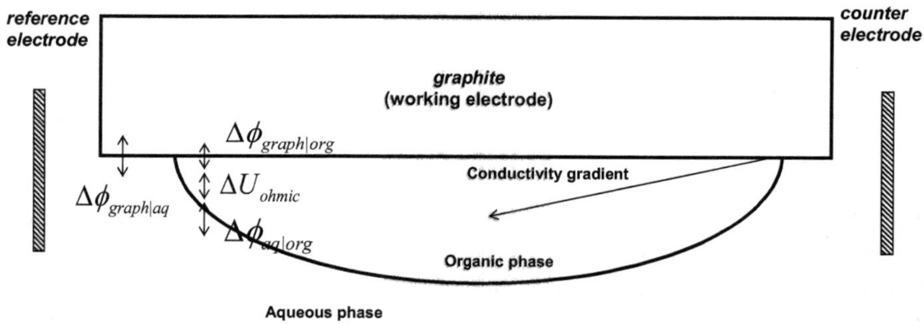

Fig. 6.23 Potential drops at three-phase electrode (details see text) [B280]. Copyright ©2004 with permission from Elsevier

The potential drop at the graphite–organic phase interface is given by the following expression:

$$\Delta\phi_{\text{graph}|\text{org}} = \Delta\phi_{\text{graph}|\text{aq}} - \Delta U_{\text{ohmic}} - \Delta\phi_{\text{aq}|\text{org}} \tag{6.8}$$

The potential drop $\Delta\phi_{\text{graph|aq}}$ at the interface graphite electrode–aqueous solution in Eq. 6.8 is controlled potentiostatically. However, the potential drop $\Delta\phi_{\text{graph|org}}$ at the interface graphite–organic solution is in fact the applied potential difference ($\Delta\phi_{\text{graph|aq}}$) diminished (i) by the value of the ohmic drop ΔU_{ohmic} inside the organic phase and (ii) by the value of the potential drop $\Delta\phi_{\text{aq|org}}$ (i.e., the potential drop due to the partition of ionic species) at the interface aqueous solution–organic solution. Indeed, the value of $\Delta\phi_{\text{aq|org}}$ will be different for different ions that are initially present in the aqueous phase. Consequently, the potential $\Delta\phi_{\text{graph|org}}$ will be altered in the same way as the nature of the ions in the aqueous phase is changed. When the organic droplet does not contain any deliberately added electrolyte, ΔU_{ohmic} will be very large all over the graphite–droplet interface with the only exception of the very edge, i.e., the three-phase junction region. A thin layer that is situated at the droplet–water interface will acquire a higher ionic conductivity due to salt partition from the aqueous phase. Therefore, an electrochemical oxidation of Red (see Fig. 6.22) can start without ohmic impediment only in that three-phase junction region. The ohmic drop will decrease within the droplet with progressing of the electrode reaction because Ox^+ and X^- are ions that provide ionic conductivity (the concentration of Red is usually rather high, e.g., higher than 10^{-2} mol L^{-1}).

The *electron transfer* process occurring at the electrode–organic phase interface is *necessarily coupled* to a *transfer of ions* across the aqueous phase–organic phase interface in order to maintain the electroneutrality of the organic phase. In voltammetric experiments, both the electron and ion transfer processes are recorded together. The nature of the ion transfer takes place, i.e., the transfer of anions from water to the organic phase or the transfer of electrochemically created cations from the organic phase to water depends primarily on the ratio of the standard Gibbs energies of ion transfer of both candidates. Generally, if the solvation of the electrochemically created cations Ox^+ in the organic phase is stronger than the hydration of the anions in the aqueous phase X^-, then a transfer of anions from the aqueous phase to the organic phase will take place, as depicted in the Fig. 6.22.

A *transfer of cations* from aqueous phase to the organic phase can be achieved if the solvation of electrochemically generated anions in the organic phase, created by reduction of a neutral electroreducible compound present in the organic phase, is stronger than the hydration of the cations in the aqueous phase.

In both cases, the standard Gibbs energies of ion transfer can be deduced from the formal potentials of cyclic or square-wave voltammograms (or of any other voltammetric technique), following the algorithm described next.

According to Fig. 6.22, the overall reaction proceeding at a microdroplet-modified electrode can be written as follows:

$$\mathrm{RED}_{(o)} + \mathrm{A}^-_{(aq)} \leftrightarrows \mathrm{Ox}^+_{(o)} + \mathrm{A}^-_{(o)} + \mathrm{e}^- \qquad \text{(XIV)}$$

If no kinetic constraints exist with respect to the electron and ion transfer, the thermodynamic treatment applied to Reaction (XIV) leads to the following form of the Nernst equation:

$$E = E^{\ominus}_{\mathrm{Ox}^+_{(o)}/\mathrm{Red}_{(o)}} + \Delta\phi^{(o)\ominus}_{(aq),\ \mathrm{A}^-} + \frac{RT}{F}\ln\frac{a_{\mathrm{Ox}^+_{(o)}}a_{\mathrm{A}^-_{(o)}}}{a_{\mathrm{Red}_{(o)}}a_{\mathrm{A}^-_{(aq)}}} \qquad (6.9)$$

In Eq. 6.9, E is the applied potential between the working and the reference electrode, $E^{\ominus}_{\mathrm{Ox}^+_{(o)}/\mathrm{Red}_{(o)}}$ is the standard potential of the redox couple Ox^+/Red in the organic solvent, $\Delta\phi^{(o)\ominus}_{(aq),\ \mathrm{A}^-}$ is the standard potential of transfer of anions from the aqueous phase to the organic phase, $a_{\mathrm{Ox}^+_{(o)}}$ and $a_{\mathrm{Red}_{(o)}}$ are the activities of the oxidized and reduced forms of the electroactive compound in the organic phase, while $a_{\mathrm{A}^-_{(o)}}$ and $a_{\mathrm{A}^-_{(aq)}}$ are the activities of anions in the organic phase and aqueous solutions, respectively. In a first approximation, the activities in the Nernst equation have been replaced by concentrations. Since the concentration of the anions in the aqueous phase does not change significantly during the experiment, Eq. 6.9 can be rewritten as

$$E = E^{\ominus}_{\mathrm{Ox}^+_{(o)}/\mathrm{Red}_{(o)}} + \Delta\phi^{(o)\ominus}_{(aq),\ \mathrm{A}^-} - \frac{RT}{F}\ln c_{\mathrm{A}^-_{(aq)}} + \frac{RT}{F}\ln\frac{c_{\mathrm{Ox}^+_{(o)}}c_{\mathrm{A}^-_{(o)}}}{c_{\mathrm{Red}_{(o)}}} \qquad (6.10)$$

The electroneutrality principle requires that

$$c_{\mathrm{Ox}^+_{(\mathrm{o})}} = c_{\mathrm{A}^-_{(\mathrm{o})}} \tag{6.11}$$

The mass conservation law in respect to the organic phase leads to

$$c_{\mathrm{Red}_{(\mathrm{o})}} + c_{\mathrm{Ox}^+_{(\mathrm{o})}} = c^*_{\mathrm{Red}_{(\mathrm{o})}} \tag{6.12}$$

where $c^*_{\mathrm{Red}_{(\mathrm{o})}}$ is the initial concentration of the oxidizable compound in the organic phase. When the applied potential equals the formal potential of the redox pair, it follows that

$$c_{\mathrm{Red}_{(\mathrm{o})}} = c_{\mathrm{Ox}^+_{(\mathrm{o})}} \tag{6.13}$$

Substitution of Eqs. 6.11, 6.12, and 6.13 in Eq. 6.10 allows calculating the formal potential ($E^{\ominus'}_{\mathrm{c}}$) of the system:

$$E^{\ominus'}_{\mathrm{c}} = E^{\ominus}_{\mathrm{Ox}^+_{(\mathrm{o})}/\mathrm{Red}_{(\mathrm{o})}} + \Delta\phi^{(\mathrm{o})\ominus}_{(\mathrm{aq}),\ \mathrm{A}^-} - \frac{RT}{F}\ln c_{\mathrm{A}^-_{(\mathrm{aq})}} + \frac{RT}{F}\ln\frac{c^*_{\mathrm{Red}_{(\mathrm{o})}}}{2} \tag{6.14}$$

Since the voltammetric systems obtained by these experiments, at least with the used redox probes, possess all features of electrochemical reversibility and the transfer coefficients are obviously near to 0.5, it is reasonable to take the mid-peak potentials of cyclic voltammograms as the formal potential of the system.

The last equation shows that the formal potential of the voltammograms depends on the *nature of the anions in aqueous phase* (via the values of $\Delta\phi^{(\mathrm{o})\ominus}_{(\mathrm{aq}),\ \mathrm{A}^-}$). Generally, the more lipophilic the anions are, the more negative is the value of $\Delta\phi^{(\mathrm{o})\ominus}_{(\mathrm{aq}),\ \mathrm{A}^-}$. Consequently, the oxidation of the compound Red in the organic phase will occur at more negative potentials (i.e., it will be easier oxidized) as the lipophilicity of the transferable anions increases. Further, the formal potential should shift by about 59 mV in negative direction for a tenfold increase of the concentration of the transferable anions in aqueous phase. This criterion, taken together with the stability of the voltammograms recorded during consecutive cycling, serves as an important indicator for the reversibility of the processes occurring at the droplet-modified electrodes.

When an *electroreducible* compound is dissolved in the organic phase, then its reduction at a three-phase electrode can provoke the transfer of cations from water to the organic phase. This can be described by the following reaction:

$$\mathrm{Ox}_{(\mathrm{o})} + \mathrm{Cat}^+_{(\mathrm{aq})} + \mathrm{e}^- \leftrightarrows \mathrm{Red}^-_{(\mathrm{o})} + \mathrm{Cat}^+_{(\mathrm{o})} \tag{XV}$$

Analogously to the previous case, the thermodynamic treatment of Reaction (XV) leads to the following Nernst equation, which is valid when cations are transferred from the aqueous to the organic phase:

$$E_{\mathrm{c}}^{\ominus\prime} = E_{\mathrm{Ox}_{(\mathrm{o})}^{+}/\mathrm{Red}_{(\mathrm{o})}}^{\ominus} + \Delta\phi_{(\mathrm{aq}),\ \mathrm{Cat}^{+}}^{(\mathrm{o})\ominus} + \frac{RT}{F}\ln c_{\mathrm{Cat}_{(\mathrm{aq})}^{+}} + \frac{RT}{F}\ln\frac{2}{c_{\mathrm{Ox}_{(\mathrm{o})}}^{*}} \tag{6.15}$$

The more lipophilic the cations present in the aqueous phase are, the more positive will be $\Delta\phi_{(\mathrm{aq}),\ \mathrm{Cat}^{+}}^{(\mathrm{o})\ominus}$. Subsequently, the reduction of the organic compound Ox in the oil phase will occur at more positive potentials as the lipophilicity of the cations in the aqueous phase increases. The formal potential of the coupled electron–ion reaction at the three-phase electrode will shift by 59 mV in positive direction for a tenfold increase of the concentration of the transferable cations in the aqueous solution.

6.2.1 The Determination of Standard Gibbs Energies of Transfer of Anions

The oxidation of a neutral lipophilic organic compound dissolved in an organic solvent is a precondition for transferring anions across the aqueous–organic interface. The compound decamethylferrocene (dmfc) [bis(pentamethylcyclopentadienyl)iron(II)] possesses that ability (see Fig. 6.24). It is a lipophilic compound, soluble in various organic solvents, but almost insoluble in water[10]. It exhibits an electrochemically reversible one-electron redox reaction in organic solvents, and it is commonly used as a reference standard for potential in measurements in nonaqueous media (see footnote 4). The oxidation of dmfc in a droplet of nitrobenzene (NB) attached to an electrode and immersed in aqueous solutions of different sodium salts gives rise to well-defined square-wave voltammograms, the formal potentials of which are sensitive to the nature and concentrations of the anions present in the aqueous solutions (see Fig. 6.25). Of course, in principle, any voltammetric measuring technique can be applied; however, square-wave voltammograms offer a simple way to measure the formal potential as it is the peak potential of the signals. The overall electrode process proceeding at the three-phase electrode can be written as follows:

$$\mathrm{dmfc}_{(\mathrm{nb})} + \mathrm{A}_{(\mathrm{aq})}^{-} \leftrightarrows \mathrm{dmfc}_{(\mathrm{nb})}^{+} + \mathrm{A}_{(\mathrm{nb})}^{-} + \mathrm{e}^{-} \tag{XVI}$$

The dependence of the formal potentials of the square-wave voltammograms (Fig. 6.25) on the standard potentials of transfer of the anions from water to nitrobenzene is linear and characterized by a slope of 0.95 and an intercept

[10] Noviandri I, Brown KN, Fleming DS, Gulyas PT, Lay PA, Masters AF, Philips L (1999) J Phys Chem B, 103:6713–6722

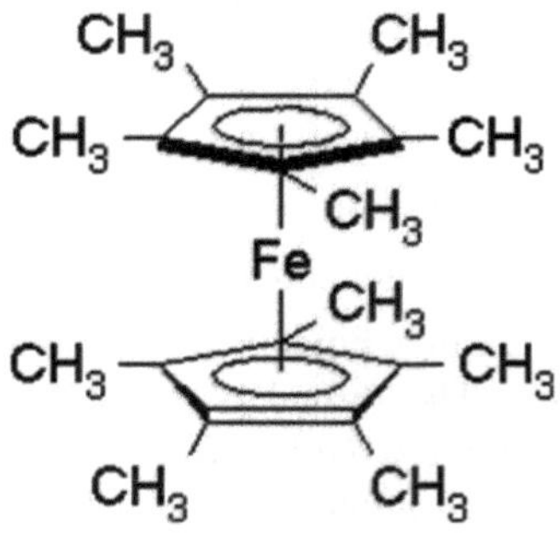

Fig. 6.24 Structure of decamethylferrocene (dmfc)

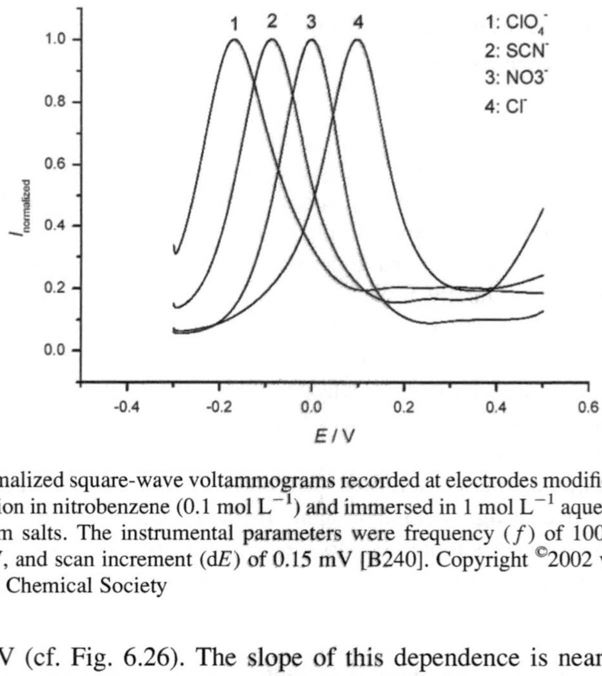

Fig. 6.25 Normalized square-wave voltammograms recorded at electrodes modified with droplets of a dmfc solution in nitrobenzene (0.1 mol L^{-1}) and immersed in 1 mol L^{-1} aqueous solutions of different sodium salts. The instrumental parameters were frequency (f) of 100 Hz, amplitude (E_{sw}) of 50 mV, and scan increment (dE) of 0.15 mV [B240]. Copyright ©2002 with permission from American Chemical Society

of –0.259 mV (cf. Fig. 6.26). The slope of this dependence is near to 1, just as predicted by Eq. 6.14. From the intercept of this dependence, one can evaluate the standard potential of the couple $dmfc^{+}$/dmfc in nitrobenzene which reads –0.184 V (vs. Ag/AgCl, saturated KCl). This value can be also determined by common measurements in the nonaqueous phase in the presence of a reference standard.

Knowing the value of the standard potential of $dmfc^{+}$/dmfc in nitrobenzene, and studying the oxidation of nitrobenzene solution of dmfc at a droplet-modified electrode which is immersed in aqueous solutions containing different anions, one can estimate the standard potentials of transfer (and thus the standard Gibbs energies of transfer) of the anions from water to nitrobenzene using Eq. 6.14. To identify the electrode system to follow Reaction (XVI), one can use the dependence

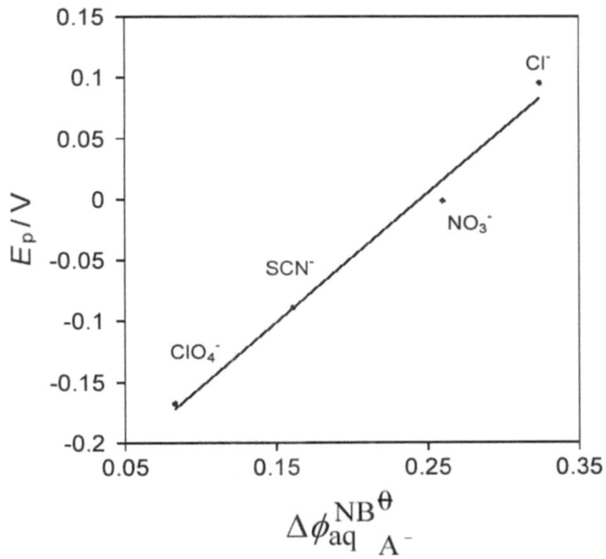

Fig. 6.26 Dependence of the formal potentials of the square-wave voltammograms depicted in Fig. 6.25 on the standard potentials of transfer of anions across the water–nitrobenzene interface [B240]. Copyright ©2002 with permission from American Chemical Society

of the formal potentials on the logarithm of concentration of anions in the aqueous phase and the stability of the voltammetric responses during consecutive cycling.

The electrochemical reversibility of the electrode process should be checked by studying the dependence of the formal potentials on the frequency (in square-wave voltammetry) or on the scan rate (in cyclic voltammetry). Representative square-wave voltammogram of all current components of the electrode process of dmfc obtained at a three-phase electrode with coupled transfer of thiocyanate anions is given in Fig. 6.27.

This approach has been utilized for the determination of the standard Gibbs energies of transfer of a large number of inorganic anions [B132, B206, B240, B259] as well as of various organic anions, such as the anions of aliphatic and aromatic carboxylic acids [B240, B259], substituted phenols [B240, B269], drugs [B269], and amino acids and peptides [B262, B266]. As organic phases, the common solvents nitrobenzene and dichloroethane [B132, B240, B259, B262, B266] have been used. Remarkably, also *n*-octanol [B206, B269], nitrophenyl octyl ether [B282], and D- and L-menthol [B221] have been exploited as organic solvents. Some of the data of lipophilicities of anions are given in Table 6.5.

It was an important achievement in the field of ion transfer studies that the determination of lipophilicities of ions across the water–*n*-octanol interface became feasible with the help of the three-phase electrode approach [B206, B269]. This is so important because the traditional measure of lipophilicity as a predictor of solute membrane partitioning is the partition coefficient in the *n*-octanol–water system log

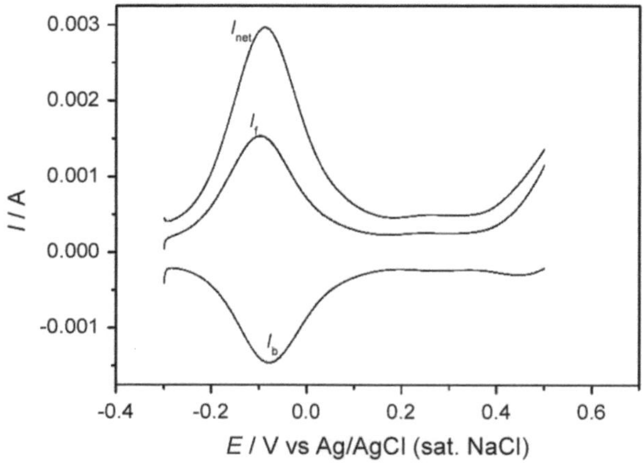

Fig. 6.27 Forward (I_f), backward (I_b), and net (I_{net}) components of a square-wave voltammetric response of a nitrobenzene droplet containing 0.1 mol L^{-1} dmfc attached to the surface of the working electrode and immersed in 1 M aqueous solution containing SCN^- anions. The experimental conditions were frequency $f = 100$ Hz, amplitude ΔEsw $= 50$ mV, and scan increment d$E = 0.15$ mV [B240]. Copyright ©2002 with permission from American Chemical Society

(P_{aq}^{n-oct}). The lipophilic chain of *n*-octanol together with the hydrophilic hydroxyl group makes *n*-octanol a good mimic for the major constituents of biological membranes (i.e., phospholipids). However, due to the nonpolarizability of the water–*n*-octanol interface in the 4-electrode ITIES technique, no lipophilicity data of ions measured in that system were accessible until now. The nonpolarizability of that interface is a result of not having yet found an appropriate lipophilic electrolyte for *n*-octanol to perform the necessary 4-electrode ITES measurements. Using the three-phase electrode approach, the determination of standard Gibbs energies of ion transfer across the water–*n*-octanol interface is quite easy [B206, B269].

The knowledge of the lipophilicities of amino acids and peptides [B262, B266] is of essential importance to understand their biological activity, peptide interactions, and peptide folding. Studying the transfer of monoanionic forms of various oligopeptides was performed to determine the individual contributions of the amino acid residues to the overall lipophilicity of the oligopeptides [B266]. The effect of the *position* of an amino acid residue in the chain of a peptide is in some cases very pronounced, especially when an aromatic amino acid residue is placed next to the terminal amino acid at which the negative charge is located in the anionic form. Therefore, it is unreasonable to approximate the entire lipophilicity of peptide anions as additive functions of the contributions of their amino acid residues, as it has been proposed in all theoretical models for neutral peptides. A first study of the lipophilicities of various peptide anions and amino acid anions has been published [B262, B266]. A part of the data is given here in Table 6.6.

Table 6.5 Formal potentials $E_c^{\ominus'}$, slope of the formal potentials versus concentration of anions in aqueous phase, standard deviation of formal potentials, and standard Gibbs energies of ion transfer of some anions across the water–NB (NB = nitrobenzene) interface [B259]

Anion	$E_c^{\ominus'}$/mV	slope $E_c^{\ominus'}$ versus log (c)/mV	$s(E_c^{\ominus'})^a$/mV	$\Delta G^{\ominus}$ /kJ mol^{-1}
ClO_3^-	2	–55.3	6.43	25.40
BrO_3^-	60	–58.7	7.17	30.90
IO_3^-	74	–54.3	8.08	32.40
IO_4^-	–132	–56.4	2.00	12.50
OCN^-	45	–50.5	2.45	29.50
$SeCN^-$	–136	–43.0	5.30	11.80
CN^-	41	–58.1	4.43	29.60
N_3^-	14	–52.1	3.44	26.80
Monofluoroacetate	44	–54.4	5.48	29.90
Difluoroacetate	34	–48.5	3.90	28.90
Trifluoroacetate	–2	–60.1	1.79	25.30
Monochloroacetate	36	–51.5	4.73	29.10
Dichloroacetate	9	–58.0	1.15	26.40
Trichloroacetate	–66	–60.1	1.97	18.80
Monobromoacetate	12	–39.3	3.44	26.70
Dibromoacetate	–7	–59.0	2.00	24.80
Tribromoacetate	–94	–59.8	1.03	16.00
Monoiodoacetate	0	–54.6	1.20	25.10
Formate	58	–56.4	2.40	30.60
Acetate	52	–58.0	1.50	30.10
Propionate	29	–54.6	0.80	27.98
Butyrate	11	–53.1	2.20	26.25
Valeriate	–31	–63.5	2.80	22.30
Capronate	–75	–60.3	1.40	18.10
Oenanthate	–115	–55.2	1.80	14.20
Caprylate	–125	–57.4	4.20	12.64
Pelargonate	–120	–52.9	3.20	13.40
Caprinate	–118	–58.4	2.50	13.60
Cyclopropane carboxylate	–20	–60.0	1.10	23.25
Cyclobutane carboxylate	–61	–57.8	1.40	19.30
Cyclopentane carboxylate	–100	–63.2	1.60	15.54
Cyclohexane carboxylate	–131	–56.8	2.80	12.54
Cycloheptane carboxylate	–155	–55.4	2.00	10.22

[a]This is the standard deviation of the measured formal potentials

The Transfer of Chiral Ions Across the Water–Chiral Organic Liquid Interface

It has been shown that the transfer of ions from water to a *chiral* organic solvent can be studied with three-phase electrodes [B450]. Thus, it was possible to measure the

Table 6.6 The standard Gibbs energies of transfer, the standard potential of transfer, and the logarithm of the partition coefficients of monoanionic forms of amino acids and peptides as determined by three-phase electrode approach [B266]

Peptide anions	$\Delta\phi^{(nb)\ominus}_{(aq),\ A^-}$/ V	$\Delta G^{(nb)\ominus}_{(aq),\ A^-}$/ kJ mol^{-1}	$\log P^{(nb)\ominus}_{(aq),\ A^-}$	slope $E^{\ominus'}_{c}$ versus log (c)/ mV
Trp$^-$	0.115	10.80	−1.90	−64
Trp-Ala$^-$	0.165	15.75	−2.75	−80
Trp-Gly$^-$	0.162	15.60	−2.73	−73
Trp-Val$^-$	0.120	11.60	−2.05	−75
Trp-Leu$^-$	0.100	9.50	−1.66	−73
Trp-Tyr$^-$	0.075	7.40	−1.30	−65
Trp-Phe$^-$	0.055	5.30	−0.93	−77
Trp-Trp$^-$	0.050	4.80	−0.85	−70
Trp-Gly-Gly$^-$	0.165	15.80	−2.75	−75
Trp-Gly-Tyr$^-$	0.165	15.00	−2.65	−74
Trp-Gly-Gly-Tyr$^-$	0.160	15.50	−2.70	−74
Leu-Leu$^-$	0.245	23.70	−4.15	−71
Leu-Leu-Ala$^-$	0.293	28.20	−4.95	−57
Leu-Leu-Gly$^-$	0.290	28.00	−4.91	−80
Leu-Leu-Leu$^-$	0.240	23.20	−4.05	−80
Leu-Leu-Tyr$^-$	0.205	19.70	−3.45	−56
Leu-Leu-Phe$^-$	0.180	17.50	−3.05	−64
Gly-Gly-Val$^-$	0.275	26.40	−4.60	−57
Gly-Gly-Leu$^-$	0.280	26.80	−4.70	−56
Gly-Gly-Phe$^-$	0.270	26.00	−4.55	−58
Gly-Trp-Gly$^-$	0.165	15.80	−2.75	−48
Gly-Gly-Trp$^-$	0.195	19.00	−3.35	−56
Lys-Tyr-Thr$^-$	0.310	30.00	−5.25	−58
Tyr-Ala-Gly$^-$	0.260	24.90	−4.40	−48

differences in Gibbs energies of transfer for *chiral* ions from water to *chiral* liquids. By attaching a single droplet of a *chiral* liquid containing decamethylferrocene on a paraffin-impregnated graphite electrode and immersing that electrode in an aqueous solution containing *chiral* anions, the Gibbs energies of all four ion–solvent combinations can be determined (see Fig. 6.28).

For symmetry reasons, it is expected that the Gibbs energies of transfer of a D-ion from water to the D-liquid is the same as that of the L-ion to the L-liquid. The same holds true for the combinations D-ion/L-liquid and L-ion/D-liquid. Square-wave voltammetric experiments conducted in aqueous solutions containing D- or L-anionic forms of some amino acids (phenylalanine, tyrosine, and lysine) performed with dmfc solutions in D- and L-2-octanol droplets allowed to determine these Gibbs energies of transfers relative to each other, i.e., the differences between $\Delta G^{\ominus,\ w\rightarrow D-liq}_{D\text{-ion}}$ and $\Delta G^{\ominus,\ w\rightarrow L-liq}_{D\text{-ion}}$, i.e., $\Delta\Delta G^{\ominus,\ w\rightarrow D-liq;\ D-liq}_{D\text{-ion}}$, and between $\Delta G^{\ominus,\ w\rightarrow L-liq}_{L\text{-ion}}$ and $\Delta G^{\ominus,\ w\rightarrow D-liq}_{L\text{-ion}}$, i.e., $\Delta\Delta G^{\ominus,\ w\rightarrow L-liq;\ D-liq}_{L\text{-ion}}$. In all studied

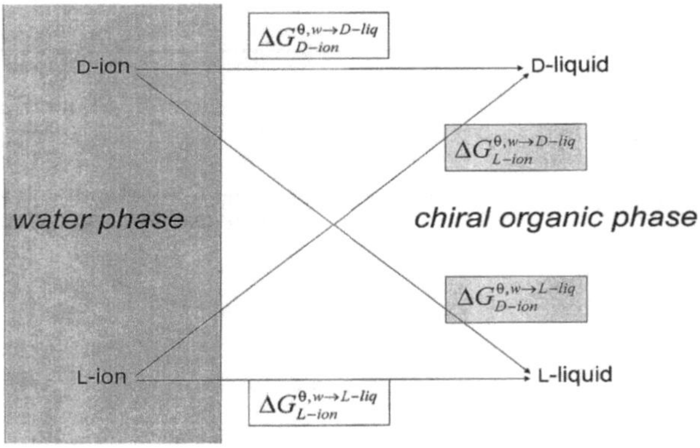

Fig. 6.28 Accessible Gibbs energies of transfer of D- and L-ions between water and D- and L-liquid [B450]. Copyright ©2005 with permission from Royal Soc. Chem.

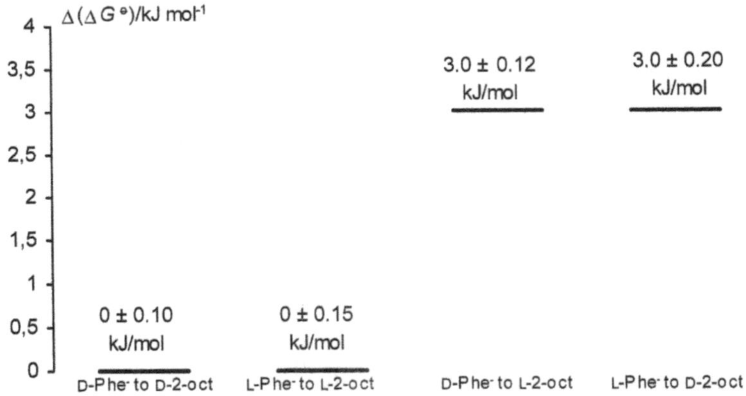

Fig. 6.29 Relations between the Gibbs energies of transfer of the anions of D- and L-phenylalanine from water to D- and L-2-octanol. The values for D-Phe$^-$ and L-Phe$^-$ have been set arbitrarily to zero. The values following the signs "±"denote to standard deviations [B450]. Copyright ©2005 with permission from Royal Soc. Chem.

combinations, symmetrical solvation behavior has been observed, which is in complete agreement with theory [B450]. Thus, the energy of interactions between D-anions and D-2-octanol was identical with those between L-anions and L-2-octanol. The same holds true for the combinations D-anions and L-2-octanol and L-anions and D-2-octanol (see the example in Fig. 6.29). This figure illustrates that the three-phase electrodes with immobilized droplets allow a reproducible measurement of the rather small energetic differences caused by the chirality of solvents and dissolved ions, which have not yet been determined by other techniques.

When the transfer of D- and L-monoanionic forms of the amino acid tryptophan was followed across the water–D-menthol and water–L-menthol interface, an

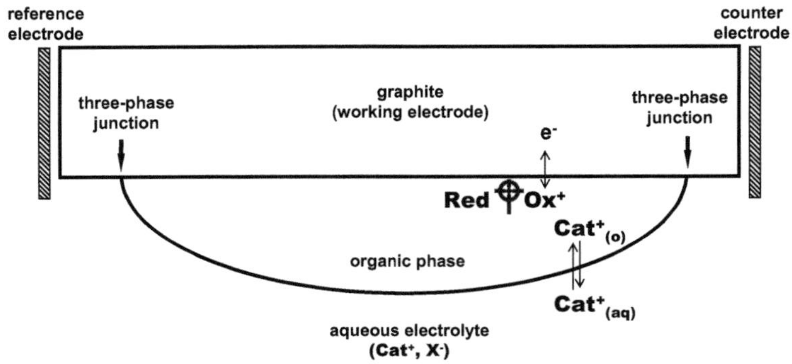

Fig. 6.30 Processes occurring at a three-phase electrode when an electroreducible compound is dissolved in the organic phase and cations are transferred across the interface water | organic liquid [B271]. Copyright ©2003 with permission from Elsevier

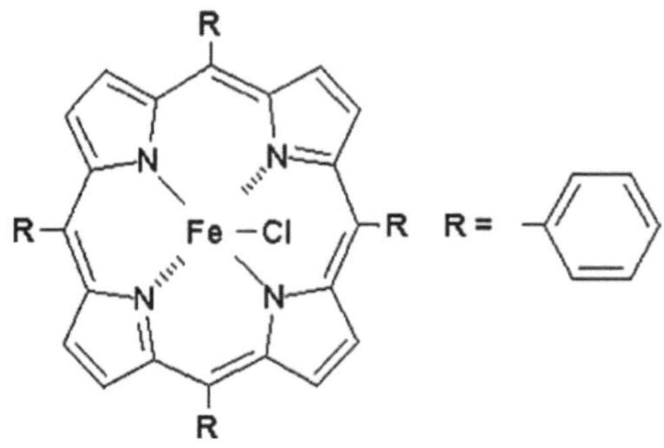

Fig. 6.31 Fe(III) tetraphenylporphyrine chloride

apparent asymmetrical solvation behavior was observed [B221]. The reasons for this asymmetry may be adsorption or precipitation at the liquid–liquid interface or even different kinetics.

6.2.2 *The Determination of Standard Gibbs Energies of Transfer of Cations*

Following the principles outlined for anion transfer with three-phase electrodes, the *reduction* of an electroreducible compound dissolved in an organic liquid is required to transfer cations from water to an organic liquid. Figure 6.30 visualizes the electrode system. The reduction of Fe(III) tetraphenylporphyrine chloride [Fe(III)TPP-Cl] (Fig. 6.31) in a nitrobenzene microdroplet in a three-phase

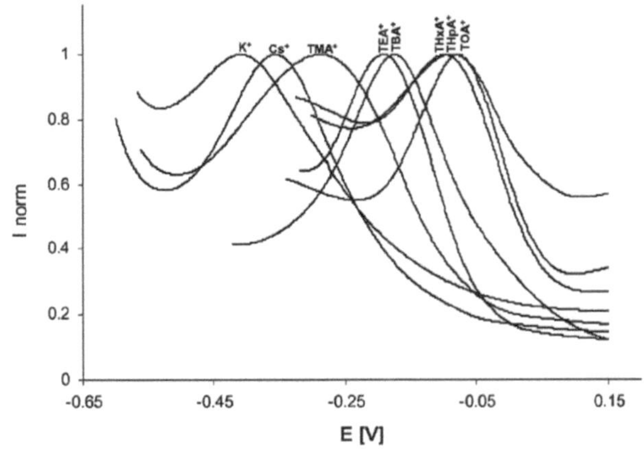

Fig. 6.32 Normalized square-wave voltammograms of the redox reaction of Fe(III)TPP-Cl in nitrobenzene droplets accompanied by the transfer of cations from aqueous solution to NB [B271]. Copyright ©2003 with permission from Elsevier

Table 6.7 Standard Gibbs energies of transfer of cations across the water–NB interface determined from the formal potentials of the voltammograms of Fe(III) TPP-Cl at three-phase electrode [B271]

Cations	$\Delta G^{(\mathrm{nb})\ominus}_{(\mathrm{aq}),\ \mathrm{Cat}^+}$/ kJ mol^{-1}
K^+	22.65
Rb^+	19.80
Tl^+	19.30
Cs^+	17.80
TMA^+	9.60
TEA^+	–0.50
TBA^+	–8.20
$THxA^+$	–8.21
$THpA^+$	–8.78
TOA^+	–10.10

arrangement gives rise to well-developed square-wave voltammetric signals with a peak potential depending on the nature of the cations present in the aqueous phase [B271]. The more lipophilic the cations in the aqueous phase are, i.e., the more *positive* the standard potentials of cation transfer from water to the organic liquid are, the easier is the reduction of [Fe(III)TPP-Cl]. As predicted by Eq. 6.15, this shifts the reduction to more positive potentials. Representative voltammograms showing the transfer of cations across the water–NB interface are shown in Fig. 6.32.

The standard Gibbs energies of transfer of several common inorganic cations as well as of some tetraalkyl ammonium cations across the water–NB interface have been determined using this approach. The data are given in Table 6.7.

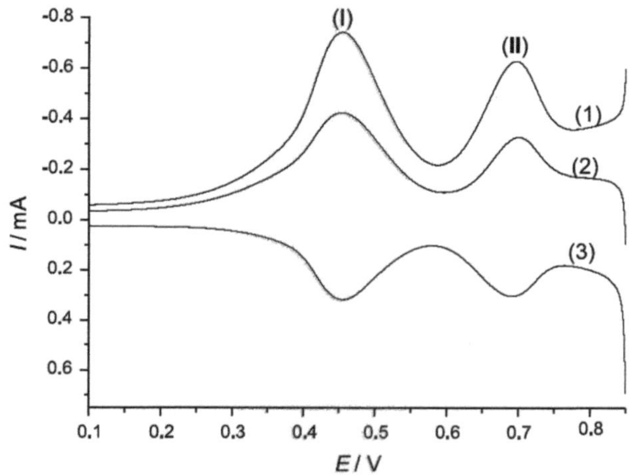

Fig. 6.33 The net (1), forward (2), and backward (3) components of the square-wave voltammetric response of a nitrobenzene droplet containing 0.1 mol L^{-1} iodine attached to the paraffin-impregnated graphite electrode and immersed in 0.25 M aqueous solution of NaCl: frequency $f = 50$ Hz, SW amplitude $E_{sw} = 50$ mV, scan increment $dE = 0.15$ mV, and starting potential $E_s = +0.90$ V [B234]. Copyright ©2002 with permission from Elsevier

Standard Gibbs Energies of Transfer of Cations Determined by Utilizing the Reduction of Iodine

The reduction of iodine dissolved in the organic droplet of a three-phase electrode is a reversible process, characterized by well-defined square-wave voltammograms. With aqueous solutions that do not contain any halide anions but different cations, the electrode reaction of iodine is accompanied by the expulsion of created iodide anions from the nitrobenzene to the aqueous phase [B215]:

$$I_{2(nb)} + 2e^- \leftrightarrows 2I^-_{(nb)} \quad \text{(XVII)}$$

$$I^-_{(nb)} \leftrightarrows I^-_{(aq)} \quad \text{(XVIII)}$$

An interesting effect is observed when the electrode with the attached NB droplet containing dissolved iodine is immersed in aqueous solutions of *chlorides* of different cations [B234]. Additionally to the previous peak [Reactions (XVII) and (XVIII); peak I in Fig. 6.33], a new reversible process is observed at more positive potentials (peak II in Fig. 6.33). The last peak shifts about 60 mV in negative direction per decade increase of chloride concentration in the aqueous phase [B215, B234], and it is attributed to the reduction of iodine in the nitrobenzene droplet followed by expulsion of chloride ions from nitrobenzene into the aqueous phase.

The electrochemical behavior of the system can be explained by assuming a preceding chemical reaction, i.e., the partition of the chloride salt between the aqueous and the nitrobenzene phases, mainly driven by the formation of I_2Cl^- ions

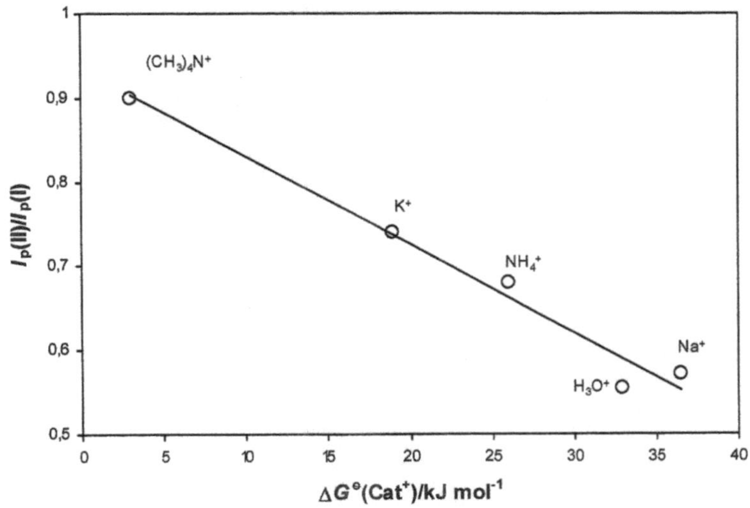

Fig. 6.34 Correlation of the peak currents ratio with the standard Gibbs energies of transfer of cations across the water–nitrobenzene interface. The other conditions are the same as in Fig. 6.33 [B234]. Copyright ©2002 with permission from Elsevier

in nitrobenzene [B234]. The ratio of currents of peaks II and I, i.e., $I_p(II)/I_p(I)$, increases linearly with increasing chloride concentration in the water phase. This is mainly due to the absolute increase of $I_p(II)$. In chloride solutions with different cations, the peak current ratio depends linearly on the standard Gibbs energies of transfer of the cations across the water–nitrobenzene interface (Fig. 6.34). The overall reaction associated with peak II is of CE type (electrochemical reaction coupled to a preceding chemical reaction) [B234] and can be formulated as follows:

$$I_{2(nb)} + Cl^-_{(aq)} + Cat^+_{(aq)} \leftrightarrows I_2Cl^-_{(nb)} + Cat^+_{(nb)} \quad \text{(IXX)}$$

$$I_{2(nb)} + 2e^- \leftrightarrows 2I^-_{(nb)} \quad \text{(XX)}$$

$$Cl^-_{(nb)} \leftrightarrows Cl^-_{(aq)} \quad \text{(XXI)}$$

The equation corresponding to the calibration line given in Fig. 6.34 can be used to determine the standard Gibbs energies of transfer of cations present in aqueous chloride solutions. This approach has been used to determine these data of several amino acid cations for the transfer across the water–nitrobenzene interface [B234, B262]. The application of the iodine method is limited by the high reactivity of iodine toward organic compounds as well as by the complexity of the entire mechanism.

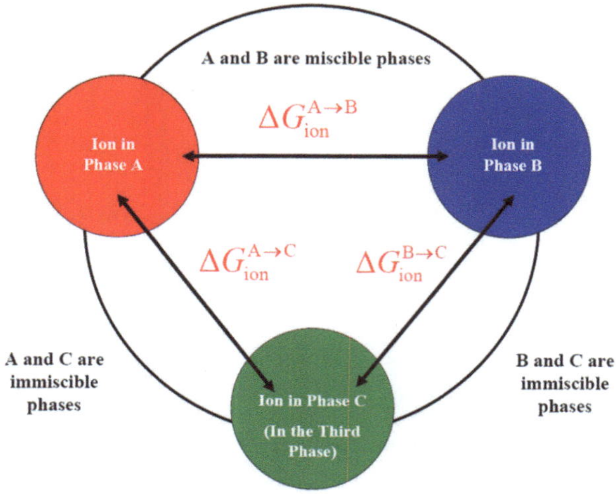

Fig. 6.35 Schematic situation for the determination of free energies of ion transfer between miscible solutions (here the phases A and B) by using a third phase C which is either a solid phase insoluble in A and B or a liquid phase immiscible with A and B

6.2.3 The Determination of Standard Gibbs Energies of Transfer of Anions and Cations Between Miscible Solvents: The "Third Phase Strategy"

In Sect. 5.6.1, the insertion electrochemistry of solid hexacyanometalates is explained, where charge compensating cations are transferred between the solid and the adjacent solution. Sections 6.2.1 and 6.2.2 present the basics of ion transfer between immiscible solutions. Here it will be shown that electrochemical measurements with immobilized particles and droplets can be also applied for the determination of ion transfer energies between miscible solutions: Figure 6.35 explains schematically the approach used in such determinations.

When the two liquid phases A and B are miscible, it is impossible to transfer ions between these phases following the simple strategy outlined in Sects. 6.2.1 and 6.2.2. However, it is easy to solve that problem by using a third phase (C), which, when liquid, needs to be immiscible with A and B and, when solid, needs to be insoluble in A and B. This may be called the "third phase strategy," and it can be well illustrated by the determination of ion transfer energies between water and heavy water, two miscible phases: Figure 6.36 shows the scheme of experiments.

Figure 6.37 shows the application of the "third phase strategy" for cases where the third phase is solid. Doménech-Carbó et al. [B360, B361, B420] have discovered that the solid heterotrimetallic Au(I)–Cu(I) compound $[\{Au_3Cu_2(C_2C_6H_4Fc)_6\}Au_3(PPh_2C_6H_4PPh_2)_3](PF_6)_2$ (=L) exhibits a well-defined insertion electrochemistry using anions for the charge compensating ion transfer between a solution and the solid. The cationic unit $[\{Au_3Cu_2(C_2C_6H_4Fc)_6\}$

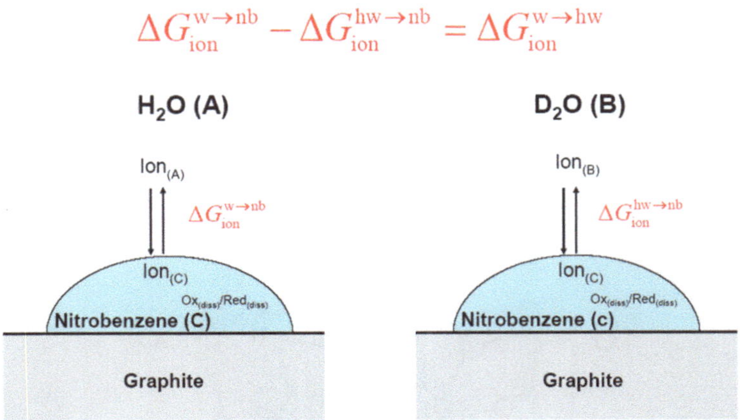

Fig. 6.36 Scheme of the experiments to determine the free energies of ion transfer between water (w) and heavy water (hw) using nitrobenzene as third phase, which is immiscible both with water and heavy water [B277]

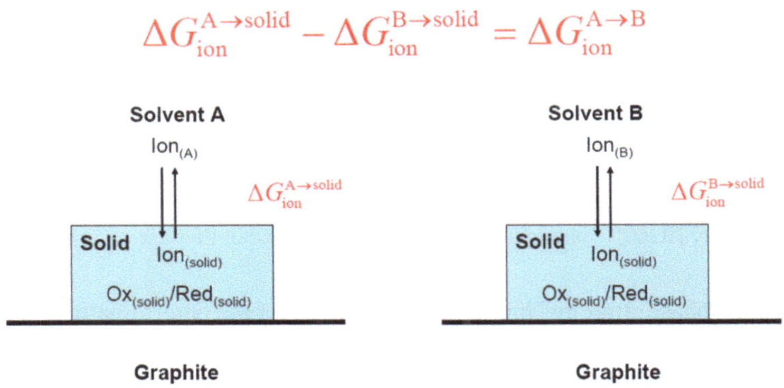

Fig. 6.37 Scheme of the experiments to determine the free energies of ion transfer between miscible solvents A and B using a solid compound as third phase

$Au_3(PPh_2C_6H_4PPh_2)_3]^{2+}$ consists of the alkynyl clusters $[Au_3Cu_2(C_2C_6H_4Fc)_6]^-$ surrounded by the cationic $[Au_3\ (PPh_2C_6H_4PPh_2)_3]^{3+}$ "belt" (cf. Fig. 6.38). The electrochemistry of L dissolved in nonaqueous solvents is dominated by the oxidation of ferrocenyl units. By using the solid compound $[\{Au_3Cu_2(C_2C_6H_4Fc)_6\}$ $Au_3(PPh_2C_6H_4PPh_2)_3](PF_6)_2$, it has been possible to determine the Gibbs free energies of ion transfer from water to methanol and acetonitrile [B421].

In Table 6.8, a compilation of ion transfer energies is given, and the comparison with literature data indicates the reliability of the "third phase strategy."

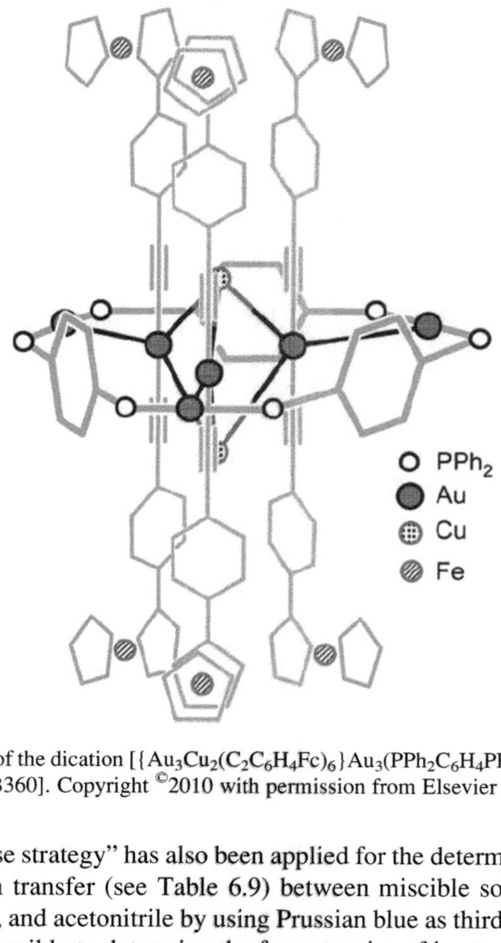

Fig. 6.38 Structure of the dication $[\{Au_3Cu_2(C_2C_6H_4Fc)_6\}Au_3(PPh_2C_6H_4PPh_2)_3]^{2+}$. Phenyl rings omitted for clarity [B360]. Copyright ©2010 with permission from Elsevier

The "third phase strategy" has also been applied for the determination of the free energies of cation transfer (see Table 6.9) between miscible solvents like water, methanol, DMSO, and acetonitrile by using Prussian blue as third phase [B422]. Of course it is also possible to determine the free energies of ion transfer between the pure solvents and solvent mixtures, as well as between different solvent mixtures. Further, solvation energies can be determined for all these systems. Figure 6.39 shows the excess free energies of ion solvation of sodium and potassium ions in water–DMSO mixtures.

The "third phase strategy" opens also a road to estimate solvent-independent redox potentials. On the basis of a theoretical modeling of the diffusion processes taking place under the conditions of voltammetry of immobilized particles, voltammetric and chronoamperometric data can be combined to obtain solvent-independent electrode potentials for the K^+-assisted one-electron reduction of Prussian blue to Berlin white [B423, B424, B425].

Table 6.8 Comparison of Gibbs free energies of anion transfer from water to methanol (MeOH) and acetonitrile (MeCN) from voltammetric experiments at alkynyl-diphosphine dinuclear Au(I) complexes and heterometallic Au(I)–Cu(I) cluster complexes[a] [B421] with literature[b, c] data

Anion	$\Delta G_t^\circ(w \rightarrow MeOH)$	$\Delta G_t^\circ(w \rightarrow MeCN)$
F^-	26 ± 2^a	65 ± 5^a
Cl^-	13.5 ± 1.2^a (12.5^b 13.7^c)	36 ± 2^a (37.6^b 31.4^c)
Br^-	11.5 ± 1.5^a (10.4^b 11.4^c)	36 ± 2^a (37.6^b 31.4^c)
NO_3^-	2.0 ± 0.5^a	12.0 ± 0.6^a
ClO_4^-	5.8 ± 0.2^a (5.7^c)	7.2 ± 0.4^a (4.0^c)
SO_4^{2-}	36 ± 2^a	52 ± 3^a
HCO_3^-	21 ± 1^a	32 ± 2^a
CO_3^{2-}	41 ± 3^a	126 ± 12^a
$H_2PO_4^-$	29 ± 3^a	
PO_4^{3-}	158 ± 12^a	

[a][B421]
[b]Kelly CP, Cramer, CJ, Truhlar DG (2007) J Phys Chem B 111:408–422
[c]Cox BG, Waghorne WE (1980) Chem Soc Rev 9:381–411

Table 6.9 Gibbs free energy of cation transfer from water to MeOH, MeCN, and DMSO $\Delta G^{\ominus}_{w \rightarrow org}$, kJ/mol) obtained for Li^+, Na^+, and K^+ ions using Prussian blue-modified electrodes immersed into different electrolytes [B422]. In parentheses are those theoretically calculated by Tissandier et al.[a]

Ion	$\Delta G^{\ominus}_{w \rightarrow MeOH}(M^+)$ (kJ/mol)	$\Delta G^{\ominus}_{w \rightarrow MeCN}(M^+)$ (kJ/mol)	$\Delta G^{\ominus}_{w \rightarrow DMSO}(M^+)$ (kJ/mol)
Li^+	5.2 ± 0.4	9.1 ± 0.6	-19.0 ± 0.4
Na^+	8.7 ± 0.4 (8.8)	8.6 ± 0.8 (8.8)	-21.4 ± 1.0 (−23.8)
K^+	11.5 ± 0.6 (11.3)	7.1 ± 0.8 (2.1)	-21.9 ± 0.4 (−25.9)

[a]Tissandier, MD, Cowen KA, Feng WY, Gundlach E, Cohen MH, Earhart AD, Coe JV (1998) The Protons Absolute Aqueous Enthalpy and Gibbs Free Energy of Solvation from Cluster-Ion Solvation Data. J Phys Chem A 102:7787–7794

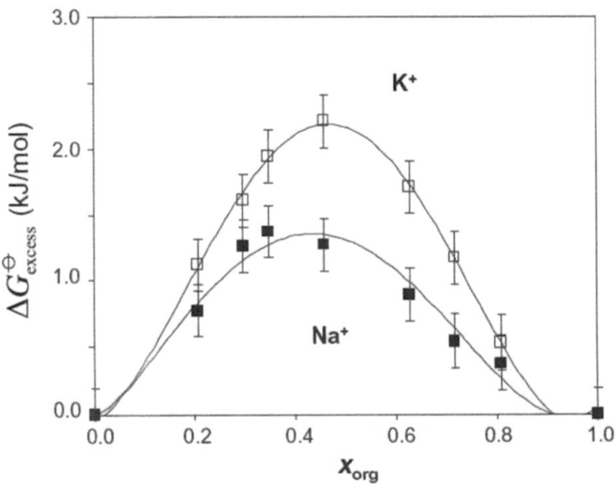

Fig. 6.39 Excess free energy of solvation calculated from electrochemical data for Na^+ and K^+ ions in water–DMSO binary mixtures [B422]. Copyright ©2011 with permission from Elsevier

6.3 Theoretical Models for the Voltammetry of Immobilized Droplets

The modeling of the processes occurring at the droplet-modified electrodes is not a simple task since several effects have to be taken into account. The mass transfer is for sure achieved by diffusion and migration. Convection may play a role, perhaps as Marangoni convection. Due to the absence of supporting electrolyte in the organic phase at the start of the experiments, the ohmic drop can play a role. These effects have been studied theoretically by different groups and some progress has been achieved.

6.3.1 Droplets of Organic Solvents with Dissolved Electroactive Compound

Semi-infinitive Planar Diffusion Model

Considering the reaction

$$\mathrm{dmfc}_{(o)} + \mathrm{A}^{-}_{(aq)} \leftrightarrows \mathrm{dmfc}^{+}_{(o)} + \mathrm{A}^{-}_{(o)} + \mathrm{e}^{-} \qquad \text{(XXII)}$$

and following the analogy with the theoretical model for electron transfer across the liquid–liquid interface developed by Stewart et al.[11], a theoretical model has been developed taking into account the diffusion of all species [B174, B188]. The model considers a mass transfer through semi-infinite planar diffusion and was solved under the following initial and boundary conditions:

$$\left.\begin{aligned} t = 0;\ x \geq 0;\ c_{\mathrm{dmfc}_{(o)}} &= c^{*}_{\mathrm{dmfc}_{(o)}} \\ c_{\mathrm{A}^{-}_{(aq)}} &= c^{*}_{\mathrm{A}^{-}_{(aq)}} \\ c_{\mathrm{A}^{-}_{(o)}} = c_{\mathrm{dmfc}^{+}_{(o)}} &= 0 \end{aligned}\right\} \qquad (6.16)$$

$$\left.\begin{aligned} t = 0;\ x \to \infty;\ c_{\mathrm{dmfc}_{(o)}} &\to c^{*}_{\mathrm{dmfc}_{(o)}} \\ c_{\mathrm{A}^{-}_{(aq)}} &\to c^{*}_{\mathrm{A}^{-}_{(aq)}} \\ c_{\mathrm{A}^{-}_{(o)}} = c_{\mathrm{dmfc}^{+}_{(o)}} &\to 0 \end{aligned}\right\} \qquad (6.17)$$

[11] Stewart AA, Campbell JA, Girault HH, Eddowes M (1990) Ber Bunsen Ges Phys Chem 94:83–87

$$\left.\begin{array}{l} t>0 \\ \dfrac{\partial c_{\mathrm{dmfc}_{(\mathrm{o})}}}{\partial t}=D_{\mathrm{dmfc}_{(\mathrm{o})}}\dfrac{\partial^2 c_{\mathrm{dmfc}_{(\mathrm{o})}}}{\partial x^2};\quad \dfrac{\partial c_{\mathrm{dmfc}^+_{(\mathrm{o})}}}{\partial t}=D_{\mathrm{dmfc}^+_{(\mathrm{o})}}\dfrac{\partial^2 c_{\mathrm{dmfc}^+_{(\mathrm{o})}}}{\partial x^2} \\ \dfrac{\partial c_{\mathrm{A}^-_{(\mathrm{aq})}}}{\partial t}=D_{\mathrm{A}^-_{(\mathrm{aq})}}\dfrac{\partial^2 c_{\mathrm{A}^-_{(\mathrm{aq})}}}{\partial x^2};\quad \dfrac{\partial c_{\mathrm{A}^-_{(\mathrm{o})}}}{\partial t}=D_{\mathrm{A}^-_{(\mathrm{o})}}\dfrac{\partial^2 c_{\mathrm{A}^-_{(\mathrm{o})}}}{\partial x^2} \end{array}\right\} \quad (6.18)$$

For $x = 0$, the fluxes are given by

$$\left.\begin{array}{l} \dfrac{\partial c_{\mathrm{dmfc}_{(\mathrm{o})}}}{\partial x}=\dfrac{\partial c_{\mathrm{dmfc}^+_{(\mathrm{o})}}}{\partial x}=\dfrac{I}{nFSD} \\ \dfrac{\partial c_{\mathrm{A}^-_{(\mathrm{aq})}}}{\partial x}=\dfrac{I}{nFSD_{\mathrm{A}^-_{(\mathrm{aq})}}} \\ \dfrac{\partial c_{\mathrm{A}^-_{(\mathrm{o})}}}{\partial x}=\dfrac{I}{nFSD} \end{array}\right\} \quad (6.19)$$

(it is assumed that the diffusion coefficients of A^- and $dmfc^+$ in the organic phase are the same, i.e., D). In this model, the distance x represents a distance to the three-phase junction, i.e., the position where the reaction can start.

The integral solutions for all species involved in Reaction (XXII), which were obtained by means of Laplace transformation, read

$$\left.\begin{array}{l} c_{\mathrm{dmfc}_{(\mathrm{o})}(x=0)}=c^*_{\mathrm{dmfc}_{(\mathrm{o})}}-\dfrac{1}{\sqrt{\pi}}\displaystyle\int_0^t\dfrac{I(\tau)}{nFS\sqrt{D}}(t-\tau)^{-0.5}\mathrm{d}\tau \\ c_{\mathrm{dmfc}^+_{(\mathrm{o})}(x=0)}=\dfrac{1}{\sqrt{\pi}}\displaystyle\int_0^t\dfrac{I(\tau)}{nFS\sqrt{D}}(t-\tau)^{-0.5}\mathrm{d}\tau \\ c_{\mathrm{A}^-_{(\mathrm{aq})}(x=0)}=c^*_{\mathrm{A}^-_{(\mathrm{aq})}}-\dfrac{1}{\sqrt{\pi}}\displaystyle\int_0^t\dfrac{I(\tau)}{nFS\sqrt{D}}(t-\tau)^{-0.5}\mathrm{d}\tau \\ c_{\mathrm{A}^-_{(\mathrm{o})}(x=0)}=\dfrac{1}{\sqrt{\pi}}\displaystyle\int_0^t\dfrac{I(\tau)}{nFS\sqrt{D}}(t-\tau)^{-0.5}\mathrm{d}\tau \end{array}\right\} \quad (6.20)$$

The solutions given in Eq. 6.20 were substituted in the below Nernst equation:

$$\left.\begin{array}{l} \dfrac{c_{\mathrm{A}^-_{(\mathrm{o})}(x=0)}c_{\mathrm{dmfc}^+_{(\mathrm{o})}(x=0)}}{c_{\mathrm{A}^-_{(\mathrm{aq})}(x=0)}c_{\mathrm{dmfc}_{(\mathrm{o})}(x=0)}}=\exp(\phi),\ \text{where} \\ \phi=\left(E-E^{\ominus}_{\mathrm{dmfc}^+_{(\mathrm{o})}/\mathrm{dmfc}_{(\mathrm{o})}}-\Delta\phi^{(\mathrm{o})\,\ominus}_{(\mathrm{aq}),\,\mathrm{A}^-}\right)\left(\dfrac{F}{RT}\right) \end{array}\right\} \quad (6.21)$$

Equation 6.21 was solved under conditions of cyclic [B188] and square-wave voltammetry [B174]. The concentration gradients of all species involved in the redox process are given in Fig. 6.40.

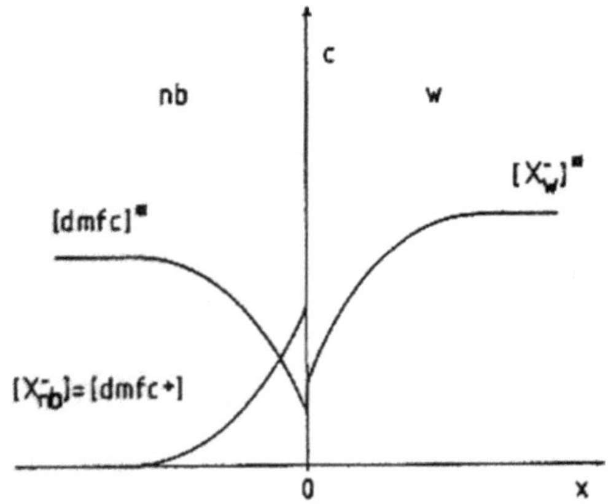

Fig. 6.40 A scheme of concentrations of active species in the organic [here nitrobenzene (NB)] and water (w) phases in the vicinity of the three-phase junction [B174]. Copyright ©2002 with permission from publisher

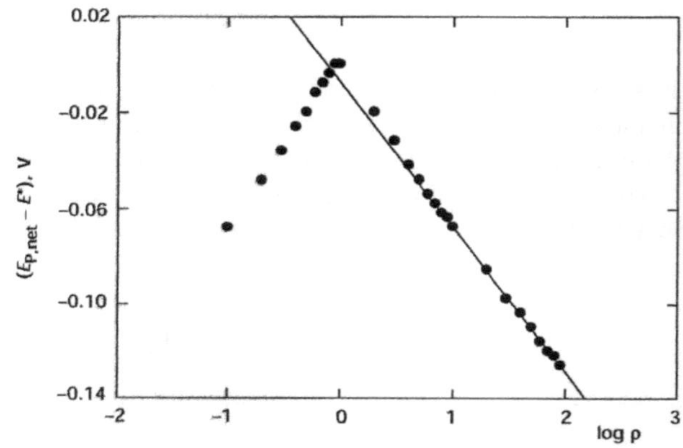

Fig. 6.41 Relationship between the theoretical square-wave voltammetric peak potentials and the logarithm of the dimensionless parameter ρ [B174]. Copyright ©2002 with permission from publisher

The formal potentials (i.e., the mid-peak potentials of the cyclic voltammograms or the peak potentials of the net square-wave voltammograms) of the simulated voltammograms depend on one dimensionless parameter ρ, defined as a ratio between the initial concentrations of anions in the aqueous phase and dmfc in the organic phase, i.e., $\rho = c^*_{\mathrm{A}^-_{(aq)}}/c^*_{\mathrm{dmfc(o)}}$. If $c^*_{\mathrm{A}^-_{(aq)}}\sqrt{D} \geq c^*_{\mathrm{dmfc(o)}}\sqrt{D}$, then the dependence of the formal potential on the logarithm of the dimensionless parameter ρ is linear, with a slope of $\mathrm{d}E^{\ominus\prime}_{\mathrm{c}}/\mathrm{dlog}(\rho) = -60$ mV (at 25 °C), which is identical with the slope in Equation 6.14, i.e., in an equation derived on a purely thermodynamic basis. This behavior has also been experimentally observed [B132, B174, B206, B240, B259, B262, B266].

However, when $c^*_{\mathrm{A}^-_{(aq)}}\sqrt{D_{\mathrm{A}^-_{(aq)}}} < c^*_{\mathrm{dmfc(o)}}\sqrt{D}$, then the slope of the dependence d $E^{\ominus\prime}_{\mathrm{c}}/\mathrm{dlog}(\rho) = +60$ mV (see Fig. 6.41). This behavior has not been observed in the experiments, which suggests that the experiments are more complicated and affected by additional effects.

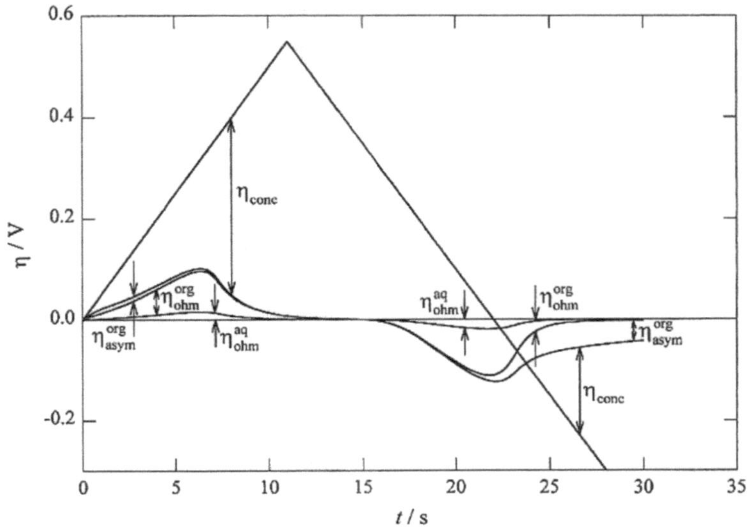

Fig. 6.42 Schematic drawing of the time dependence of all overpotentials affecting the voltammetric response in the Oldham model [B237]. Copyright ©2002 with permission from Elsevier

A Convolutive Model for Cyclic Voltammetry in a Thin Organic Layer

Considering the reaction

$$N_{(o)} + A^{-}_{(aq)} \leftrightarrows P^{+}_{(o)} + A^{-}_{(o)} + e^{-} \qquad \text{(XXIII)}$$

occurring in an insulating organic layer with surface area A and thickness Z, that is, sandwiched between the working electrode and an aqueous electrolyte solution, Oldham et al. [B237] have developed a rather robust model, comprising several effects which affect the voltammetric features of Reaction (XXIII). By assuming that the ion and electron transfer in Reaction (XXIII) occurs without kinetic constraints, i.e., there is no activation polarization, the flow of current in such a setup may induce at least three different kinds of polarization: the concentration polarization, the ohmic polarization (in both organic and aqueous phase), and an asymmetry polarization due to the different diffusivities in the organic and aqueous phases. Each polarization can generate an overvoltage which will be reflected throughout the features of the obtained theoretical responses. A schematic drawing of all the overpotentials considered in the model [B237] is given in Fig. 6.42.

It is a complicate task to predict the behavior of such complex systems under potentiodynamic conditions. The transport of four species, i.e., $N_{(o)}$, $A^{-}_{(aq)}$, $P^{+}_{(o)}$, and $A^{-}_{(o)}$, must be modeled, three of them migrating and all of them diffusing. Another major problem is the existence of a significant time-dependent resistance in the organic layer and the likely existence of a diffusion potential difference

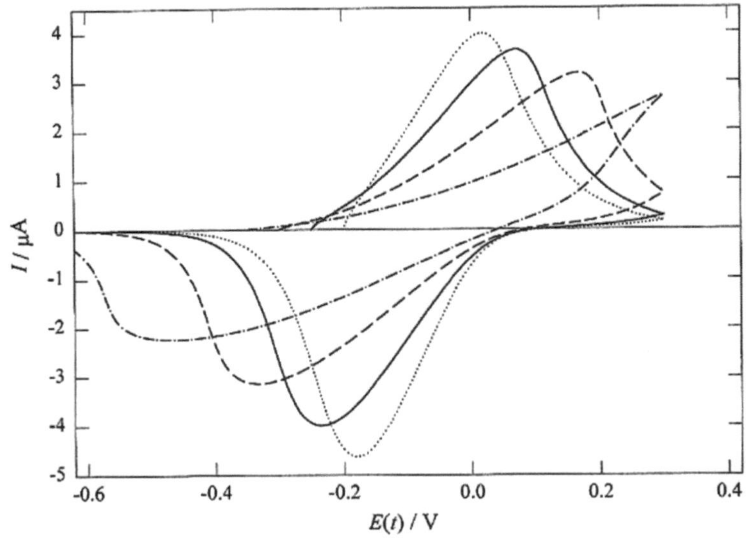

Fig. 6.43 Cyclic voltammograms simulated for the thin-film voltammetry obtained by varying the starting potentials of $E_{in} = -200$ mV (*dotted line*), -250 mV (*solid line*), -300 mV (*dashed line*), and -350 mV (*chained line*). The other simulation conditions are $E^{\ominus} = 0$ V, $E_{reverse} = 300$ mV, $v = 50$ mVs^{-1}, $A = 10^{-5}$ m^2, $Z = 20$ μm, $c^{salt}_{(aq)} = 0.01$ mol L^{-1}, $c_{N_{(o)}} = 0.001$ mol L^{-1}, $D_N = D_{P^+} = 10^{-10}$ m^2 s^{-1}, $D_{A^-_{(o)}} = 8 \times 10^{-10}$ m^2 s^{-1}, $D_{A^-_{(aq)}} = 10^{-9}$ m^2 s^{-1} [B237]. Copyright ©2002 with permission from Elsevier

(asymmetry overvoltage) across the organic layer. By developing a rather complicated algorithm, the authors succeeded to elucidate all these effects theoretically.

On the contrary to common cyclic voltammetry experiments, where the choice of the rest (starting) potential that is imposed on the working electrode is innocuous, in this theoretical model the choice of the rest potential is the most important parameter. It determines the initial ionic content in the organic layer and hence strongly affects the ohmic polarization. For example, the concentration of P^+A^- in the organic phase, created by choosing a rest potential of -250 mV more negative than $E^{\ominus}_{N/P^+}$, will be about 5 mm (if $c_{N_{(o)}} = 0.05$ M and $c^{salt}_{(aq)} = 0.5$ M), which is far from being negligible. Indeed, the question whether the center of the thin film will be active or not depends on the time of electrolysis as well as of the film thickness. Several theoretical voltammograms simulated for different values of the starting potential are depicted in Fig. 6.43. Obviously, the obtained theoretical voltammograms differ from the conventional ones. The separation of the peak potentials is much greater than the 59 mV typical for reversible systems in cyclic voltammetry under semi-infinite planar diffusion. This is attributed mainly to the ohmic resistance in the organic layer. Figure 6.43 shows that the more positive the starting potential is, the lower will be the resistance in the organic phase due to the higher ionic content there, and consequently the simulated voltammograms are

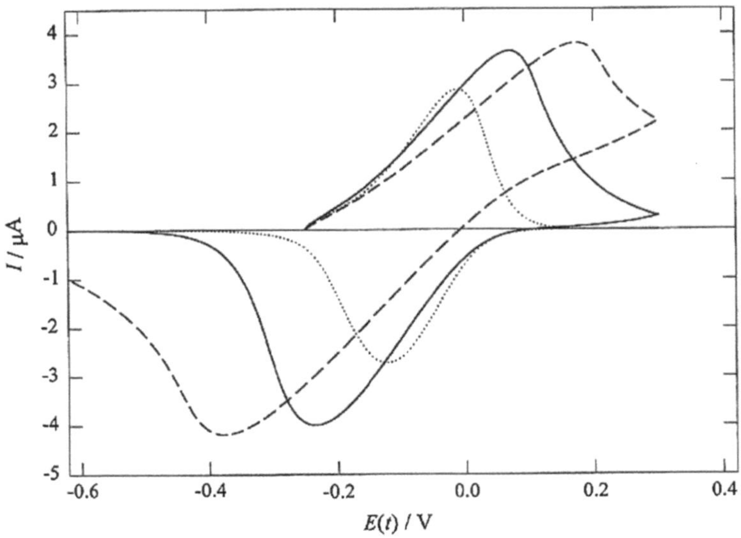

Fig. 6.44 Theoretical cyclic voltammograms obtained in thin-film voltammetry by varying the thickness of the organic film: $Z = 10$ μm (*dotted line*), 20 μm (*solid line*), and 40 μm (*dashed line*). The rest potential was $E_{in} = -250$ mV. All other conditions were the same as in Fig. 6.38 [B237]. Copyright ©2002 with permission from Elsevier

getting closer to the conventional forms. The curve in Fig. 6.43 corresponding to an electrolysis potential of 350 mV more negative than $E^{\ominus}$ shows such high resistance that the voltammogram is dominated by ohmic polarization.

Voltammograms simulated for conditions that differ only in the width of the organic layer are shown in Fig. 6.44. As expected, the distortion due to the resistance becomes more pronounced as the layer becomes thicker. The features of the simulated voltammograms are additionally affected by the scan rate, by the concentration of the electroactive compound N in the organic phase, as well as by the concentration of the salt in the aqueous solutions. A detailed overview of all contributing overvoltages is given in reference [B237]. The interplay of the various factors makes the result of this type of experiment complicate. In such an experimental setup, without an existing three-phase junction line, caution is needed in drawing conclusions from CV data.

Modeling Cyclic Voltammograms at a Conic Drop Three-Phase Electrode

A model that more closely approaches the processes occurring at a three-phase electrode has been developed using the geometry of a conic "drop" (cf. Fig. 6.45)

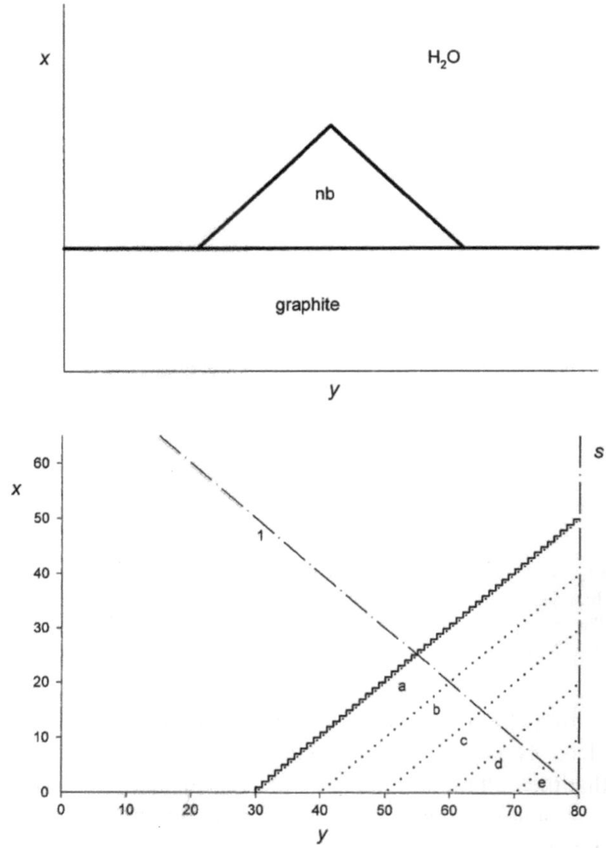

Fig. 6.45 Scheme of a conic drop of an organic solvent (nitrobenzene) attached to a graphite electrode surface and immersed in an aqueous electrolyte (**a**) and a radial section of the conic drop electrode in the Cartesian plane (**b**). The axis of radial symmetry passing through the top of the cone is denoted by *s* [B257]. Copyright ©2003 with permission from Elsevier

[B257]. Considering Reaction (XVI), i.e.,

$$\mathrm{dmfc}_{(\mathrm{nb})} + \mathrm{A}^-_{(\mathrm{aq})} \leftrightarrows \mathrm{dmfc}^+_{(\mathrm{nb})} + \mathrm{A}^-_{(\mathrm{o})} + \mathrm{e}^-$$

occurring at a three-phase electrode (where dmfc is dissolved in nitrobenzene that is devoid of any electrolyte), the major obstacle is the origin of the initial conductivity in the organic phase, which is needed for initiating the reaction. Contrary to the model of Oldham et al. [B237], where the authors have assumed that the initial conductivity of the organic film is established as a result of the electrode Reaction (XVI) proceeding at potentials below the formal potential, the authors of [B257] have recognized in their model the *natural partition of the salt* between the aqueous solution and the organic drop as the reason for supplying the initial conductivity in the organic phase.

Before the onset of the CV experiments, when the nitrobenzene-modified electrode is immersed in the aqueous solution, the partition of the salt can occur under open circuit conditions. Certainly, the delay time before applying the potential will be a major factor determining how far the partitioned ions can penetrate the nitrobenzene phase. Assuming that concentrations of the salts in the aqueous solution are 1 M, from the partition constant across the water–nitrobenzene interface, one can estimate that in some cases the concentration of partitioned salts in nitrobenzene phase can exceed some mmol L^{-1}. This gives the model creditability.

The second parameter that affects the voltammetric features of Reaction (XVI) in this model is certainly the resistance in the organic phase. In this model, the ohmic overvoltage has been calculated as a product of the dimensionless current and the equivalent of the resistance ρ, defined as

$\rho = \frac{RT}{2dF}\sum_{i=1}^{n}\left(1/c_{A^-_{(NB)}}\right)$, where $d = D\frac{\Delta t}{\Delta x^2}$, $n = \frac{L}{\Delta x}$. Δt and Δx are the time and space increments, respectively; L is the thickness of the film; and D is the diffusion coefficient (assumed to be equal for all species in the organic film).

The effect of the thickness of the conic film on the shape of the simulated CV curves and on the equivalents of the resistance is shown in Figs. 6.46a–c. The components of Fig. 6.46b correspond to an increase of the thickness of the conic film from 10 to 40 space increments in the direction parallel to the axis of symmetry s. Obviously, for higher space increments (which correspond to higher values of the equivalent resistance), the ohmic effect is dominating the shape of the simulated voltammograms (see curve 4 in Fig. 6.46b). Moreover, this also influences the mid-peak potentials of the square-wave voltammograms. The equivalence of the resistance ρ corresponding to the same parts of the organic film as the currents in Fig. 6.46b as a function of the applied potential is given in Fig. 6.46c. The dependence of ρ on the electrode potential shows that the ingress of anions from water starts at the edge of the cone. During the anodic polarization (forward scan), the equivalence of the resistance ρ decreases with increasing applied potential. In the reverse scan of the cyclic voltammograms, the equivalence of the resistance does not reach the initial value (see Fig. 6.46c). This is probably caused by the lateral diffusion of the created decamethylferrocenium cations from the center of the cone toward its edge. After the first cycle of the potential, the conductivity of the cone is significantly enhanced. Consequently, the second and further cyclic voltammograms should be less affected by the ohmic drop than the first scan. Three successive cyclic voltammograms of the conic drop simulated under

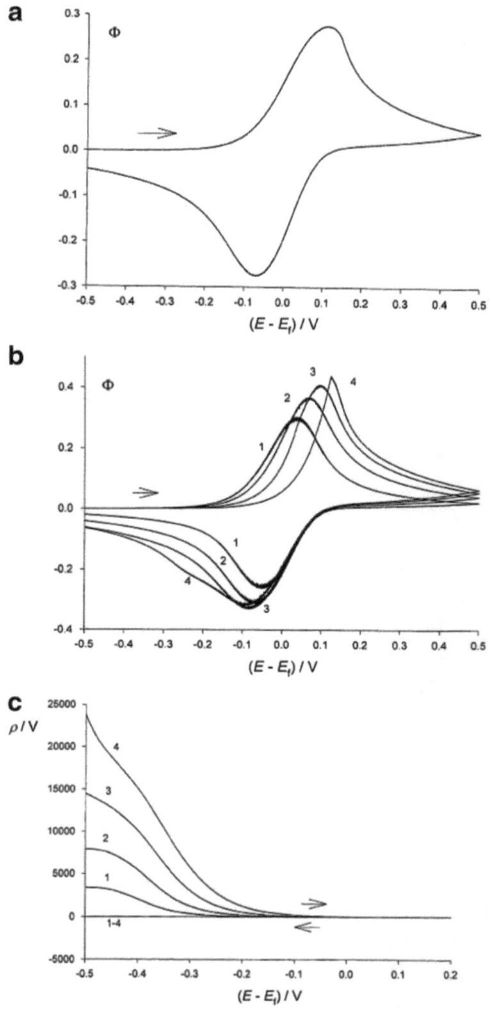

Fig. 6.46 (**a**) Cyclic voltammogram of Reaction (XVI) in a conic drop of nitrobenzene, (**b**) the components of the average current, and (**c**) the equivalents of the resistance that correspond to the parts of the conic drop with a thickness of 10 (1), 20 (2), 30 (3), and 40 (4) space increments in the direction of x axis. The duration of the initial contact was 1,000 time increments, and the partition constant of the salt MX was $K_{MX}^{w \rightarrow nb} = 2 \times 10^{-4}$ [B257]. Copyright ©2003 with permission from Elsevier

conditions of Fig. 6.46 are depicted in Fig. 6.47. The second and third cycles in Fig. 6.42 are identical, which confirms the former statement. The anodic peak potential of the first scan is more positive than that of the 2nd and 3rd scans due to the influence of the ohmic drop in the first cycle. This is a very important

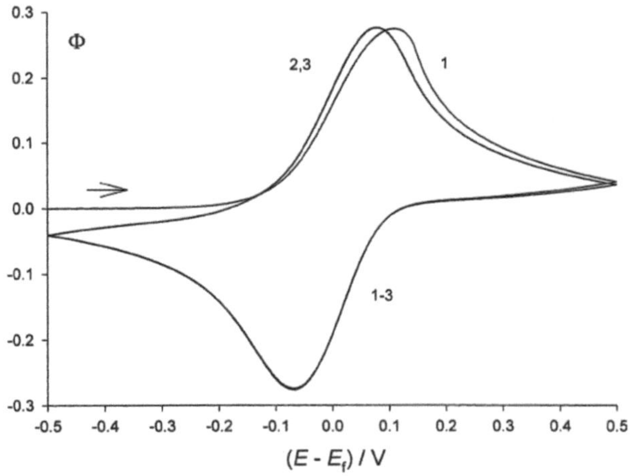

Fig. 6.47 Three consecutive cyclic voltammograms of the redox Reaction (XVI) considering a conic drop of nitrobenzene simulated with 10 space increments. All other conditions are the same as in Fig. 6.46 [B257]. Copyright ©2003 with permission from Elsevier

observation since it shows that the ohmic overvoltage does not have an influence on the mid-peak potential of the second and all other subsequent cycles in multiple cyclic voltammograms of reaction XVI. This influence was significantly pronounced in the model of Oldham et al. [B237].

Uncompensated Resistance and Charge Transfer Kinetics

The theoretical model of a conic drop three-phase electrode as described above implies that the electrode process is not confined to the three-phase junction line, where it starts, but that it advances into the droplet. Depending on the contact time of the two adjacent liquid phases under open circuit condition, as well as depending on the scan rate, the thickness of the reaction zone in the organic droplet will vary. An immobilized droplet can be regarded as a film with a thickness varying from zero to rather large values. The fact that the droplet has zero thickness at the three-phase junction line enables the electron transfer reaction to be started even with lowest conductivities of the organic phase, which is a big advantage over the conventional thin-film techniques[12,13]. The droplet can be treated theoretically as a thin-film electrode with a low conductivity of the film. When a fast technique like square-wave voltammetry (SWV) is used, the reaction will be restricted to a "thin-film" adjacent to the three-phase junction line. With this in mind, the effect of the uncompensated resistance on the voltammetric properties of a simple electrode reaction occurring in a limited diffusion space under conditions of SWV has been studied theoretically as well as experimentally [B281].

[12] Shi C, Anson FC (1999) J Phys Chem B 103:6283–6289

[13] Shi C, Anson FC (1998) Anal Chem 70:3114–3118

Table 6.10 Minimal values of the resistance parameter ρ_{min} that affect the SW voltammetric response of a reversible electrode reaction for different thicknesses of the film. The conditions of the simulations were diffusion coefficient $D = 10^{-5}$ cm^2 s^{-1}, SW frequency $f = 100$ Hz, SW amplitude $E_{sw} = 50$ mV, and scan increment d$E = 10$ mV [B281]

L/μm	ρ_{min}
1	0.1
2	0.03
5	0.04
10	2
50	3
500	4

Considering the reaction

$$\text{Ox} + ne^- \leftrightarrows \text{Red} \qquad \text{(XXIV)}$$

proceeding in a limited diffusion space ($0 \leq x \leq L$), the voltammetric properties of a reversible redox reaction in the absence of ohmic polarization are a function of one dimensionless thickness parameter Λ, defined as $\Lambda = L\sqrt{(f/D)}$, where L is the thickness of the film, f is the SW frequency, and D is the common diffusion coefficient. The dimensionless peak current is a sigmoid function of Λ, while the peak potential does not depend on Λ.

In the presence of a resistance in the thin film, the ohmic overvoltage of the system, owing to the low conductivity of the film, is represented by a complex dimensionless resistance parameter ρ, defined as $\rho = R_{\Omega}\left(n^2F^2/RT\right)Sc_{\text{Ox}}^{*}\sqrt{Df}$, where R_{Ω} the resistance in the film, S is the electrode surface area, and the meaning of other symbols is as usual. The influence of the resistance parameter ρ upon the voltammetric features of the reaction depends on the thickness of the film. The thicker the film, the smaller is the effect of the resistance parameter.

The minimal values of the resistance parameter that exhibit an influence on the voltammetric response of Reaction (XXIV) in the thin film are given in Table 6.10. For films with lower thicknesses, the increase of the resistance parameter ρ influences all the components of the SW voltammetric response (see Fig. 6.48).

By increasing the resistance parameter ρ, the symmetry of forward and backward components is lost, and the forward (cathodic) branch of the SW voltammograms shifts toward more positive potentials, whereas the backward (anodic) branch shifts toward more negative potentials. In this range of the values of ρ, the overall net current increases. This is due to the specific chronoamperometric properties of a thin-film electrode reaction. During each potential pulse of the SW voltammetric excitation signal, the current diminishes severely with time, as the equilibrium of Reaction (XXIV) is rapidly established between the species of the reversible couple that diffuse in a limiting space. Consequently, a low current remains for measuring at the end of each potential pulse. However, when a significant resistance is present in the thin film, the equilibrium cannot be

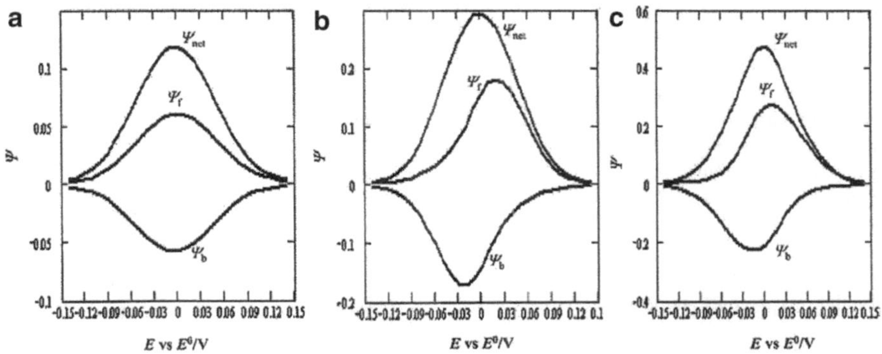

Fig. 6.48 The effect of the uncompensated resistance on the dimensionless SW response of a reversible electrode reaction. The resistance parameter was: $\rho = 0$ (**a**), 1 (**b**), and 3 (**c**). The other conditions of the simulations were: $\Lambda = 0.362$, $E_{sw} = 50$ mV, $D = 10^{-5}$ cm^2 s^{-1}, d$E = 10$ mV. Ψ_f, Ψ_b, and Ψ_{net} are symbols for forward, backward and net components of the SW voltammetric responses, respectively [B281]. Copyright ©2004 with permission from Elsevier

established so rapidly, causing the redox reaction to proceed to a large extent even at the end of the pulse and thus increasing the current compared with the situation without resistance (compare the net currents of Fig. 6.48c and b with that of a). In other words, the observed feature in the presence of uncompensated resistance is a consequence of the coupled properties of the studied reaction, of the chronoamperometric properties of the used technique (SWV), and of the delay of the mass transfer (caused by increased resistance of the thin film). The shifting of the cathodic branch toward positive potentials and of the anodic branch toward negative potentials with increasing ρ is opposite to the influence of the resistance parameter on the voltammetric features of the same reaction occurring under semi-infinite planar diffusion[14]. The effect of ρ on the dimensionless net-peak currents studied for films with different thicknesses is given in Fig. 6.49. For thicker films, a parabolic dependence exists between the dimensionless net-peak currents $\Delta\Psi_p$ and the logarithm of ρ. In the same direction, increasing ρ causes a growing of the half-peak width of the net square-wave voltammograms; however, the peak potentials remain unaltered. This effect of ρ in a reversible reaction occurring in thin-film voltammetry resembles completely the effect of charge transfer kinetics in a thin film without resistance (see Fig. 6.50). The kinetic parameter k in Fig. 6.50 is defined as $k = k_s/\sqrt{Df}$, where k_s (cm s^{-1}) is the standard rate constant of electron transfer. Considering a quasireversible reaction in a thin film under conditions of SWV, the so-called quasireversible maximum (see curves a, b, and c in Fig. 6.50) appears in films with a thickness parameter $\Lambda \leq 0.949$. Within the same range of Λ values ($\Lambda \leq 0.949$), a parabolic dependence of $\Delta\Psi_p$ on logρ appears (see Fig. 6.49). Since, in thin-film voltammetric experiments, the frequency of the SW voltammetric signal simultaneously influences both ρ and k, criteria must be

[14] Mirčeski V, Lovrić M (2001) J Electroanal Chem 497:114–124

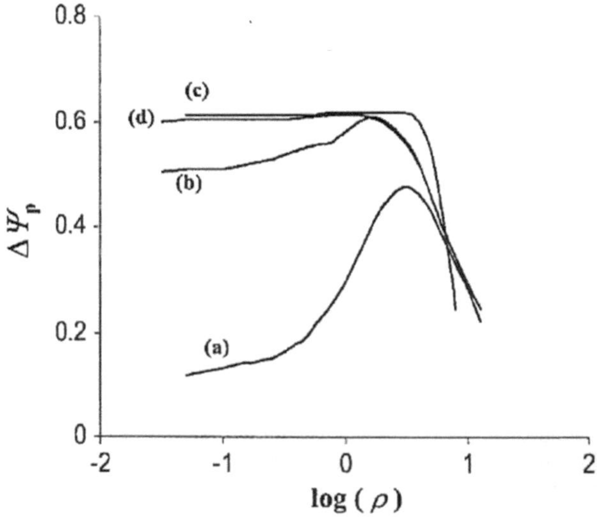

Fig. 6.49 Influence of the resistance parameter ρ on the dimensionless net-peak current $\Delta\Psi_p$ of a reversible electrode reaction for different thickness parameters. The thickness parameter was $\Lambda = 0.632$ (a), 0.948 (b), 1.581 (c), and 15.811 (d). The other conditions of the simulations were the same as in Fig. 6.48 [B281]. Copyright ©2004 with permission from Elsevier

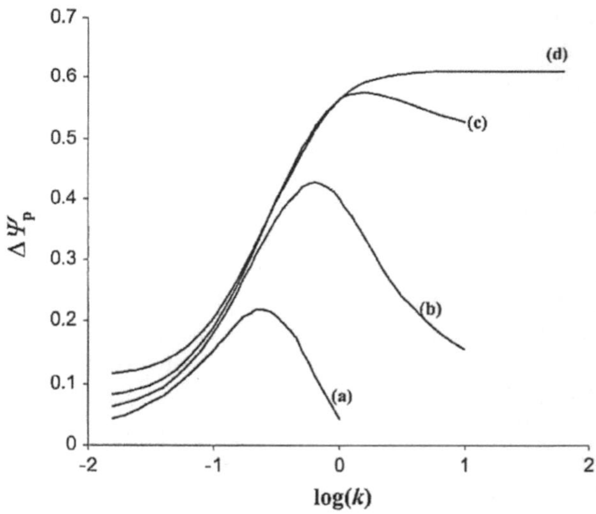

Fig. 6.50 Influence of the kinetic parameter k on the dimensionless net-peak current $\Delta\Psi_p$ of a quasireversible electrode reaction for different thickness parameters. The thickness parameter was $\Lambda = 0.316$ (a), 0.632 (b), 0.948 (c), and 1.581 (d). The other conditions of the simulations were the same as in Fig. 6.48 [B281]. Copyright ©2004 with permission from Elsevier

found to distinguish between the effect of ohmic overvoltage and that of the kinetics of charge transfer. *The shape of the net voltammograms and the half-peak width can serve as simple and safe criteria for this purpose.* The uncompensated resistance effect is associated with an increase of the half-peak width [B281] and, under extreme conditions, by a strong distortion of the SW voltammograms. On the contrary, the effect of charge transfer kinetics does not influence the half-peak width significantly. Both effects have been experimentally observed for two different types of reactions proceeding at three-phase electrodes [B281].

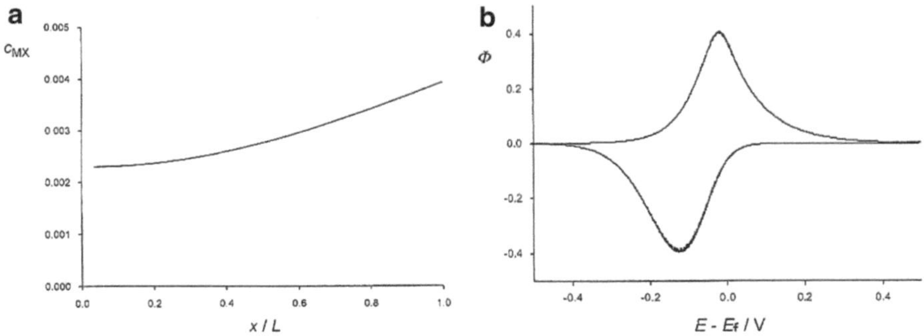

Fig. 6.51 (**a**) The dimensionless concentration of the electrolyte MX in the film at the end of the initial contact with aqueous solution and (**b**) a simulated cyclic voltammogram of decamethylferrocene in the film electrode $c^*_{MX,(NB)}/c^*_{dmfc} = 10^9$ (see footnote 15). Copyright ©2003 with permission from Elsevier

The Effect of Migration

Studying Reaction (XVI), i.e., $dmfc_{(nb)} + A^-_{(aq)} \leftrightarrows dmfc^+_{(nb)} + A^-_{(o)} + e^-$, in a thin-film arrangement, one fundamental question concerns the type of involved mass transfer. In the models described before, the diffusion was considered to be the major mass transport process (except in the model of Oldham et al. [B237] where migration was also taken into account). However, since, *in thin-film experiments*, the flux of $dmfc^+$ cations at the electrode surface and the flux of the anions at the liquid–liquid interface are separated by the thickness of the film, it must be migration of ionic species through the film that ensures the electroneutrality in the film. Recently, the effect of migration and distribution of all active species in the thin film have been theoretically analyzed[15]. Two models of thin-film voltammetry have been considered, one for the presence of a supporting electrolyte and one for the absence of supporting electrolyte in the organic film. Here, the second one is described since it is closer to the reality of the experiments considered in this book. In the model, the initial ionic conductivity of the film is achieved by free partition of the aqueous electrolyte MX between water and the film. The concentration of the partitioned salt in the film is rather small, since the partition constant is taken to be $K_{MX}^{w\rightarrow nb} = 4 \times 10^{-4}$. Figure 6.51a shows the profile of the dimensionless concentration of the electrolyte MX in the thin film after a short contact with the aqueous electrolyte. Figure 6.51b is a theoretical cyclic voltammogram of dmfc obtained in this setup. Dimensionless concentrations of dmfc, $dmfc^+$, X^-, and M^+ in the thin film, estimated at four different potentials corresponding to the Fig. 6.51b are shown in Fig. 6.52a–d. Contrary to the situation shown in Fig. 6.51a, where $c_{X^-} = c_{M^+}$ in the organic film under the conditions of open circuit, when a potential is applied, the oxidation of dmfc to $dmfc^+$ at a potential of E versus $E_c^{\ominus\prime} = -0.30$ V

[15] Lovrić M, Komorsky-Lovrić Š (2003) Electrochem Commun 5:637–643

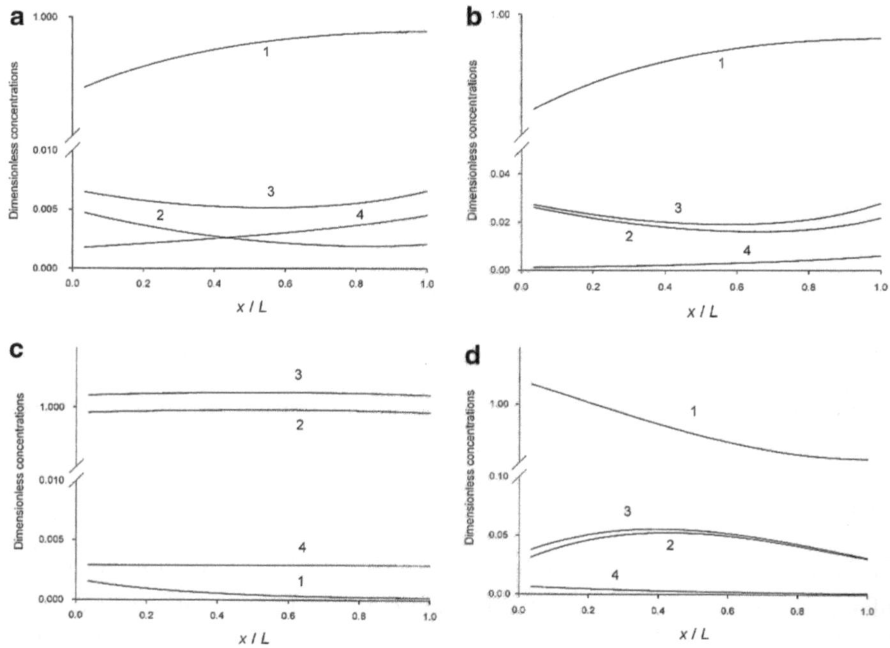

Fig. 6.52 Dimensionless concentrations of decamethylferrocene (1), $dmfc^+$ (2), X^- anion (3), and M^+ cation (4) in the film at the potentials $E = -0.3$ V (**a**) and -0.2 (**b**) on the anodic branch and 0.1 (**c**) and -0.3 (**d**) on the cathodic branch of cyclic voltammogram shown in Fig. 6.51b (see footnote 15). Copyright ©2003 with permission from Elsevier

(see Fig. 6.52a) causes a change of distributions of M^+ and X^- ions in order to compensate the charge of $dmfc^+$ cations. However, at higher potentials, the current is carried almost completely by $dmfc^+$ and X^- ions due to the low concentrations of M^+ ions. Thus, the concentrations of $dmfc^+$ and X^- are almost equal (see Fig. 6.52b). The curves in Fig. 6.52c correspond to the cathodic (backward) branch of the cyclic voltammogram in Fig. 6.51b estimated at a potential of 0.1 V. At this potential, the film is rapidly exhausted of dmfc. As the current approaches zero in this potential range (see Fig. 6.51b), the gradients of concentrations of ionic species in the thin film are very small. However, the reduction of $dmfc^+$ at potential of -0.3 V in the cathodic branch of the cyclic voltammogram causes migration of M^+ ions toward the electrode surface and the flux of X^- anions from the film toward the water, as can be seen in Fig. 6.52d.

The present model shows that in simple thin-film voltammetry, the current of the redox Reaction (XVI) mainly depends on the diffusion and migration of decamethylferrocenium cations and anions of the aqueous electrolyte, although the separation of charges appearing in the very beginning of the oxidation of dmfc can be compensated only by the migration of cations of the electrolyte in the film. The single contributions of the migration and diffusion currents of all species can be found in reference (see footnote 15).

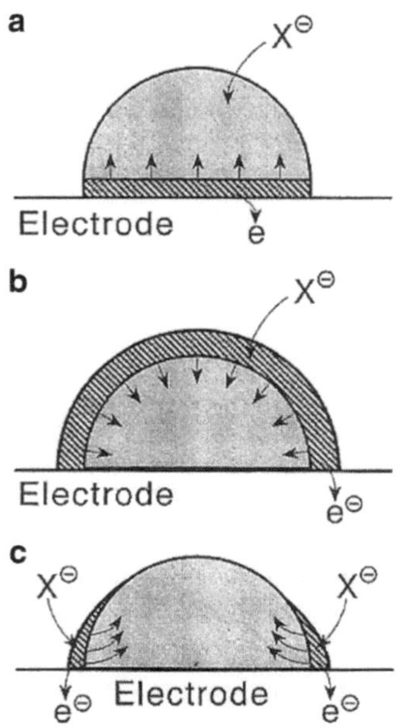

Fig. 6.53 Scheme of three mechanistic models for processes occurring at a three-phase electrode with immobilized droplets [B149]. Copyright ©2000 with permission from Wiley-VCH

6.3.2 *Droplets of Electroactive Liquids*

Numerical Modeling of Three Possible Mechanisms

The mathematical modeling of the processes occurring in microdroplet experiments conducted with electroactive liquids is similar to the theoretical considerations described previously in this chapter. Considering the reaction of the type

$$B_{(droplet)} + X^-_{(aq)} \leftrightarrows A_{(droplet)} + e^- \qquad \text{(XXV)}$$

Compton et al. [B149] have developed theoretical models for three different types of three-phase electrode voltammetry with droplets, considering different geometries and corresponding to different electroactive regions (see Fig. 6.53):

(a) The electrochemical reactions start from the oil–electrode surface interface.
(b) A rapid charge conduction over the droplet surface is assumed so that ion insertion occurs from the droplet–aqueous electrolyte interface.
(c) The electrochemical reaction occurs only at the three-phase junction electrode-electroactive droplet–aqueous electrolyte.

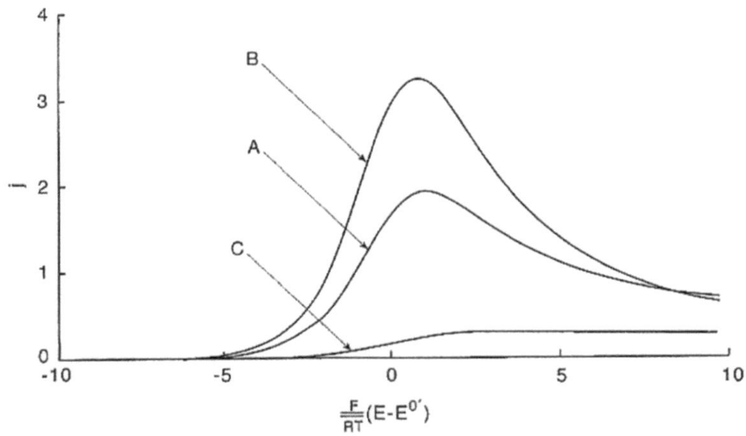

Fig. 6.54 Superimposed linear sweep voltammograms for the three models shown in Fig. 6.48 obtained with intermediate scan rates [B149]. Copyright ©2000 with permission from Wiley-VCH

In all cases, the diffusion is taken as the only way of mass transport and the solutions of Fick's equations corresponding to each separate case were solved numerically using the dual reciprocity finite element method. In model A, the planar diffusion is considered, while in the models B and C, concave and convex diffusions have been taken into account, respectively. The influence of the scan rate on the shape of the linear sweep voltammograms, on the potential separations, as well as on the normalized peak currents for all three geometries has been examined theoretically.

At low scan rates, identical linear sweep voltammograms are obtained in all cases. This behavior is due to the complete conversion to the product during the time scale of the potential sweep so that no effect of mass transport or geometry is evident. Unlike in the case of low scan rates, three different types of voltammograms are obtained by faster scan rates (see Fig. 6.54). While the voltammograms corresponding to the models A and B have conventional diffusion forms, the result for model C exhibit an entirely different feature. A steady-state voltammogram was obtained similar to that known for microelectrodes.

The dimensionless peak currents for all considered models depend linearly on the scan rates (with slope of 1) in the regions of low scan rates (see Fig. 6.55). Such behavior is typical for thin-film voltammetry. For the faster scan rates, the peak currents for models A and B increase in proportion to the square root of the scan rate, with a slope of 0.5. However, the normalized peak currents corresponding to the model C are almost independent on the scan rate for faster scan rates, like for microband electrodes.

The potential separation between the peak potentials and the formal potential of Reaction (XXV) for all three models is characterized by sigmoid dependences

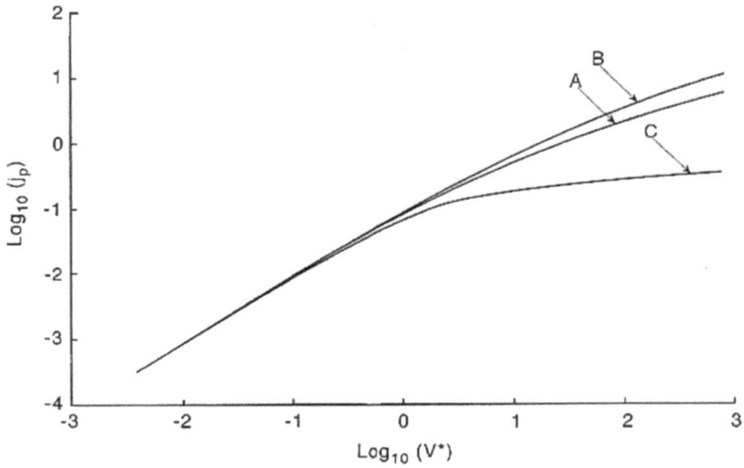

Fig. 6.55 Dependence of the dimensionless normalized peak currents on the dimensionless scan rate for the three models shown in Fig. 6.53 [B149]. Copyright ©2000 with permission from Wiley-VCH

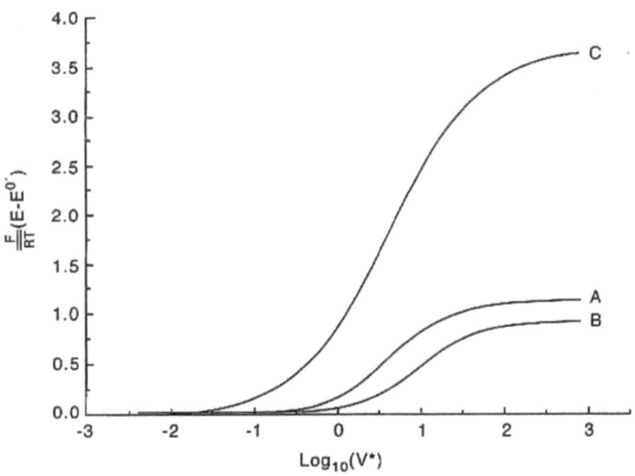

Fig. 6.56 Dependence of the separation between the peak potential and the formal potential on dimensionless scan rate [B149]. Copyright ©2000 with permission from Wiley-VCH

(see Fig. 6.56). By increasing the scan rates, the magnitude of the peak-to-peak separation of the model A converges to the value of 56 mV as typical for planar diffusion. The hemispherical model B gives a smaller potential separation converging to 49 mV. The most pronounced peak-to-peak separation was observed in model C, where that value converges toward 140 mV (at 298 K).

Conducting the experiments with electroactive liquid *para*-tetrahexylendiamine (THPD) microdroplets deposited on a pyrolytic graphite electrode and immersed in SCN^- solution, Compton et al. [B150] have shown that the obtained experimental

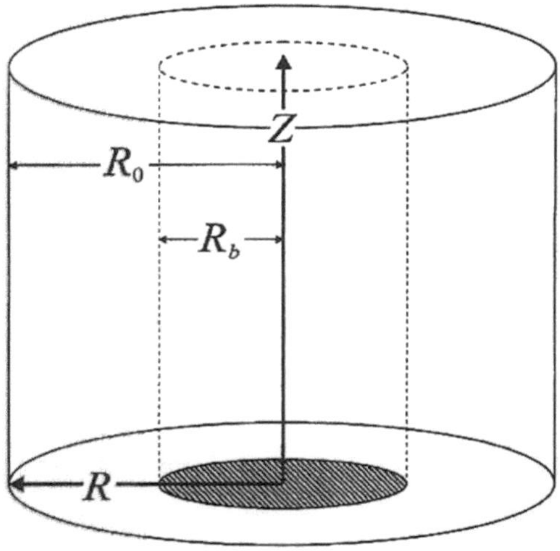

Fig. 6.57 The cylindrical polar coordinate system used to model the blocked electrode surface [B288]. Copyright ©2003 with permission from American Chemical Society

results fit best the theoretical results obtained with model C, i.e., for convex diffusion. However, the authors have determined experimentally the values for the diffusion coefficient of SCN^- that are three orders of magnitude lower than that estimated theoretically. It means that the mass transport process under fast scan rate conditions is orders of magnitude faster than possible by diffusion! This feature was ascribed to additional convection processes like that of Marangoni type. Probably, the neglecting of migration in the modeling is a major reason for the poor fit of experimental results.

Cyclic Voltammetry Response of Partially Blocked Electrodes

In the last few decades, significant attempts have been undertaken to quantify the electrochemistry of partially blocked electrodes, primarily to produce a reliable method for interrogating the characteristics of surface blocking. Compton et al. [B288] have recently developed a theoretical model in order to characterize the voltammetric responses of a partially blocked electrode (PBE) for which the blocks are distributed at random intervals.

The first mathematical model considers a circular inert disk of radius R_b situated at the bottom of a cylinder of radius R_o (see Fig. 6.57) and the following electrode reaction $A \leftrightarrows B + e^-$.

The authors developed a protocol that allowed calculating the dimensionless peak current, Ψ_{max}, from the dimensionless scan rate, ν_{dl}; microscopic coverage, Θ_{micro}; and dimensionless rate constant, k^0_{dl}, using the relation $\Psi_{max} = f(\nu_{dl}, \Theta_{micro}, k^0_{dl})$. The three variables on which the dimensionless peak current depends are given by

$$\left.\begin{aligned} \nu_{\mathrm{dl}} &= \frac{nFR_0^2}{DRT}\nu \\ \Theta_{\mathrm{micro}} &= \left(\frac{R_{\mathrm{b}}}{R_0}\right)^2 \\ k_{\mathrm{dl}}^0 &= \frac{k^0 R_0}{D} \end{aligned}\right\} \tag{6.22}$$

where ν is the scan rate, k^0 is the heterogeneous rate constant for the redox reaction, while the other symbols have their usual meaning. In the theoretical model, random distribution of inert disks and inert microdroplets has been considered. Here we shall consider the model of randomly distributed inert microdroplets.

For modeling a random distribution of inert microdroplets, it has been assumed that each droplet is hemispherical and that it is attached to the electrode surface with a circular base of surface radius R_{b}. The diffusion around each droplet is symmetric and the volume of the blocking material, V_{block}, is given by

$$V_{\mathrm{block}} = \frac{2}{3} N_{\mathrm{block}} \pi R_{\mathrm{b}}^3 \tag{6.23}$$

where N_{block} is the number of the droplet locks. The global coverage, Θ_{global}, is given by

$$\Theta_{\mathrm{global}} = \frac{N_{\mathrm{block}} \pi R_{\mathrm{b}}^2}{A_{\mathrm{elec}}} \tag{6.24}$$

In the model of a *monodisperse distribution of diffusion domains* (model A), it is assumed that all droplets are regularly dispersed on the electrode surface, i.e., every diffusion domain is identical. Consequently, R_0 is constant for a certain value R_{b}, and the dimensionless peak current for one diffusion domain will be equal to that for the whole electrode ($\Theta_{\mathrm{micro}} = \Theta_{\mathrm{global}}$). In this model, no overlap of the droplets occurs. In such a case, one needs to calculate a value of R_0 for particular value of R_{b}. This can be done by using the relation $R_0 = R_{\mathrm{b}}/\sqrt{\Theta_{\mathrm{global}}}$, where the global coverage Θ_{global} is given by Eq. 6.24. However, in the case of microdroplets, one has no prior knowledge of Θ_{global} or N_{block}, so there is an equation with two unknowns. Thus, rearranging Eq. 6.23 to obtain an expression for N_{block}, one can use Eq. 6.24 to obtain an expression for Θ_{global}. The resulting equations contain just one unknown (V_{block} can be measured experimentally, and it is assumed a certain value of R_{b}):

$$N_{\mathrm{block}} = \frac{3V_{\mathrm{block}}}{2\pi R_{\mathrm{b}}^3} \tag{6.25}$$

$$\Theta_{\mathrm{global}} = \frac{3V_{\mathrm{block}}}{2R_{\mathrm{b}} A_{\mathrm{elec}}} \tag{6.26}$$

For a certain volume of blocking material, using Eq. 6.26 with the relationship

$R_0 = R_b/\sqrt{\Theta_{global}}$, one can calculate values of ν_{dl}, Θ_{micro}, and k^0_{dl} for a given R_b. Hence, Ψ_{max} can be determined for a range of R_b values assuming that V_{block} is known.

In the model of a *polydisperse distribution of diffusion domains* (model B_1), it is assumed that when the droplets overlap, their geometry at the surface does not change, whereas their shape above the surface is undefined. Thus, one can use the following equation for calculating the Ψ_{max} for a range of R_b values:

$$\Psi_{max} = \frac{\pi N_{block}}{A_{elec}} \int_0^{3R_0} R_0^2 \Psi_{max}(R_0) P(R_0) dR_0 \tag{6.27}$$

where $P(R_0)$ is the probability of domains with radii greater than $3R_0$.

In the additional model B_2, the authors [B288] have approximated overlapping droplets of model B_1 as *larger droplets that do not overlap*. These larger droplets have a surface radius of R_b'. R_b' is related to the real coverage Θ_{real}, in exactly the same way as R_b is related to Θ_{global} , i.e.,

$$\Theta_{real} = \frac{3V_{block}}{2R_b' A_{elec}} \tag{6.28}$$

The relation between the real coverage, Θ_{real}, and the total coverage, Θ_{global}, is given by $\Theta_{real} = 1 - e^{-\Theta_{global}}$. For each value of R_b, one can calculate an average radius that accounts for overlapping, R_b', by using Eq. 6.28. Because the value of Θ_{real} is always less than Θ_{global}, R_b' is always larger than R_b.

Figures 6.58a–f illustrate the dependence of the dimensionless peak current on the droplet size, R_b, for the models A, B_1, and B_2 over a range of six block volumes at a scan rate of 50 mV s^{-1}. The effect of taking an overlapping into account is apparent in the differences between the models A and B. As R_b increases, there is a decrease in both Θ_{global} and N_{block}. Thus, the spread of probable R_0 values in models B_1 and B_2 increases, and one observes a gentle increase to the limiting value of Ψ_{max}. At large values of R_b, models A and B converge to one limiting peak dimensionless current. These theoretical models have been applied to experimental systems with electrodes modified by THPD in order to deduce the most "realistic case" [B288].

Conclusions

The theoretical modeling of the processes occurring at three-phase electrodes is an extraordinarily difficult task since the mathematical models have to take into account various effects. A major issue is the influence of the time-dependent ohmic resistance in the organic phase on the voltammetric properties of the studied systems. The origin of the initial conductivity in the

(continued)

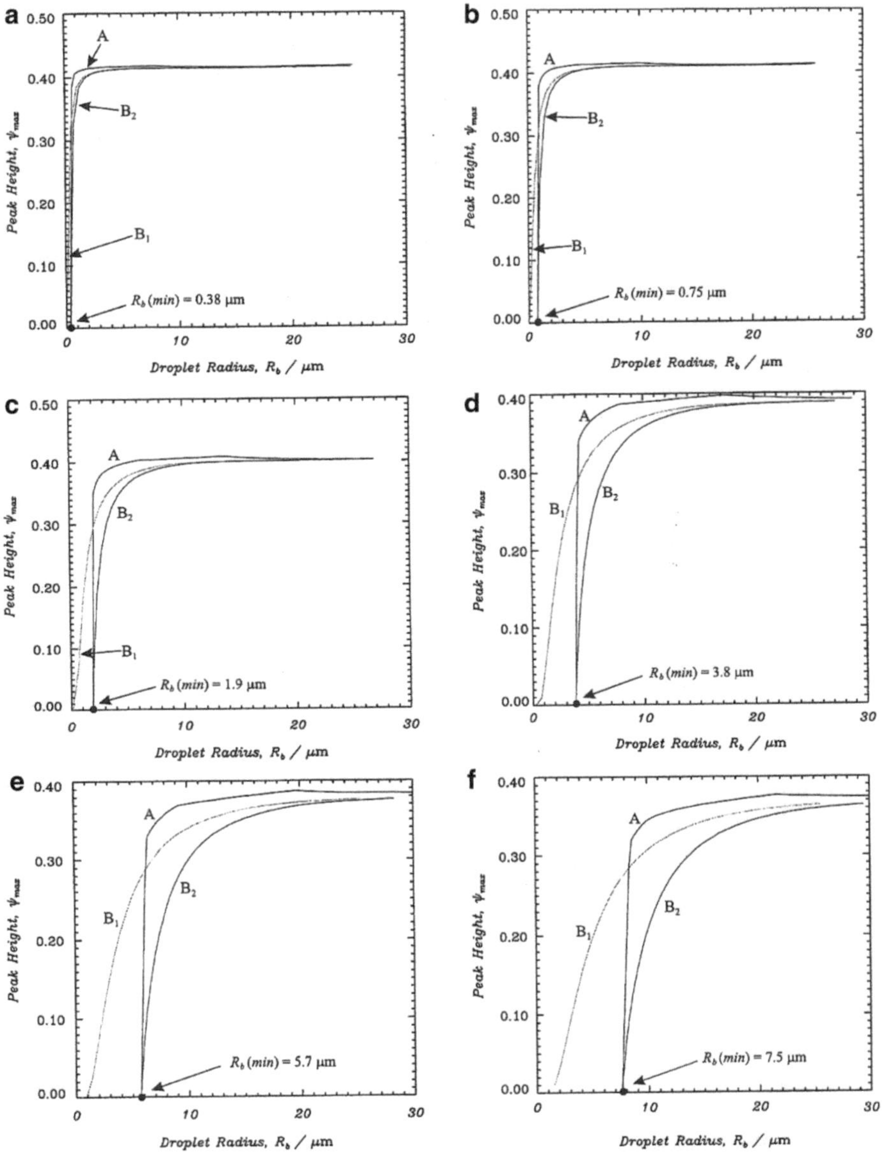

Fig. 6.58 Simulation of the dimensionless peak current response over a range of R_b values to models A, B_1, and B_2 for a microdroplet partially blocked electrode. In all cases, $v = 50$ mV s^{-1}, $D = 6.7 \times 10^{-6}$ cm^2 s^{-1}, $k^0 = 7 \times 10^{-3}$ cm s^{-1}, $A_{elec} = 0.19$ cm^2, and $T = 298$ K. The volume of blocking material on electrode is (**a**) 4.77×10^{-6}, (**b**) 9.55×10^{-6}, (**c**) 2.39×10^{-5}, (**d**) 4.77×10^{-5}, (**e**) 7.16×10^{-5}, and (**f**) 9.55×10^{-5} cm^3 [B288]. Copyright ©2003 with permission from American Chemical Society

organic phase that enables the start of the electrochemical reaction is another important point. Further, the modeling of the mass transfer through the organic phase is difficult because of its geometry.

The influence of the ohmic overvoltage on the voltammetric characteristics has been considered theoretically by several groups. It could be shown that the ohmic overvoltage influences mainly the current and shape of the voltammograms, but not the formal potentials, which is of importance for using these data to determine thermodynamic quantities. The ohmic drop effect in square-wave voltammetry resembles the effect of electron transfer kinetics, and criteria for distinguishing both effects in SWV have been proposed [B281]. All the features owing to the ohmic overvoltage have been experimentally confirmed [B281]. It has been also confirmed that the initial conductivity in the organic phase which is necessary for initiating the electrochemical reaction can be provided by the natural partition of the electrolyte from the adjoined aqueous phase. The mass transfer in the organic phase has been shown to be achieved by diffusion and migration. The effect of the migration on the overall current should not be underestimated, as shown in a recent publication (see footnote 15).

The question of *the three-phase junction line* is probably the most exciting one. In order to follow the development of the reaction zone inside the organic droplets, experiments have been performed to probe the dmfc or ferrocene (fc) concentration in the immobilized NB droplets with a Pt microdisk electrode [B207]. The authors have shown that in the three-phase electrode arrangement, the oxidation of dmfc and fc starts at the electrode–NB droplet–aqueous electrolyte three-phase junction line, where the anions from the aqueous phase can balance the excess charge of the electrochemically created cations in the organic phase. From that three-phase junction line, the ferrocenium and decamethylferrocenium cations diffuse into the bulk of the NB droplet, and they can be detected at the Pt microelectrode. The delay in detecting these cations was longer as the distance between the microdisk surface and the three-phase junction line increases [B207]. The detection time of these cations at a Pt microelectrode depends additionally on the droplet volume, as well as on the electrolyte concentration in the aqueous phase. Placing the microelectrode in the center of the NB droplet at a distance of some 25 μm from the carbon surface, the authors observed delay times for ferrocenium and decamethylferrocenium cations ranging from 10 to 30 s, depending on the experimental conditions (see Fig. 6.59). However, one can easily estimate that the time necessary to pass a distance of 25 μm by taking into account the diffusion only is more than 100 s. Similar results regarding the progress of the reaction zone have been obtained also by conducting experiments in a specially designed spectroelectrochemical cell [B280]. By an in situ spectroelectrochemical study of the simultaneous transfer of

(continued)

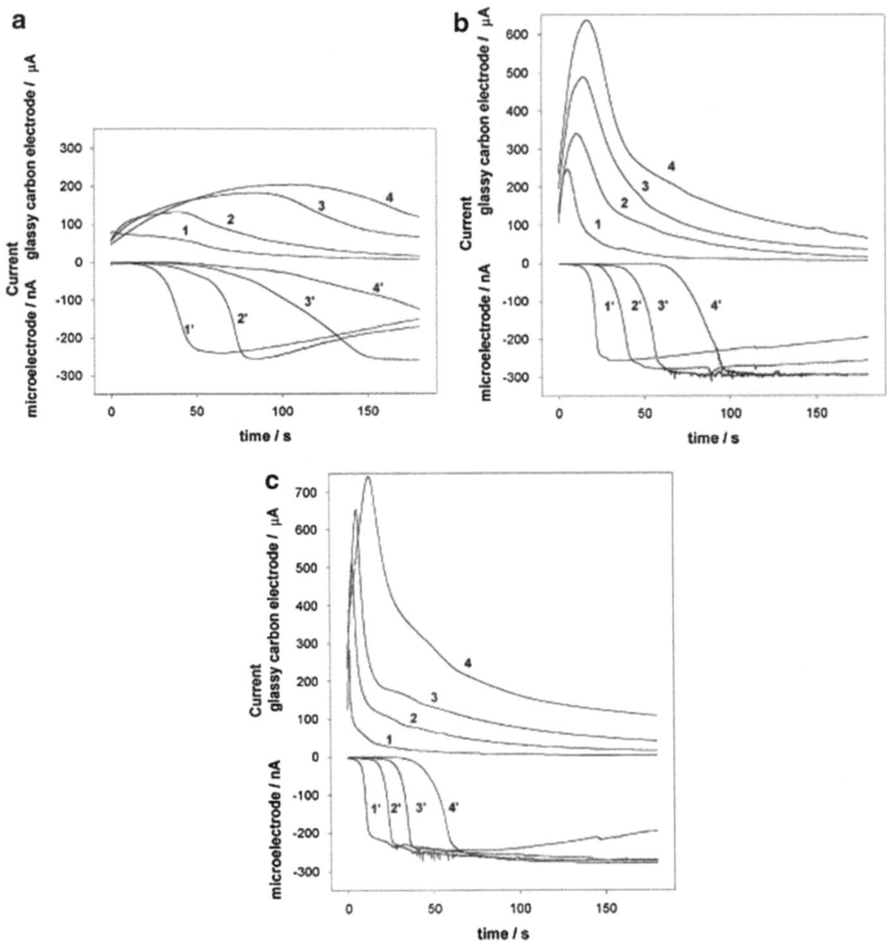

Fig. 6.59 Chronoamperometric oxidation of ferrocene at GC electrode-top, and reduction of Fc^+ at Pt microelectrode in NB drop-bottom. nitrobenzene droplets contain 0.1 mol L^{-1} ferrocene. Volume of nitrobenzene drops: 1 μL (1, 1'), 2 μL (2, 2'), 4 μL (3, 3'), and 8 μL (4, 4'). Water phase contains $LiClO_4$ at 0.01 mol L^{-1} (**a**), 0.1 mol L^{-1} (**b**), and 1 M (**c**) [B207]. Copyright ©2002 with permission from Elsevier

electrons and ions at a three-phase electrode, the authors [B280] observed a rather fast propagation of the reaction of dmfc oxidation that cannot be explained by a simple diffusion only. These results imply that the diffusion is not the only way of mass transport through the NB droplet. Indeed, migration and/or the Marangoni effect [B150] may play a role in the mass transport, too. These studies clearly showed that the reactions occurring at

(continued)

three-phase droplet-modified electrodes are not exclusively confined to the three-phase junction line, but they advance toward the center of the droplet.

The questions addressing to the structure of the liquid–liquid interface are also attractive owing to the importance of such interfaces in organic synthesis, chemical separations, and ion transfer processes. An issue of particular interest is whether the interfacial region is molecularly sharp or whether the interface is comprised of a diffuse mixed interfacial region. By using the vibrational sum-frequency spectroscopy as a probe of the structure, orientation, and bonding of the interfacial water, Richmond et al.[16,17,18] have recently shown that the interfaces between water and nonpolar organic liquids, like alkanes (C_6–C_8) or CCl_4, are very sharp on a molecular level. Beside these interfaces, the interface of water and polar 1,2-dichlorethane was proven to be molecularly disordered with properties similar to a mixed-phase interfacial region[19].

References

[B13] Bond AM, Scholz F (1991) Calculation of thermodynamic data from voltammetry of solid lead and mercury dithiocarbamate complexes mechanically attached to a graphite electrode. J Phys Chem 95:7460–7465

[B74] Marken F, Webster RD, Bull SD, Davies SG (1997) Redox processes in microproplets studied by voltammetry, microscopy, and ESR spectroscopy: oxidation of *N,N,N′,N′*-tetrahexylphenylene diamine deposited on solid electrode surfaces and immersed in aqueous electrolyte solution. J Electroanal Chem 437:209–218

[B97] Marken F, Compton RG, Goeting CH, Foord JS, Bull SD, Davies SG (1998) Anion detection by electro-insertion into *N,N,N′,N′*-tetrahexyl-phenylene-diamine (THPD) microdroplets studied by voltammetry, EQCM, and SEM techniques. Electroanal 10:821–826

[B132] Scholz F, Komorsky-Lovrić Š, Lovrić M (2000) A new access to Gibbs free energies of transfer of ions across liquid-liquid interfaces and a new method to study electrochemical processes at well-defined three-phase junctions. Electrochem Commun 2:112–118

[B149] Fulian Q, Ball JC, Marken F, Compton R, Fisher AC (2000) Voltammetry of electroactive oil droplets. Part I: Numerical modelling for three mechanistic models using the dual reciprocity finite element method. Electroanalysis 12:1012–1016

[B150] Ball JC, Marken F, Fulian Q, Wadhawan JD, Blythe AN, Schröder U, Compton R, Bull SD, Davies SG (2000) Voltammetry of electroactive oil droplets. Part II: Comparison of experimental and simulation data for coupled ion and electron insertion processes and evidence for microscale convection. Electroanalysis 12:1017–1025

[16] Scatena LF, Richmond GL (2001) J Phys Chem B 105:11240–11250

[17] Brown MG, Walker DS, Raymond EA, Richmond GL (2003) J Phys Chem B 107:237–244

[18] Scatena LF, Brown MG, Richmond GL (2001) Science 292:908–912

[19] Walker DS, Brown MG, McFearin CL, Richmond GL (2004) J Phys Chem B 108:2111–2114

[B166] Marken F, Blythe AN, Wadhawan JD, Compton RG, Bull SD, Aplin RT, Davies SG (2001) Voltammetry of electroactive liquid redox systems: anion insertion and chemical reactions in microdroplets of para-tetrakis(6-methoxyhexyl) phenylenediamine, para- and meta-tetrahexylphenylenediamine. J Solid State Electrochem 5:17–22

[B168] Marken F, Compton RG, Goeting CH, Foord JS, Bull SD, Davies SG (2001) Fast electrochemical triple-interface processes at boron-doped diamond electrodes. J Solid State Electrochem 5:88–93

[B171] Schröder U, Compton RG, Marken F, Bull AD, Davies SG, Gilmour S (2001) Electrochemically driven ion insertion processes across liquid/liquid boundaries: neutral versus ionic redox liquids. J Phys Chem B 105:1344–1350

[B174] Komorsky-Lovrić Š, Lovrić M, Scholz F (2001) Square-wave voltammetry of decamethylferrocene at the three-phase junction organic liquid|aqueous solution|graphite. Coll Czech Chem Commun 66:434–444

[B180] Ulmeanu S, Lee HJ, Fermin DJ, Girault HH, Shao Y (2001) Voltammetry at a liquid-liquid interface supported on a metallic electrode. Electrochem Commun 3:219–223

[B182] Wadhawan JD, Compton RG, Marken F, Bull SD, Davies SG (2001) Photoelectrochemically driven processes at the *N,N,N′,N′*-tetrahexylphenylenediamine microdroplet/electrode/aqueous electrolyte triple interface. J Solid State Electrochem 5:348–354

[B188] Komorsky-Lovrić Š, Lovrić M, Scholz F (2001) Cyclic voltammetry of decamethylferrocene at the three-phase junction organic liquid|aqueous solution|graphite. J Electroanal Chem 508:129–137

[B197] Gergely A, Inzelt G (2001) Electropolymerization of 3-methylthiophene at liquid 3-methylthiphene/aqueous solution/graphite three-phase junction. Electrochem Commun 3:753–757

[B206] Gulaboski R, Mirčeski V, Scholz F (2002) An electrochemical method for the determination of the standard Gibbs energy of anion transfer between water and *n*-octanol. Electrochem Commun 4:277–283

[B207] Donten M, Stojek Z, Scholz F (2002) Electron transfer – ion insertion electrochemistry at an immobilised droplet: probing the three-phase electrode-reaction zone with a Pt disk microelectrode. Electrochem Commun 4:324–329

[B213] Marken F, Hayman CM, Bulman Page PC (2002) Phosphate and arsenate electro-insertion processes into a *N,N,N′,N′*-tetraoctylphenylenediamine redox liquid. Electrochem Commun 4:462–467

[B214] Marken F, Hayman CM, Bulman Page PC (2002) Chromate and dichromate electro-insertion process into a *N,N,N′,N′*--tetraoctylphenylenediamine redox liquid. Electroanalysis 14:1–5

[B215] Mirčeski V, Scholz F (2002) Reduction of iodine at the organic liquid/aqueous solution/graphite electrode three-phase arrangement. J Electroanal Chem 522:189–198

[B221] Scholz F, Gulaboski R, Mirčeski V, Langer P (2002) Quantification of the chiral recognition in electrochemically driven ion transfer across the interface water|chiral liquid. Electrochem Commun 4:659–662

[B230] Wadhawan JD, Evans RG, Banks CE, Wilkins SJ, France RR, Oldham NJ, Fairbanks AJ, Wood B, Walton DJ, Schröder U, Compton RG (2002) Voltammetry of electroactive oil droplets: electrochemically-induced ion insertion, expulsion and reaction processes at microdroplets of *N,N,N′,N′*-tetraalkyl-para-phenylenediamines (TRPD, R = *n*-butyl, *n*-hexyl, *n*-heptyl and n-nonyl). J Phys Chem B 106:9619–9632

[B234] Mirčeski V, Gulaboski R, Scholz F (2002) Determination of the standard Gibbs energies of transfer of cations across the nitrobenzene/water interface utilizing the reduction of iodine in an immobilized nitrobenzene droplet. Electrochem Commun 4:813–818

[B237] Myland JC, Oldham KB (2002) A model of cyclic voltammetry for a thin organic layer sandwiched between an electrode and an aqueous solution. Convolutive modelling in the absence of supporting electrolyte. J Electroanal Chem 530:1–9

[B240] Komorsky-Lovrić Š, Riedl K, Gulaboski R, Mirčeski V, Scholz F (2002) Determination of standard Gibbs energies of transfer of organic anions across the water|nitrobenzene interface. Langmuir 18:8000–8005 (and Komorsky-Lovrić Š, Riedl K, Gulaboski R, Mirčeski V, Scholz F (2003) Langmuir 19:3090)

[B241] Wadhawan JD, Evans RG, Compton RG (2002) Voltammetric characteristics of graphite electrodes modified with microdroplets of n-butylferrocene. J Electroanal Chem 533:71–84

[B254] Schröder U, Wadhawan J, Evans RG, Compton RG, Wood B, Walton DJ, France RR, Marken F, Bulman Page PC, Hayman CM (2002) Probing thermodynamic aspects of electrochemically driven ion-transfer processes across liquid|liquid interfaces: pure versus diluted redox liquids. J Phys Chem B 106:8697–8704

[B257] Lovrić M, Scholz F (2003) Modeling cyclic voltammograms of simultaneous electron and ion transfer reactions at a conic film three-phase electrode. J Electroanal Chem 540:89–96

[B259] Gulaboski R, Riedl K, Scholz F (2003) Standard Gibbs energies of transfer of halogenate and pseudohalogenate ions, halogen substituted acetates, and cycloalkyl carboxylate anions at the water|nitrobenzene interface. PCCP 5:1284–1289

[B262] Gulaboski R, Mirčeski V, Scholz F (2003) Determination of the standard Gibbs energies of transfer of cations and anions of amino acids and small peptides across the water|nitrobenzene interface. Amino Acids 24:149–154

[B266] Gulaboski R, Scholz F (2003) The lipophilicity of peptide anions – an experimental data set for lipophilicity calculations. J Phys Chem B 107:5650–5657

[B269] Bouchard G, Galland A, Carrupt PA, Gulaboski R, Mirčeski V, Scholz F, Girault HH (2003) Standard partition coefficients of anionic drugs in the *n*-octanol/water system determined by voltammetry at three-phase electrodes. Phys Chem Chem Phys 5:3748–3751

[B271] Scholz F, Gulaboski R, Caban K (2003) The determination of standard Gibbs energies of transfer of cations across the nitrobenzene|water interface with the help of a three-phase electrode. Electrochem Commun 5:929–934

[B273] Wain AJ, Wadhawan JD, Compton RG (2003) Electrochemical studies of vitamin K_1 microdroplets: electrocatalytic hydrogen evolution. Chem Phys Chem 4:974–982

[B277] Gulaboski R, Caban K, Stojek Z, Scholz F (2004) The determination of the standard Gibbs energies of ion transfer between water and heavy water by using the three-phase electrode approach. Electrochem Commun 6:215–218

[B280] Komorsky-Lovrić Š, Mirčeski V, Kabbe C, Scholz F (2004) An *in situ* microscopic spectroelectrochemical study of a three-phase electrode where an ion transfer at the water|nitrobenzene interface is coupled to an electron transfer at the interface ITO|nitrobenzene. J Electroanal Chem 566:371–377

[B281] Mirčeski V, Gulaboski R, Scholz F (2004) Square-wave thin-film voltammetry in the presence of uncompensated resistance. Electroanal Chem 566:351–360, Electrochem Commun 12:206–209

[B282] Gulaboski R, Galland A, Bouchard G, Caban K, Kretschmer A, Carrupt PA, Stojek Z, Girault HH, Scholz F (2004) A comparison of the solvation properties of 2-nitrophenyloctyl ether, nitrobenzene, and *n*-octanol as assessed by ion transfer experiments. J Phys Chem B 108:4565

[B284] Wadhawan JD, Wain AJ, Kirkham AN, Walton DJ, Wood B, France RR, Bull SD, Compton RG (2003) Electrocatalytic reactions mediated by *N,N,N′,N′*-tetraalkyl-1,4-phenylenediamine redox liquid microdroplet-modified electrodes: chemical and photochemical reactions in, and at the surface of femtoliter droplets. J Am Chem Soc 125:11418–11429

[B285] Banks CE, Davies TJ, Evans RG, Hignett G, Wain AJ, Lawrence NS, Wadhawan JD, Marken F, Compton RG (2003) Electrochemistry of immobilised redox droplets: concepts and applications. Phys Chem Chem Phys 5:4053–4069

[B286] Wain AJ, Lawrence NS, Greene PR, Wadhawan JD, Compton RG (2003) Reactive chemistry via the redox switching of microdroplets of 4-nitrophenyl nonyl ether in the presence of aqueous electrolytes. Phys Chem Chem Phys 5:1867–1875

[B287] Wadhawan JD, Wain AJ, Compton RG (2003) Electrochemical probing of photochemical reactions inside femtolitre droplets confined to electrodes. Chem Phys Chem 4:1211–1215

[B288] Davies TJ, Brooks BA, Fisher AC, Yunus K, Wilkins AJ, Greene PR, Wadhawan JD, Compton RG (2003) A computational and experimental study of the cyclic voltammetry response of partially blocked electrodes. Part II: Randomly distributed and overlapping blocking systems. J Phys Chem B 107:6431–6444

[B289] Marken F, Blythe A, Compton RG, Bull SD, Davies SG (1999) Sulfide accumulation and sensing based on electrochemical processes in microdroplets of N^1-[4-(dihexylamino) phenyl]-N^1,N^4,N^4-trihexyl-1,4-phenylenediamine. Chem Commun 1823–1824

[B360] Doménech-Carbó A, Koshevoy IO, Montoya N, Pakkanen TA (2010) Electrochemically assisted anion insertion in Au(I)-Cu(I) heterotrimetallic clusters bearing ferrocenyl groups: application to the flouride/chloride discrimination in aqueous media. Electrochem Commun 12:206–209

[B361] Doménech-Carbó A, Koshevoy IO, Montoya N, Pakkanen TA (2010) Electrochemical anion sensing using electrodes chemically modified with Au(I)-Cu(I) heterotrimetallic alkynyl cluster complexes containing ferrocenyl groups. Anal Bioanal Chem 397:2013–2022

[B420] Doménech-Carbó A, Koshevoy IO, Montoya N, Pakkanen TA (2011) Estimation of free energies of anion transfer from solid state electrochemistry of alkynyl-based Au (I) dinuclear and Au(I)-Cu(I) cluster complexes containing ferrocenyl groups. Electrochem Commun 13:96–98

[B421] Doménech-Carbó A, Koshevoy IO, Montoya N, Karttunen AJ, Pakkanen TA (2011) Determination of individual Gibbs energies of anion transfer and excess Gibbs energies using an electrochemical method based on insertion electrochemistry of solid compounds. J Chem Eng Data 56:4577–4586

[B422] Doménech-Carbó A, Montoya N, Scholz F (2011) Estimation of individual Gibbs energies of cation transfer employing the insertion electrochemistry of solid Prussian blue. J Electroanal Chem 657:117–122

[B423] Doménech-Carbó A (2012) Solvent-independent electrode potentials of solids undergoing insertion electrochemical reactions: Part I. Theory. J Phys Chem C 116:25977–25983

[B424] Doménech-Carbó A, Koshevoy IO, Montoya N, Pakkanen TA, Doménech-Carbó MT (2012) Solvent-independent electrode potentials of solids undergoing insertion electrochemical reactions: Part II. Experimental data for alkynyl-diphosphine dinuclear Au (I) complexes undergoing electron exchange coupled to anion exchange. J Phys Chem C 116:25984–25992

[B425] Doménech-Carbó A, Scholz F, Montoya N (2012) Solvent-independent electrode potentials of solids undergoing insertion electrochemical reactions: Part III. Experimental data for Prussian blue undergoing electron exchange coupled to cation exchange. J Phys Chem C 116:25993–25999

[B450] Scholz F, Gulaboski R (2005) Gibbs energies of transfer of chiral anions across the interface water|chiral organic solvent determined with the help of three-phase electrodes. Faraday Discuss 129:169–177

Bibliography

[B1] Scholz F, Nitschke L, Henrion G (1989) A new procedure for fast electrochemical analysis of solid materials. Naturwiss 76:71–72

[B2] Scholz F, Nitschke L, Henrion G, Damaschun F (1989) A new technique to study the electrochemistry of minerals. Naturwiss 76:167–168

[B3] Scholz F, Nitschke L, Henrion G (1989) Identification of solid materials with a new electrochemical technique – the abrasive stripping voltammetry. Fresenius Z Anal Chem 334:56–58

[B4] Scholz F, Nitschke L, Henrion G, Damaschun F (1989) Abrasive stripping voltammetry – the electrochemical spectroscopy for solid state: application for mineral analysis. Fresenius Z Anal Chem 335:189–194

[B5] Scholz F, Nitschke L, Kemnitz E, Olesch T, Henrion G, Hass D, Bagchi RN, Herrmann R, Pruss N, Wilde W (1989) A simple and convenient solid state microanalytical technique for identification and characterization of the high temperature superconductor $YBa_2Cu_3O_{7-x}$. Fresenius Z Anal Chem 335:571–572

[B6] Nitschke L (1989) Neue methoden in der voltammetrischen analytik. Ph.D. thesis, Humboldt-Universität, Berlin

[B7] Scholz F, Nitschke L, Henrion G (1990) Abrasive stripping voltammetric analysis of tin-bismuth. Electroanalysis 2:85–87

[B8] Scholz F, Müller WD, Nitschke L, Rabi F, Livanova L, Fleischfresser C, Thierfelder C (1990) Fast and nondestructive identification of dental alloys by abrasive stripping voltammetry. Fresenius J Anal Chem 338:37–40

[B9] Scholz F, Lange B (1990) High-performance abrasive stripping voltammetry. Fresenius J Anal Chem 338:293–294

[B10] Björnbom P, Bursell M (1990) A method for studying microelectrodes by means of micromanipulators as applied to carbon agglomerates from oxygen reduction electrode catalyst. J Electrochem Soc 137:363–365

[B11] Scholz F, Lange B, Jaworski A, Pelzer J (1991) Analysis of powder mixtures with the help of abrasive stripping voltammetry and coulometry. Fresenius J Anal Chem 340:140–144

[B12] Bond AM, Bobrowski A, Scholz F (1991) Electrochemically generated copper(II), tin(II), bismuth(III) and zinc(II) complexes in dichloromethane containing tetrabuthylammonium perchlorate, tetrafluoroborate or hexafluorophosphate as the electrolyte. J Chem Soc Dalton Trans 1991:411–416

[B13] Bond AM, Scholz F (1991) Calculation of thermodynamic data from voltammetry of solid lead and mercury dithiocarbamate complexes mechanically attached to a graphite electrode. J Phys Chem 95:7460–7465

F. Scholz et al., *Electrochemistry of Immobilized Particles and Droplets*,
DOI 10.1007/978-3-319-10843-8

[B14] Bond AM, Scholz F (1991) Electrochemical, thermodynamic and mechanistic data derived from voltammetric studies on insoluble metallocenes, mercury halide and sulfide compounds, mixed silver halide crystals and other metal complexes following their mechanical transfer to a graphite electrode. Langmuir 7:3197–3204

[B15] Lovrić M, Komorsky-Lovrić Š, Bond AM (1991) Theory of square-wave stripping voltammetry and chronoamperometry of immobilized reactants. J Electroanal Chem 319:1–18

[B16] Büchi FN, Bond AM (1991) Interpretation of the electrochemistry of cytochrome c at macro and micro sized carbon electrodes using a microscopic model based on partially blocked surfaces. J Electroanal Chem 314:191–206

[B17] Bond AM, Scholz F (1992) Field-based identification of minerals using a battery-operated electrochemical measuring system and mechanical transfer of the solid to a graphite electrode. J Geochem Explor 42:227–235

[B18] Scholz F, Rabi F, Müller WD (1992) The anodic dissolution of dental amalgams studied with abrasive stripping voltammetry. Electroanalysis 4:339–346

[B19] Scheurell S, Scholz F, Olesch T, Kemnitz E (1992) Electrochemical evidence for Cu^{3+}-Cu^{2+}-Cu^{+} transitions in the orthorhombic $YBa_2Cu_3O_{7-x}$ phase. Supercond Sci Technol 5:303–305

[B20] Scholz F (1992) Abrasive stripping voltammetry – a new method for the study of the electrochemistry of solid materials. GIT Fachz Lab 36:917–918

[B21] Scholz F, Lange B (1992) Abrasive stripping voltammetry – an electrochemical solid state spectroscopy of wide applicability. Trends Anal Chem 11:359–367

[B22] Komorsky-Lovrić Š, Lovrić M, Bond AM (1992) Comparison of the square-wave stripping voltammetry of lead and mercury following their electrochemical or abrasive deposition onto a paraffin impregnated graphite electrode. Anal Chim Acta 258:299–305

[B23] Dueber RE, Bond AM, Dickens PG (1992) Investigation of the mechanism of formation of insertion compounds of uranium oxides by voltammetric reduction of the solid phase after mechanical transfer to a carbon electrode. J Electrochem Soc 139:2363–2371

[B24] Lange B, Scholz F, Bautsch HJ, Damaschun F, Wappler G (1993) Thermodynamics of the xanthoconite-proustite and pyrostilpnite-pyrargyrite transitions as determined by abrasive stripping voltammetry. Phys Chem Miner 19:486–491

[B25] Lange B, Scholz F, Weiß A, Schwedt G, Behnert J, Raezke KP (1993) Abrasive stripping voltammetry – the electrochemical alternative for pigment analysis. Int Lab 23:23–26

[B26] Jaworski A, Stojek Z, Scholz F (1993) A comparison of simulated and experimental abrasive stripping voltammetric curves of ionic crystals. Reversible case. J Electroanal Chem 354:1–9

[B27] Scholz F, Düssel H, Meyer B (1993) A new pH sensor based on quinhydrone. Fresenius J Anal Chem 347:458–459

[B28] Bond AM, Colton R, Daniels F, Fernando DR, Marken F, Nagaosa Y, Van Steveninck RFM, Walter JN (1993) Voltammetry, electron microscopy, and X-ray electron probe microanalysis at the electrode-aqueous electrolyte interface of solid microcrystalline *cis*- and *trans*-$Cr(CO)_2(dpe)_2$ and *trans*-$[Cr(CO)_2(dpe)_2]^+$ complexes (dpe = $Ph_2PCH_2CH_2PDh_2$) mechanically attached to carbon electrodes. J Am Chem Soc 115:9556–9562

[B29] Bond AM, Marken F (1994) Mechanistic aspects of the electron and ion transport processes across the electrode|solid solvent (electrolyte) interface of microcrystalline decamethylferrocene attached mechanically to a graphite electrode. J Electroanal Chem 372:125–135

[B30] Scholz F, Meyer B (1994) Electrochemical solid state analysis – state of the art. Chem Soc Rev 23:341–347

[B31] Bond AM, Colton R, Marken F, Walter JN (1994) Mechanistic study of the voltammetry of nonconducting microcrystalline *cis*- and *trans*-$Cr(CO)_2(dpe)_2$ complexes (dpe = $Ph_2PCH_2CH_2PDh_2$) mechanically attached to a graphite electrode and immersed in different aqueous electrolyte media: identification by infrared spectroscopy of *cis*-[Cr $(CO)_2(dpe)_2]^+$ stabilized at the electrode-solid-solution interface. Organometallics 13:5122–5131

[B32] Zhang S, Meyer B, Moh G, Scholz F (1995) Development of analytical procedures based on abrasive stripping coulometry and voltammetry for solid state phase microanalysis of natural and synthetic tin-, arsenic-, and antimony bearing sulfosalts and sulfides of thallium, tin, lead, and silver. Electroanalysis 7:319–328

[B33] Dostal A, Schröder U, Scholz F (1995) Electrochemistry of chromium (II) hexacyanochromate(III)) and electrochemically induced isomerization of solid iron (II) hexacyanochromate(III)) mechanically imobilized on the surface of a graphite electrode. Inorg Chem 34:1711–1717

[B34] Scholz F, Schröder U, Meyer S, Brainina KZ, Zakharchuk NF, Sobolev NV, Kozmenko OA (1995) The electrochemical response of radiation defects of non-conducting materials. An electrochemical access to age determinations. J Electroanal Chem 385:139–142

[B35] Dostal A, Meyer B, Scholz F, Schröder U, Bond AM, Marken F, Shaw SJ (1995) Electrochemical study of microcrystalline solid Prussian blue particles mechanically attached to graphite and gold electrodes: electrochemically induced lattice reconstruction. J Phys Chem 99:2096–2103

[B36] Meyer B, Ziemer B, Scholz F (1995) In situ X-ray diffraction study of the electrochemical reduction of tetragonal lead oxide and orthorhombic Pb(OH)Cl mechanically immobilized on a graphite electrode. J Electroanal Chem 392:79–83

[B37] Grygar T, Šubrt J, Boháček J (1995) Electrochemical dissolution of goethite by abrasive stripping voltammetry. Coll Czech Chem Commun 60:950–959

[B38] Bond AM (1994) Mechanistic aspects of electron transfer and mass transport processes associated with the voltammetry of microcrystalline solids mechanically attached to electrode surfaces. In: 45th annual meeting of the international society of electrochemistry, Porto, Portugal, Book of Abstracts, p IV

[B39] Bond AM, Cooper JB, Marken F, Way DM (1995) Redox and electroinsertion processes associated with the voltammetry of microcrystalline forms of Dawson molybdate anion salts mechanically attached to graphite electrodes and immersed in aqueous electrolyte media. J Electroanal Chem 396:407–418

[B40] Grygar T (1995) Kinetics of electrochemical reductive dissolution of iron(III) hydroxyoxides. Coll Czech Chem Commun 60:1261–1273

[B41] Scholz F, Dostal A (1995) The formal potentials of the solid metal hexacyanometalates. Angew Chem 107:2876–2878 (Angew Chem Int Ed Engl 34(1995):2685–2687)

[B42] Zakharchuk NF, Meyer B, Hennig H, Scholz F, Jaworski A, Stojek Z (1995) A comparative study with Prussian blue modified graphite paste electrodes and solid graphite electrodes with mechanically immobilized Prussian blue. J Electroanal Chem 398:23–35

[B43] Komorsky–Lovrić S (1995) Voltammetry of microcrystals of cobalt and manganese phthalocyanines. J Electroanal Chem 397:211–215

[B44] Madigan NA, Murphy TJ, Fortune JM, Hagan CRS, Coury LA (1995) Sonochemical stripping voltammetry. J Anal Chem 67:2781–2786

[B45] Downard AJ, Bond AM, Hanton LR, Heath GA (1995) A comparison of solution, microcrystalline solid, and thin film phase voltammetry of $[Co(mtas)_2](X)_n$ (mtas = bis (2–(dimethylarsino)–phenyl)methylarsine; $X = BF_4^-$, n = 3; $X = ClO_4^-$, n = 2, 3; $X = BPh_4^-$, $n = 2$). Inorg Chem 34:6387–6395

[B46] Meyer B (1995) Entwicklung von festkörperelektrochemischen Methoden für die Mikroanalyse von Mineralen und synthetischen Festkörpern. Ph.D. thesis, Humboldt-Universität, Berlin

[B47] Ura H, Nishina T, Uchida I (1995) Electrochemical measurements of single particles of Pd and $LaNi_5$ with a microelectrode technique. J Electroanal Chem 396:169–173

[B48] Blum D, Leyffer W, Holze R (1996) Pencil-leads as new electrodes for abrasive stripping voltammetry. Electroanalysis 8:296–297

[B49] Reddy SJ, Dostal A, Scholz F (1996) Solid state electrochemical studies of mixed nickel-iron hexacyanoferrates with the help of abrasive stripping voltammetry. J Electroanal Chem 403:209–212
[B50] Grygar T (1996) Electrochemical dissolution of iron(III) hydroxy-oxides: more information about the particles. Coll Czech Chem Commun 61:93–106
[B51] Dostal A, Kauschka G, Reddy SJ, Scholz F (1996) Lattice contractions and expansions which accompany the electrochemical conversion of Prussian blue and the reversible and irreversible insertion of rubidium and thallium ions. J Electroanal Chem 406:155–163
[B52] Grygar T (1996) The electrochemical dissolution of iron(III) and chromium(III) oxides and ferrites under conditions of abrasive stripping voltammetry. J Electroanal Chem 405:117–125
[B53] Shaw SJ, Marken F, Bond AM (1996) Simultaneous electrochemical and quartz crystal microbalance studies of nonconducting microcrystalline particles of *trans*-$Cr(CO)_2(dpe)_2$ and *trans*-$[Cr(CO)_2(dpe)_2]^+$ (dpe = $Ph_2PCH_2CH_2PDh_2$) attached to gold electrodes. J Electroanal Chem 404:227–235
[B54] Shaw SJ, Marken F, Bond AM (1996) Detection of new features associated with the oxidation of microcrystalline tetrathiafulvalene attached to gold electrodes by the simultaneous application of electrochemical and quartz crystal microbalance techniques. Electroanalysis 8:732–741
[B55] Dostal A, Hermes M, Scholz F (1996) The formation of bilayered nickel-iron, cadmium-iron and cadmium-silver hexacyanoferrates by an electrochemically driven insertion-substitution mechanism. J Electroanal Chem 415:133–141
[B56] Bond AM, Fletcher S, Marken F, Shaw SJ, Symons PG (1996) Electrochemical and X-ray diffraction study of the redox cycling of nanocrystals of 7,7,8,8-tetracyanoquinodimethane. Observation of a solid phase transformation controlled by nucleation and growth. J Chem Soc Faraday Trans 92(20):3925–2933
[B57] Meyer B, Zhang S, Scholz F (1996) The quantitative analysis of mixed crystals CuS_xSe_{1-x} with abrasive stripping voltammetry and a redetermination of the solubility product of CuSe and the standard potential of the Cu/CuSe electrode. Fresenius J Anal Chem 356:267–270
[B58] Schröder U, Meyer B, Scholz F (1996) A cell for in-situ incident-light microscopy for the study of electrochromism of solid state electrochemical reactions. Fresenius J Anal Chem 356:295–298
[B59] Reddy SJ, Hermes M, Scholz F (1996) The application of abrasive stripping voltammetry for a simple and rapid screening of pesticides. Electroanalysis 8:955–958
[B60] Lange B, Lovrić M, Scholz F (1996) The catalytic action of adsorbed thiocyanate ions and thiourea in the electron transfer from glassy carbon to solid copper(I) selenide and copper (I) sulfide particles. J Electroanal Chem 418:21–28
[B61] Komorsky-Lovrić S, Bartoll J, Stößer R, Scholz F (1997) Abrasive stripping voltammetry and ESR spectroscopy of manganese in carbonates. Croat Chem Acta 70:563–583
[B62] Meyer B, Scholz F (1997) Redetermination of the transformation enthalpies of the xanthoconite-proustite, pyrostilpnite-pyrargyrite and trechmannite-smithite phase transitions. Phys Chem Miner 24:50–52
[B63] Komorsky-Lovrić Š, Lovrić M, Scholz F (1997) Sulfide ion electrooxidation catalysed by cobalt phthalocyanine microcrystals. Mikrochim Acta 127:95–99
[B64] Lovrić M, Komorsky-Lovrić Š, Scholz F (1997) Staircase voltammetry with finite diffusion space. Electroanalysis 9:575–577
[B65] Bond AM, Colton R, Mahon PJ, Tan WT (1997) Tetrabutylammonium cation expulsion versus perchlorate electrolyte anion uptake in the electrochemical oxidation of microcrystals of $[(C_4H_9)_4N][Cr(CO)_5I]$ mechanically attached to a gold electrode: a voltammetric and quartz crystal microbalance study. J Solid State Electrochem 1:53–61

[B66] Grygar T (1997) Dissolution of pure and substituted goethites controlled by the surface reaction under conditions of abrasive stripping voltammetry. J Solid State Electrochem 1:77–82

[B67] Schröder U, Scholz F (1997) Microscopic in situ diffuse reflectance spectroelectrochemistry of solid state electrochemical reactions of particles immobilized on electrodes. J Solid State Electrochem 1:62–67

[B68] Komorsky-Lovrić Š (1997) Voltammetry of azobenzene microcrystal. J Solid State Electrochem 1:94–99

[B69] Lovrić M, Scholz F (1997) A model for the propagation of a redox reaction through microcrystals. J Solid State Electrochem 1:108–113

[B70] Bond AM, Fiedler DA (1997) *In situ* electrochemical and electron spin resonance studies of microcrystals mechanically attached to an electrode surface. J Electrochem Soc 144:1566–1574

[B71] Marken F, Leslie WM, Compton RG, Moloney MG, Sanders E, Davies SG, Bull SD (1997) Homogeneous and heterogeneous catalytic redox processes: solution and solid state voltammetry of lead complexes at carbon electrodes. J Electroanal Chem 424:25–34

[B72] Bond AM, Marken F, Hill E, Compton RG, Hügel H (1997) The electrochemical reduction of indigo dissolved in organic solvents and as a solid mechanically attached to a basal plane pyrolytic graphite electrode immersed in aqueous electrolyte solution. J Chem Soc Perkin Trans 2:1735–1742

[B73] Fiedler DA, Besenhard JO, Fooken MH (1997) Rapid electrochemical characterization of battery electrode materials in the solid state. J Power Sources 69:157–160

[B74] Marken F, Webster RD, Bull SD, Davies SG (1997) Redox processes in microproplets studied by voltammetry, microscopy, and ESR spectroscopy: oxidation of *N,N,N′,N′*-tetrahexylphenylene diamine deposited on solid electrode surfaces and immersed in aqueous electrolyte solution. J Electroanal Chem 437:209–218

[B75] Bond AM, Colton R, Marken F, Walter JN (1997) Voltammetric, specular reflectance infrared, and X-ray electron probe characterization of redox and isomerization processes associated with the $[Mn(CO)_2(\eta^3{-}P_2P')Br]^{+/0}$ ($P_2P' = \{Ph_2P(CH_2)_2\}PPh$), $[Mn(CO)_2(\eta^3{-}P_3P')Br]^{+/0}$, ($P_3P' = \{Ph_2PCH_2\}_3P$), and $[\{Mn(CO)(\eta^2{-}dpe)Br\}_2(\mu{-}dpe)]^{2+/0}$ (dpe = $Ph_2P(CH_2)_2PPh_2$) solid state systems. Organometallics 16:5006–5014

[B76] Gálova M, Markušová K, Lux L, Zezula I, Orináková R (1997) Determination of the surface activity of powders by electrochemical methods. Chem Pap 51:5–10

[B77] Bond AM, Colton R (1997) Electrochemical studies of metal carbonyl compounds. Coord Chem Rev 166:161–180

[B78] Uchida I, Fujiyoshi H, Waki S (1997) Microvoltammetric studies on single particles of battery active materials. J Power Sources 68:139–144

[B79] Scholz F, Lovrić M, Stojek Z (1997) The role of redox mixed phases $\{ox_x(C_n red)_{1-x}\}$ in solid state electrochemical reactions and the effect of miscibility gaps in voltammetry. J Solid State Electrochem 1:134–142

[B80] Perdicakis M, Grosselin N, Bessière J (1997) Voltammetric behaviour of single particles of silver chalcogenides in the micrometric range. Comparison with measurements at modified carbon electrodes. Electrochim Acta 42:3351–3358

[B81] Grygar T (1998) Phenomenological kinetics of irreversible electrochemical dissolution of metal-oxide microparticles. J Solid State Electrochem 2:127–136

[B82] Bond AM, Colton R, Mahon PJ, Snook GA, Tan WT (1998) Voltammetric oxidation of solution and solid phases of salts of $[V(CO)_6]^-$ in aqueous (electrolyte) media. J Phys Chem B 102:1229–1234

[B83] Friedrich A, Hefele H, Mickler W, Mönner A, Uhlemann E, Scholz F (1998) Voltammetric and potentiometric studies of the stability of vanadium(IV) complexes. A comparison of solution phase voltammetry with the voltammetry of the microcrystalline solid compounds. Electroanalysis 10:244–248

[B84] Komorsky-Lovrić Š, Scholz F (1998) Stripping chronopotentiometry of immobilized microparticles. J Electroanal Chem 445:81–87

[B85] Scholz F, Meyer B (1998) Voltammetry of solid microparticles immobilized on electrode surfaces. In: Bard AJ, Rubinstein I (eds) Electroanalytical chemistry, a series of advances, vol 20. Marcel Dekker, New York, NY, pp 1–86

[B86] Fiedler DA (1998) Rapid evaluation of the rechargeability of γ-MnO_2 in alkaline media by abrasive stripping voltammetry. J Solid State Electrochem 2:315–320

[B87] Lovrić M, Hermes M, Scholz F (1998) The effect of electrolyte concentration in the solution on the voltammetric response of insertion electrodes. J Solid State Electrochem 2:401–404

[B88] Oldham KB (1998) Voltammetry at a three-phase junction. J Solid State Electrochem 2:367–377

[B89] Fiedler DA, Albering JH, Besenhard JO (1998) Characterization of strontium and barium manganates by abrasive stripping voltammetry. J Solid State Electrochem 2:413–419

[B90] Bond AM, Colton R, Humphrey DG, Mahon PJ, Snook GA, Tedesco V, Walter JN (1998) Systematic studies of 17-electron rhenium(II) carbonyl phosphine complexes. Organometallics 17:2977–2985

[B91] Komorsky–Lovrić S (1998) A simple method for the detection of manganese in marine sediments. Croat Chim Acta 71:263–269

[B92] Grygar T, Bezdička P (1998) Electrochemical dissolution of CrIII and CrIV oxides. J Solid State Electrochem 3:31–38

[B93] Mocellin E, Goscinska T (1998) Modified carbon electrodes for microscale electrochemistry. J Chem Educ 75:771–772

[B94] Doménech-Carbó A, Ribera A, Cervilla A, Llopis E (1998) Electrochemistry of hydrotalcite-supported bis(2-mercapto-2,2-diphenyl-ethanolate) dioxomolybdate complexes. J Electroanal Chem 458:31–41

[B95] Kahlert H, Retter U, Lohse H, Siegler K, Scholz F (1998) On the determination of the diffusion coefficients of electrons and potassium ions in copper(II) hexacyanoferrate(II) composite electrodes. J Phys Chem B 102:8757–8765

[B96] Kulesza PJ, Malik MA, Berrettoni M, Giorgetti M, Zamponi S, Schmidt R, Marassi R (1998) Electrochemical charging, countercation accomodation, and spectrochemical identity of microcrystalline solid cobalt hexacyanoferrate. J Phys Chem B 102:1870–1876

[B97] Marken F, Compton RG, Goeting CH, Foord JS, Bull SD, Davies SG (1998) Anion detection by electro-insertion into *N,N,N′,N′*-tetrahexyl-phenylene-diamine (THPD) microdroplets studied by voltammetry, EQCM, and SEM techniques. Electroanal 10:821–826

[B98] Nalini B, Sriman Narayanan S (1998) Ruthenium(III) diphenyldithiocarbamate as mediator for the electrocatalytic oxidation of sulfhydryl compounds at graphite electrode. Bull Electrochem 14:241–245

[B99] Ravi Shankaran D, Sriman Narayanan S (1998) Studies on copper hexacyanoferrate modified electrode and its application for the determination of cysteine. Bull Electrochem 14:267–270

[B100] Nalini B, Sriman Narayanan S (1998) Electrocatalytic oxidation of ascorbic acid at a graphite electrode modified with a new Ruthenium(III) dithiocarbamate complex. Bull Electrochem 14:267–270

[B101] Bond AM, Fletcher S, Symons PG (1998) The relationship between the electrochemistry and the crystallography of microcrystals – the case of TCNQ (7,7,8,8-tetracyanoquinodimethane). Analyst 123:1891–1904

[B102] Bond M, Deacon GB, Howitt J, MacFarlane DR, Spiccia L, Wolfbauer G (1998) Voltammetric determination of the reversible redox potential for the oxidation of the highly active polypyridyl ruthenium photovoltaic sensitizer *cis*–$Ru^{(II)}(dcbpy)_2(NCS)_2$. J Electrochem Soc 146:648–656

[B103] Dostal A (1998) Festkörperreaktionen an metallhexacyanometallat-modifizierten Elektroden. Ph.D. thesis, Humboldt–Universität, Berlin

[B104] Grygar T (1998) Elektrochemické rozpouštění práškových oxidů kovů. Ph.D. thesis, Řež

[B105] Albering JH, Grygar T (1999) A note regarding the rate-determining step of reduction of solid manganate(V). J Solid State Electrochem 3:117–120

[B106] Bond M, Fiedler DA, Lamprecht A, Tedesco V (1999) Electron-, anion-, and proton-transfer processes associated with the redox chemistry of $Fe(\eta^5-C_5Ph_5)(\eta^6-C_6H_5)C_5Ph_4)$ and its protonated form $[Fe(\eta^5-C_5Ph_5)(\eta^6-_6H_5)C_5Ph_4H)]BF_4$ at microcrystal-electrode-solvent (electrolyte) interfaces. Organometallics 18:642–649

[B107] Lovrić M, Scholz F (1999) A model for the coupled transport of ions and electrons in redox conductive microcrystals. J Solid State Electrochem 3:172–175

[B108] Zezula I, Galova M (1999) Application of abrasive stripping voltammetry in corrosion science. I. Determination of the corrosion potential of metals. J Solid State Electrochem 3:231–233

[B109] Komorsky–Lovrić S, Galic I, Penovski R (1999) Voltammetric determination of cocaine microparticles. Electroanalysis 11:120–123

[B110] Lovrić M (1999) Diffusion from a three-phase junction into a hemispherical droplet. Electrochem Commun 1:207–212

[B111] Zakharchuk NF, Naumov N, Stösser R, Schröder U, Scholz F, Mehner H (1999) Solid state electrochemistry, X-ray powder diffraction, magnetic susceptibility, electron spin resonance, Mössbauer and diffuse reflectance spectroscopy of mixed iron(III)-cadmium (II) hexacyanoferrates. J Solid State Electrochem 3(5):264–276

[B112] Lux L, Gálová M, Hezelová M, Markusová K (1999) Investigation of the reactivity of powder surfaces by abrasive voltammetry. J Solid State Electrochem 3(5):288–292

[B113] Narayanan SS, Scholz F (1999) A comparative study of the electrocatalytic activities of some metal hexacyanoferrates for the oxidation of hydrazine. Electroanalysis 11:465–469

[B114] Zhuang Q, Scholz F, Pragst F (1999) The voltammetric behaviour of solid 2,2-diphenyl-1-picrylhydrazyl (DPPH) microparticles. Electrochem Commun 1:406–4103

[B115] Shankaran DR, Narayanan SS (1999) Characterization and application of an electrode modified by mechanically immobilized copper hexacyanoferrate. Fresenius J Anal Chem 364:686–689

[B116] Jantscher W, Binder L, Fiedler DA, Andreaus R, Kordesch K (1999) Synthesis, characterization and application of doped electrolytic manganese dioxides. J Power Sources 79:9–18

[B117] Grygar T (1999) Electrochemical reactions of $La(Ni,Cr)O_3$ and $La(Ni,Fe)O_3$ in acid aqueous solution. J Solid State Electrochem 3:412–416

[B118] Grygar T, Bezdička P, Caspary EG (1999) Electrochemical dissolution of immobilized alpha–$(Fe_xCr_{1-x})_2O_3$ microparticles. J Electrochem Soc 146:3234–3238

[B119] Suárez MF, Marken F, Compton RG, Bond AM, Miao WJ, Raston CL (1999) Evidence for nucleation-growth, redistribution, and dissolution mechanisms during the course of redox cycling experiments on the C-60/NBu_4C_{60} solid-state redox system – voltammetric, SEM, and in-situ AFM studies. J Phys Chem 103:5637–5644

[B120] Shankaran DR, Narayanan SS (1999) Chemically-modified sensor for amperometric determination of sulfur-dioxide. Sens Actuators B Chem 55:191–194

[B121] Scaboo KM, Grover WH, Chambers JQ (1999) Inverse derivative chronopotentiometriy/ quartz crystal microgravimetry of solid conducting films. Anal Chim Acta 380:47–54

[B122] Mohan Rao M, Jayalakshmi M, Schäf O, Wulff H, Scholz F (1999) Electrochemical behaviour of solid lithium nickelate ($LiNiO_2$) in an aqueous electrolyte. J Solid State Electrochem 4:17–23

[B123] Manivannan A, Tryk DA, Fujishima A (1999) Abrasive stripping voltammetry at polycrystalline diamond electrodes. Chem Lett 851–852

[B124] Suárez MF, Bond AM, Compton RG (1999) Significance of redistribution reactions detected by in situ atomic force microscopy during early stages of fast scan rate redox cycling experiments at a solid 7,7,8,8-tetracyanoquinodimethane-glassy carbon electrode-aqueous (electrolyte) interface. J Solid State Electrochem 4:24–33

[B125] Bezdička P, Grygar T, Klapste B, Vondrak J (1999) MnO_x/C composites as electrode materials. I. Synthesis, XRD and cyclic voltammetric investigation. Electrochim Acta 45:913–920

[B126] Komorsky-Lovrić Š, Mirčeski V, Scholz F (1999) Voltammetry of organic microparticles. Mikrochim Acta 132:67–77

[B127] Nishizawa M, Uchida I (1999) Microelectrode-based characterization systems for advanced materials in battery and sensor applications. Electrochim Acta 44:3629–3637

[B128] Eklund JC, Bond AM (1999) Photocatalytic reactions at microcrystalline *fac*-$Mn(CO)_3(\eta^2$-$Ph_2PCH_2PPh_2)Cl$-electrode-aqueous (electrolyte) interfaces. J Am Chem Soc 121:8306–8312

[B129] Bond AM, Lamprecht A, Tedesco V, Marken F (1999) Novel features associated with the electrochemically driven bis(eta(5)-pentaphenyl-cyclopentadienyl)iron(II)-iron(III) redox transformation at an electrode-microcrystal-solvent (electrolyte) interface. Inorg Chim Acta 291:21–31

[B130] Shankaran DR, Narayanan SS (1999) Evaluation of mechanically immobilised nickel hexacyanoferrate electrode as an amperometric sensor for thiosulphate determination. Fresenius J Anal Chem 365:663

[B131] Perdicakis M, Grosselin N, Bessière J (1999) Interaction of pyrite pulps with Ag^+ and Hg^{2+} ions. Electrochemical characterization of micrometric grains. Anal Chim Acta 385:467–485

[B132] Scholz F, Komorsky-Lovrić Š, Lovrić M (2000) A new access to Gibbs free energies of transfer of ions across liquid-liquid interfaces and a new method to study electrochemical processes at well-defined three-phase junctions. Electrochem Commun 2:112–118

[B133] Scholz F, Schädel S, Schultz A, Schauer F (2000) Chronopotentiometric study of laccase catalysed oxidation of quinhydrone microcrystals immobilised on a gold electrode surface and of the oxidation of a phenol derivatised graphite electrode surface. J Electroanal Chem 480:241–248

[B134] Ramaraj R, Kabbe C, Scholz F (2000) Electrochemistry of microparticles of tris (2,2′-bipyridine)ruthenium(II) hexafluorophosphate. Electrochem Commun 2:190–194

[B135] Zhuang QK, Scholz F (2000) Electrochemical driven introduction of copper ions into the ring of 5,10,15,20-tetraphenyl-21H, 23H-porphyrin mechanically attached as solid microparticles to a graphite electrode. J Porphyr Phthalocyanines 4:202–208

[B136] Schröder U, Scholz F (2000) The solid state electrochemistry of metal octacyanomolybdates, octacyanotungstates, and hexacyanoferrates explained on the basis of dissolution and reprecipitation reactions, their lattice structures and crystallinities. Inorg Chem 39:1006–1015

[B137] Schwudke D, Stößer R, Scholz F (2000) Solid-state electrochemical, X-ray and spectroscopic characterization of substitutional solid solutions of iron-copper hexacyanoferrates. Electrochem Commun 2:301–306

[B138] Jiang J, Kucernak A (2000) The electrochemistry of platinum phthalocyanine microcrystals. I. Electrochemical behaviour in acetonitrile electrolytes. Electrochim Acta 45:2227–2239

[B139] Bond AM, Marken F, Williams CT, Beattie DA, Keyes TE, Foster RJ, Vos JG (2000) Unusually fast electron and anion transport processes observed in the oxidation of "electrochemically open" microcrystalline $[\{M(bipy)_2\}\{M'(bipy)_2\}(\text{-}L)](PF_6)_2$ complexes ($M, M' = Ru, Os$; bipy = 2,2′–bipyridyl; L = 1,4-dihydroxy-2,5-bis(pyrazol-1-yl)benzene. J Phys Chem 104:1977–1983

[B140] Myland JC, Oldham KB (2000) Modelling reversible cyclic voltammetry at the one-dimensional junction established by three phases: an organic liquid containing a neutral electroactive solute, an aqueous electrolyte solution, and an electronic conductor. Electrochem Commun 2:541–546

[B141] Bakardjieva S, Bezdička P, Grygar T, Vorm P (2000) Reductive dissolution of microparticulate manganese oxides. J Solid State Electrochem 4:306–313

[B142] Schröder U, Oldham KB, Myland JC, Mahon PJ, Scholz F (2000) Modelling of solid-state voltammetry of immobilized microcrystals assuming an initiation of the electrochemical reaction at a three-phase junction. J Solid State Electrochem 4:314–324

[B143] Doménech-Carbó A, Doménech-Carbó MT, Gimeno-Adelantado JV, Moya-Moreno M, Bosch-Reig F (2000) Voltammetric identification of lead(II) and (IV) in mediaeval glazes in abrasion-modified carbon paste and polymer film electrodes. Application to the study of alterations in archaeological ceramics. Electroanalysis 12:120–127

[B144] Kalvoda R (2000) Netradiční použití polarografie/voltametrie. Chem Listy 94:2–9

[B145] Kulesza PJ, Malik MA, Schmidt R, Smolinska A, Miecznikowski K, Zamponi S, Czerwinski A, Berrettoni M, Marassi R (2000) Electrochemical preparation and characterization of electrodes modified with mixed hexacyanoferrates of nickel and palladium. J Electroanal Chem 487:57–65

[B146] Doménech-Carbó A, Doménech-Carbó MT, Moya-Moreno M, Gimeno-Adelantado JV, Bosch-Reig F (2000) Identification of inorganic pigments from paintings and polychromed sculptures immobilized into polymer electrodes by stripping differential pulse voltammetry. Anal Chim Acta 407:275–289

[B147] Kucernak AR, Chowdhury PB, Wilde CP, Kelsall GH, Zhu YY, Williams DE (2000) Scanning electrochemical microscopy of a fuel-cell electrocatalyst deposit onto highly oriented pyrolytic graphite. Electrochim Acta 45:4483–4491

[B148] Zakharchuk N, Meyer S, Lange B, Scholz F (2000) A comparative study of lead oxyde modified graphite paste electrodes and solid graphite electrodes with mechanically immobilised lead oxides. Croat Chem Acta 73:667–704

[B149] Fulian Q, Ball JC, Marken F, Compton R, Fisher AC (2000) Voltammetry of electroactive oil droplets. Part I: Numerical modelling for three mechanistic models using the dual reciprocity finite element method. Electroanalysis 12:1012–1016

[B150] Ball JC, Marken F, Fulian Q, Wadhawan JD, Blythe AN, Schröder U, Compton R, Bull SD, Davies SG (2000) Voltammetry of electroactive oil droplets. Part II: Comparison of experimental and simulation data for coupled ion and electron insertion processes and evidence for microscale convection. Electroanalysis 12:1017–1025

[B151] Nartey VK, Binder L, Huber A (2000) Production and characterisation of titanium doped electrolytic manganese dioxide for use in rechargeable alkaline zinc/manganese dioxide batteries. J Power Sources 87:205–211

[B152] Wadhawan JD, Schröder U, Neudeck A, Wilkins SJ, Compton RG, Marken F, Consorti CS, de Souza RF, Dupont J (2000) Ionic liquid modified electrodes. Unusual partitioning and diffusion effects of $Fe(CN)_6^{4-/3-}$ in droplet and thin layer deposits of 1-methyl-3-(2,6-(*S*)-dimethylocten-2-yl)-imidazolium tetrafluoroborate. J Electroanal Chem 493:75–83

[B153] Terui N, Nakatani K, Kitamura N (2000) Kinetic analysis of electrochemically induced ion transfer across a single microdroplet/water interface. J Electroanal Chem 494:41–46

[B154] Bond AM, Miao W, Raston CL (2000) Identification of processes that occur after reduction and dissolution of C_{60} adhered to gold, glassy carbon, and platinum electrodes placed in acetonitrile (electrolyte) solution. J Phys Chem B 104:2320–2329

[B155] Foster RJ, Keyes TE, Bond AM (2000) Protonation effects on the structure and homogeneous charge transport dynamics of solid state osmium bis(bipyridyl)tetrazine chloride films. J Phys Chem B 104:6389–6396

[B156] Bond AM, Miao W, Raston CL, Sandoval CA (2000) Electrochemical, EPR, and magnetic studies on microcrystals of the $[C_{60}\subset(\text{p-benzyl-calix[5]arene})_2]\cdot 8$toluene and its one-electron-reduced encapsulation complex. J Phys Chem B 104:8129–8137

[B157] Hermes M, Scholz F (2000) The electrochemical oxidation of white phosphorus at a three-phase junction. Electrochem Commun 2:845–850

[B158] Lovrić M, Hermes M, Scholz F (2000) Solid state electrochemical reactions in systems with miscibility gaps. J Solid State Electrochem 4:394–401

[B159] Zhuang QK (2000) The mechanistic studies of the electrochemical in situ association of copper ions with 5,10,15,20-tetraphynyl-21H,23H-porphyrin. Microchem J 65:333–340

[B160] Shankaran DR, Narayanan SS (2000) Amperometric determination of sulphur dioxide at a nickel hexacyanoferrate modified electrode. Res J Chem Environ 6:13–16

[B161] Minovic A, Milošev I, Komorsky-Lovrić Š (2000) Voltametrija trdnih mikro delcev, imobiliziranih na površini grafitne electrode (Voltammetry of solid microparticles immobilized on graphite electrode surface). Materiali in Tehnologije 34:23–25

[B162] Schröder U (2000) Methodische, stofflich-mechanistische und theoretische Aspekte insertions-elektrochemischer Reaktionen. Ph.D. thesis, Humboldt-Universität, Berlin

[B163] Bárcena Soto M (2000) Thermodynamische Untersuchungen zur Insertionselektrochemie von Metallhexacyanoferraten. Ph.D. thesis, Humboldt-Universität, Berlin

[B164] Girerd JJ, Verdaguer M (2000) Electrochimie à une particule individuelle de taille micrometrique. Une technique por la charactèrisation grain par grain des matériaux conducteurs divisés. Lett Sci Chim 73:22–28

[B165] Mohan Rao M, Jayalakshmi M, Schäf O, Wulff H, Scholz F (2001) Electrochemical behaviour of solid lithium cobaltate ($LiCoO_2$) and lithium manganate ($LiMn_2O_4$) in an aqueous electrolyte System. J Solid State Electrochem 5:50–56

[B166] Marken F, Blythe AN, Wadhawan JD, Compton RG, Bull SD, Aplin RT, Davies SG (2001) Voltammetry of electroactive liquid redox systems: anion insertion and chemical reactions in microdroplets of para-tetrakis(6-methoxyhexyl) phenylenediamine, para- and meta-tetrahexylphenylenediamine. J Solid State Electrochem 5:17–22

[B167] Jiang J, Kucernak A (2001) The electrochemistry of platinum phthalocynine microcrystals. II. A microelectrode observation of nucleation-growth controlled solid-solid phase transformation in non-aqueous solvent. Electrochim Acta 46:1223–1231

[B168] Marken F, Compton RG, Goeting CH, Foord JS, Bull SD, Davies SG (2001) Fast electrochemical triple-interface processes at boron-doped diamond electrodes. J Solid State Electrochem 5:88–93

[B169] Takeda N, Stawasz ME, Parkinson BA (2001) Electrochemical oxidation and ex-situ STM observation of bis(4-dimethylamino-2-dihydroxyphenyl)squarine dye layers on HOPG electrodes. J Electroanal Chem 498:19–33

[B170] Bond M, Miao W, Raston CL, Ness TJ, Barnes MJ, Atwood JL (2001) Electrochemical and structural studies on microcrystals of the $(C_{60})_x$(CTV) inclusion complexes ($x = 1, 1.5$; CTV = cyclotriveratrylene). J Phys Chem B 105:1687–1695

[B171] Schröder U, Compton RG, Marken F, Bull AD, Davies SG, Gilmour S (2001) Electrochemically driven ion insertion processes across liquid/liquid boundaries: neutral versus ionic redox liquids. J Phys Chem B 105:1344–1350

[B172] Bond AM, Feldberg SW, Miao W, Oldham KB, Raston CL (2001) Modelling of solid-state, dissolution and solution-phase reactions at adhered solid-electrode-solvent (electrolyte) interfaces: electrochemistry of microcrystals of C_{60} adhered to an electrode in contact with dichloromethane (Bu_4NclO_4). J Electroanal Chem 501:22–32

[B173] Hermes M, Lovrić M, Hartl M, Retter U, Scholz F (2001) On the electrochemically driven formation of bilayered systems of solid Prussian-blue-type metal hexacyanoferrates: a model for Prussian blue cadmium hexacyanoferrate supported by finite difference simulations. J Electroanal Chem 501:193–204

[B174] Komorsky-Lovrić Š, Lovrić M, Scholz F (2001) Square-wave voltammetry of decamethylferrocene at the three-phase junction organic liquid|aqueous solution|graphite. Coll Czech Chem Commun 66:434–444

[B175] Doménech-Carbó MT, Casas-Catalán MJ, Doménech-Carbó A, Mateo-Castro R, Gimeno-Adelantado JV, Bosch-Reig F (2001) Analytical study of canvas painting collection from the *Basilica de la Virgen de los Desaparados* using SEM/EDX, FT–IR, GC and electrochemical techniques. Fresenius J Anal Chem 369:571–575

[B176] Doménech-Carbó A, Doménech-Carbó MT, Gimeno-Adelantado V, Bosch-Reig F, Saurí-Peris MC, Casas-Catálan MJ (2001) Electrochemical analysis of the alterations in copper pigments using charge transfer coefficient/peak potential diagrams. Application to microsamples of baroque wall paintings attached to polymer film electrodes. Fresenius J Anal Chem 369:576–581

[B177] Cepriá G, Abadias O, Pérez-Arantegui J, Castillo JR (2001) Electrochemical behavior of silver-copper alloys in voltammetry of microparticles: a simple method for screening purposes. Electroanalysis 13:477–483

[B178] Zhuang QK, Scholz F (2001) In-situ electrochemical synthesis of the metalloporphyrins with help of the voltammetry of immobilized microparticles. Chem J Chin Univ (Gaodeng Xuexiao Huaxue Xuebao) 22:171–174

[B179] Keyes TE, Foster RJ, Bond AM, Miao W (2001) Electron self-exchange in the solid-state: cocrystals of hydroquinone and bipyridyl triazole. J Am Chem Soc 123:2877–2884

[B180] Ulmeanu S, Lee HJ, Fermin DJ, Girault HH, Shao Y (2001) Voltammetry at a liquid-liquid interface supported on a metallic electrode. Electrochem Commun 3:219–223

[B181] Grygar T, Bakardjieva S, Bezdička P, Vorm P (2001) Electrochemical dissolution of mixed oxides of Mn and Fe: the relationship between phase composition and reactivity. Ceramics-Silikaty 45:55–61

[B182] Wadhawan JD, Compton RG, Marken F, Bull SD, Davies SG (2001) Photoelectrochemically driven processes at the *N,N,N′,N′*-tetrahexylphenylenediamine microdroplet/electrode/aqueous electrolyte triple interface. J Solid State Electrochem 5:348–354

[B183] Rao M, Liebenow C, Jayalakshmi M, Wulff H, Guth U, Scholz F (2001) High temperature combustion synthesis and electrochemical characterisation of $LiNiO_2$, $LiCoO_2$ and $LiMn_2O_4$ for Li-ion secondary batteries. J Solid State Electrochem 5:348–354

[B184] Hasse U, Scholz F (2001) In situ atomic force microscopy of the reduction of lead oxide nanocrystals immobilized on an electrode surface. Electrochem Commun 3:429–434

[B185] Brown RJC, Kucernak AR (2001) The electrochemistry of platinum phthalocyanine microcrystals. III. Electrochemical behaviour in aqueous electrolytes. Electrochim Acta 46:2573–2582

[B186] Frangini S, Carewska M, Passerini S, Scaccia S (2001) Intercalation behaviour of $LiCoO_2$ electrodes in aqueous alkaline electrolytes by microparticle cyclic voltammetry. J New Mater Electrochem Syst 4:83–88

[B187] Scholz F, Schwudke D, Stößer R, Boháček J (2001) The interaction of Prussian blue and dissolved hexacyanoferrate ions with goethite (α-FeOOH) studied to assess the chemical stability and physical mobility of Prussian blue in soils. Ecotoxicol Environ Saf Environ Res 49:245–254

[B188] Komorsky-Lovrić Š, Lovrić M, Scholz F (2001) Cyclic voltammetry of decamethylferrocene at the three-phase junction organic liquid|aqueous solution|graphite. J Electroanal Chem 508:129–137

[B189] Doménech-Carbó A, Doménech-Carbó MT, Osete-Cortina L (2001) Identification of manganese(IV) centers in archaeological glass using microsample coatings attached to polymer film electrodes. Electroanalysis 13:927–935

[B190] Jiang J, Kucernak AR (2001) The electrochemistry of platinum phthalocyanine microcrystals. IV. Temperature dependence of the electrochemical behaviour in non-aqueous solution. Electrochim Acta 46:3445–3456

[B191] Grygar T, Bezdička P, Piszora P, Wolska E (2001) Electrochemical reactivity of Li-Mn-O and Li-Fe-Mn-O spinels. J Solid State Electrochem 5:487–494

[B192] Galova M, Orinakova R, Grygar T, Lux L, Hezelova M (2001) Relation between the dissolution reactivity of powder substances and efficiency of their electroplating. Particulate Sci Technol 19:85–94

[B193] Cepriá G, Aranda C, Pérez-Arantegui J, Lacueva F, Castillo JR (2001) Voltammetry of immobilized microparticles: a powerful analytical technique to study the physical and chemical composition of brass. J Electroanal Chem 513:52–58

[B194] Grygar T, Bezdička P, Vorm P, Jordanova N, Krtil P (2001) Spinel solid solutions in the Li-Fe-Mn-O system. J Solid State Chem 161:152–160

[B195] Doménech-Carbó A, Doménech-Carbó MT, Gimeno-Adelantado JV, Bosch-Reig F, Saurí-Peris MC, Sánchez-Ramos S (2001) Electrochemistry of iron oxide pigments (earths) from pictorial microsamples attached to graphite-polyester composite electrodes. Analyst 126:1764–1772

[B196] Wooster TJ, Bond AM, Honeychurch MJ (2001) Resistance transitions detected by analysis of the voltammetry of tetrathiafulvalene microparticles adhered to electrode surfaces under conditions of dynamic resistance compensation. Electrochem Commun 3:746–752

[B197] Gergely A, Inzelt G (2001) Electropolymerization of 3-methylthiophene at liquid 3-methylthiphene/aqueous solution/graphite three-phase junction. Electrochem Commun 3:753–757

[B198] van Oorschot IHM, Grygar T, Dekkers MJ (2001) Detection of low concentrations of fine-grained iron oxides by voltammetry of microparticles. Earth Planet Sci Lett 193:631–642

[B199] Shankaran DR, Narayanan SS (2001) Mechanically immobilized copper hexacyanoferrate modified electrode for electrocatalysis and amperometric determination of glutathione. Bull Korean Chem Soc 22:816–820

[B200] Shankaran DR, Narayanan SS (2001) Electrochemical determination of hydrazine based on chemically modified electrode. Res J Chem Environ 5:21

[B201] Shankaran DR, Narayanan SS (2001) Amperometric sensor for thiols based on mechanically immobilised nickel hexacyanoferrate modified electrode. Bull Electrochem 17:277–280

[B202] Shankaran DR, Narayanan SS (2001) Amperometric sensor for hydrazine determination based on mechanically immobilized nickel hexacyanoferrate modified electrode. Russ J Electrochem 37:1149–1153

[B203] McKenzie KJ, Marken F (2001) Direct electrochemistry of nanoparticulate Fe_2O_3 in aqueous solution and adsorbed onto tin-doped indium oxide. Pure Appl Chem 37:1885–1894

[B204] Grygar T, Marken F, Schröder U, Scholz F (2002) Electrochemical analysis of solids. Coll Czech Chem Commun 67:163–208

[B205] Dokko K, Horikoshi S, Itoh T, Nishizawa M, Abe T, Umeda M, Uchida I (2002) Rapid evaluation of charge/discharge properties for lithium manganese oxide particles at elevated temperature. J Solid State Electrochem 6:188–193

[B206] Gulaboski R, Mirčeski V, Scholz F (2002) An electrochemical method for the determination of the standard Gibbs energy of anion transfer between water and *n*-octanol. Electrochem Commun 4:277–283

[B207] Donten M, Stojek Z, Scholz F (2002) Electron transfer – ion insertion electrochemistry at an immobilised droplet: probing the three-phase electrode-reaction zone with a Pt disk microelectrode. Electrochem Commun 4:324–329

[B208] Jiang J, Kucernak A (2002) Nanostructured platinum as an electrocatalyst for the electrooxidation of formic acid. J Electroanal Chem 520:64–70

[B209] Bárcena Soto M, Scholz F (2002) The thermodynamics of the insertion electrochemistry of solid metal hexacyanometallates. J Electroanal Chem 521:183–189

[B210] Wang J, Kawde AN (2002) Magnetic-field stimulated DNA oxidation. Electrochem Commun 4:349–352

[B211] Inzelt G (2002) Cyclic voltammetry of solid diphenylamine crystals immobilized on an electrode surface and in the presence of an aqueous solution. J Solid State Electrochem 6:265–271

[B212] Shankaran DR, Narayanan SS (2002) Amperometric sensor for glutathione based on mechanically immobilised cobalt hexacyanoferrate modified electrode. Bull Chem Soc Jpn 75:501–505

[B213] Marken F, Hayman CM, Bulman Page PC (2002) Phosphate and arsenate electro-insertion processes into a *N,N,N′,N′*-tetraoctylphenylenediamine redox liquid. Electrochem Commun 4:462–467

[B214] Marken F, Hayman CM, Bulman Page PC (2002) Chromate and dichromate electro-insertion process into a *N,N,N′,N′*--tetraoctylphenylenediamine redox liquid. Electroanalysis 14:1–5

[B215] Mirčeski V, Scholz F (2002) Reduction of iodine at the organic liquid/aqueous solution/ graphite electrode three-phase arrangement. J Electroanal Chem 522:189–198

[B216] Wang J, Xu D, Polsky R (2002) Magnetically-induced solid state electrochemical detection of DNA hybridization. J Am Chem Soc 124:4208–4209

[B217] Doménech-Carbó A, Doménech-Carbó MT, Osete-Cortina L, Gimeno-Adelantado JV, Bosch-Reig F, Mateo-Castro R (2002) Electrochemical identification of metal ions in archaeological ceramic glazes by stripping voltammetry at graphite/polyester composite electrodes. Talanta 56:161–174

[B218] Fehér K, Inzelt G (2002) Electrochemical quartz crystal microbalance study of formation and redox transformation of poly(diphenylamine). Electrochim Acta 47:3551–3559

[B219] Bárcena Soto M, Kubsch G, Scholz F (2002) Cyclic voltammetry of immobilized microparticles with *in situ* calorimetry. Part I: The thermistor electrode. J Electroanal Chem 528:18–26

[B220] Bárcena Soto M, Scholz F (2002) Cyclic voltammetry of immobilized microparticles with *in situ* calorimetry. Part II: Application of a thermistor electrode for in situ calorimetric studies of the electrochemistry of solid metal hexacyanoferrates. J Electroanal Chem 528:27–32

[B221] Scholz F, Gulaboski R, Mirčeski V, Langer P (2002) Quantification of the chiral recognition in electrochemically driven ion transfer across the interface water|chiral liquid. Electrochem Commun 4:659–662

[B222] Grygar T, Dedecek J, Hradil D (2002) Analysis of low concentrations of ferric oxides in clays by diffuse reflectance spectroscopy and voltammetry. Geologica Carpathica 53:1–7

[B223] Long JW, Ayers KE, Rolison DR (2002) Electrochemical characterization of high-surface-area catalysts and other nanoscale electroactive materials at sticky-carbon electrodes. J Electroanal Chem 522:58–65

[B224] Chatterjee A, Wiltshire R, Holt KB, Compton RG, Foord JS, Marken F (2002) Abrasive stripping voltammetry of silver and tin at boron-doped diamond electrodes. Diamond Relat Mater 11:646–650

[B225] Doménech-Carbó A, Pérez-Ramírez J, Ribera A, Kapteijn F, Mul G, Moulijn JA (2002) Characterization of iron species in ex-framework FeZSM-5 by electrochemical methods. Catal Lett 78:303–312

[B226] Doménech-Carbó A, Pérez-Ramírez J, Ribera A, Mul G, Kapteijn F, Arends IWCE (2002) Electrochemical characterization of iron sites in ex-framework FeZSM-5. J Electroanal Chem 519:72–84

[B227] Doménech-Carbó A, Alarcón J (2002) Determination of hydrogen peroxide using glassy carbon and graphite/polyester composite electrodes modified by vanadium-doped zirconias. Anal Chim Acta 452:11–22

[B228] Doménech-Carbó A, Sánchez-Ramos S, Doménech-Carbó MT, Gimeno-Adelantado JV, Bosch-Reig F, Yusá-Marco DJ, Saurí-Peris MC (2002) Electrochemical determination of the Fe(III)/Fe(II) ratio in archaeological ceramic materials using carbon paste and composite electrodes. Electroanalysis 14:685–696

[B229] Grygar T, Bezdička P, Hradil D, Doménech-Carbó A, Marken F, Pikna L, Cepriá G (2002) Voltammetric analysis of iron oxide pigments. Analyst 127:1100–1107

[B230] Wadhawan JD, Evans RG, Banks CE, Wilkins SJ, France RR, Oldham NJ, Fairbanks AJ, Wood B, Walton DJ, Schröder U, Compton RG (2002) Voltammetry of electroactive oil droplets: electrochemically-induced ion insertion, expulsion and reaction processes at microdroplets of *N,N,N′,N′*-tetraalkyl-para-phenylenediamines (TRPD, R = *n*-butyl, *n*-hexyl, *n*-heptyl and n-nonyl). J Phys Chem B 106:9619–9632

[B231] Shankaran DR, Narayanan SS (2002) Amperometric sensor for thiosulphate based on cobalt hexacyanoferrate modified electrode. Sens Actuators B 86:180–184

[B232] Fiedler DA, Scholz F (2002) Electrochemical studies of solid compounds and materials, Chap. II.8. In: Scholz F (ed) Electroanalytical methods – guide to experiments and applications. Springer, Berlin, 331 p, ISBN 3-540-42229-3

[B233] Doménech-Carbó A, Alarcón J (2002) Electrochemistry of vanadium-doped tetragonal and monoclinic ZrO_2 attached to graphite/polyester composite electrodes. J Solid State Electrochem 6:443–450

[B234] Mirčeski V, Gulaboski R, Scholz F (2002) Determination of the standard Gibbs energies of transfer of cations across the nitrobenzene/water interface utilizing the reduction of iodine in an immobilized nitrobenzene droplet. Electrochem Commun 4:813–818

[B235] McKenzie KJ, Asogan D, Marken F (2002) Adsorption and reactivity of hydrous iron nanoparticles on boron-doped diamond. Electrochem Commun 4:819–823

[B236] Wang J, Liu G, Polsky R, Merkoci A (2002) Electrochemical stripping detection of DNA hybridization based on cadmium sulfide nanoparticle tags. Electrochem Commun 4:722–726

[B237] Myland JC, Oldham KB (2002) A model of cyclic voltammetry for a thin organic layer sandwiched between an electrode and an aqueous solution. Convolutive modelling in the absence of supporting electrolyte. J Electroanal Chem 530:1–9

[B238] Yuan Y, Gao Z, Guo J, Shao Y (2002) Study of facilitated potassium ion transfer across a water vertical bar 1,2-dichloroethane interface using different phase volume ratio systems. J Electroanal Chem 526:85–91

[B239] McKenzie KJ, Marken F (2002) Electrochemical characterization of hydrous ruthenium oxide nanoparticle decorated boron-doped diamond electrodes. Electrochem Solid State Lett 5:E47–E50

[B240] Komorsky-Lovrić Š, Riedl K, Gulaboski R, Mirčeski V, Scholz F (2002) Determination of standard Gibbs energies of transfer of organic anions across the water|nitrobenzene interface. Langmuir 18:8000–8005 (and Komorsky-Lovrić Š, Riedl K, Gulaboski R, Mirčeski V, Scholz F (2003) Langmuir 19:3090)

[B241] Wadhawan JD, Evans RG, Compton RG (2002) Voltammetric characteristics of graphite electrodes modified with microdroplets of n-butylferrocene. J Electroanal Chem 533:71–84

[B242] Tasakorn P, Chen J, Aoki K (2002) Voltammetry of a single oil droplet on a large electrode. J Electroanal Chem 533:119–126

[B243] Jinag J, Kucernak A (2002) Electrooxidation of small organic molecules on mesoporous precious metal catalysts. I: CO and methanol on platinum. J Electroanal Chem 533:153–165

[B244] Sánchez-Ramos S, Bosch-Reig F, Gimeno-Adelantado JV, Yusá Marco DJ, Doménech-Carbó A (2002) Application of XRF, XRD, thermal analysis, and voltammetric techniques to the study of ancient ceramics. Anal Bioanal Chem 373:893–900

[B245] Lawrence NS, Thompson M, Davis J, Jiang L, Jones TGJ, Compton RG (2002) Carbon-epoxy electrodes: unambiguous identification of authentic triple-phase (insultor/solution/electrode) processes. Chem Commun 1028–1029

[B246] Pérez-Ramírez J, Mul G, Kapteijn F, Moulijn JA, Overweg AR, Doménech-Carbó A, Ribera A, Arends IWCE (2002) Physicochemical characterization of isomorphously substituted FeZSM-5 during activation. J Catal 207:113–126

[B247] Widmann A, Kahlert H, Petrovic-Prelevic I, Wulff H, Yakhmi JV, Bagkar N, Scholz F (2002) Structure, insertion electrochemistry and magnetic properties of a new type of substitutional solid solutions of copper, nickel and iron hexacyano-ferrates/hexacyanocobaltates. Inorg Chem 42:5706–5715

[B248] Almeida CMVB, Giannetti BF (2002) A new and practical carbon electrode for insoluble and ground samples. Electrochem Commun 4:985–988

[B249] Yang M, Li HL (2002) Determination of trace nitrite by differential pulse voltammetry using magnetic microspheres. Coll Czech Chem Commun 67:1173–1180

[B250] Doménech-Carbó A, Doménech-Carbó MT, Gimeno-Adelantado JV, Bosch-Reig F (2002) Electrochemical solid state analysis: analytical applications of electrode modification by solid microparticles. Rec Res Dev Pure Appl Anal Chem 4:107–121

[B251] Keane L, Hogan C, Forster RJ (2002) Dynamics of charge transport through osmium tris dimethoxy bipyridyl solid deposits. Langmuir 18:4826–4833

[B252] Walsh DA, Keyes TE, Forster RJ (2002) Redox switching in solid deposits: triazole bridged osmium dimers. J Electroanal Chem 538(539):75–85

[B253] Shankaran DR, Narayanan SS (2002) Cobalt hexacyanoferrate-modified electrode for amperometric assay of hydrazine. Russ J Electrochem 38:987–991

[B254] Schröder U, Wadhawan J, Evans RG, Compton RG, Wood B, Walton DJ, France RR, Marken F, Bulman Page PC, Hayman CM (2002) Probing thermodynamic aspects of electrochemically driven ion-transfer processes across liquid|liquid interfaces: pure versus diluted redox liquids. J Phys Chem B 106:8697–8704

[B255] Wooster TJ, Bond AM, Honeychurch MJ (2003) An analogy of an ion-selective electrode sensor based on the voltammetry of microcrystals of tetracyanoquinodimethane or tetrathiafulvalene adhered to an electrode surface. Anal Chem 75:586–592

[B256] Cepriá G, Usón A, Pérez-Arantegui J, Castillo JR (2003) Identification of iron(III) oxides and hydroxyl-oxides by voltammetry of immobilised microparticles. Anal Chim Acta 477:157–168

[B257] Lovrić M, Scholz F (2003) Modeling cyclic voltammograms of simultaneous electron and ion transfer reactions at a conic film three-phase electrode. J Electroanal Chem 540:89–96

[B258] Aoki K, Tasakorn P, Chen J (2003) Electrode reaction at sub-micron oil|water|electrode interfaces. J Electroanal Chem 542:51–60

[B259] Gulaboski R, Riedl K, Scholz F (2003) Standard Gibbs energies of transfer of halogenate and pseudohalogenate ions, halogen substituted acetates, and cycloalkyl carboxylate anions at the water|nitrobenzene interface. PCCP 5:1284–1289

[B260] Marken F, Cromie S, McKee V (2003) Electrochemically driven reversible solid state metal exchange processes in polynuclear copper complexes. J Solid State Electrochem 7:141–146

[B261] Cepriá G, Bolea E, Laborda F, Castillo JR (2003) Quick, easy, and inexpensive way to detect small metallic particles in suspension using voltammetry of immobilized microparticles. Anal Lett 36:921–929

[B262] Gulaboski R, Mirčeski V, Scholz F (2003) Determination of the standard Gibbs energies of transfer of cations and anions of amino acids and small peptides across the water|nitrobenzene interface. Amino Acids 24:149–154

[B263] Tan WT, Lim EB, Bond AM (2003) Voltammetric studies on microcrystalline C_{60} adhered to an electrode surface by solvent casting and mechanical transfer methods. J Solid State Electrochem 7:134–140

[B264] Grygar T, Hradilova J, Hradil D, Bezdička P, Bakardjieva S (2003) Analysis of earthy pigments in grounds of baroque paintings. Anal Bioanal Chem 375:1154–1160

[B265] Doménech-Carbó A, Dménech-Carbó MT, Saurí-Peris MC, Gimeno-Adelantado JV, Bosch-Reig F (2003) Electrochemical identification of anthraquinone-based dyes in solid microsamples by square wave voltammetry using graphite/polyester composite electrodes. Anal Bioanal Chem 375:1169–1175

[B266] Gulaboski R, Scholz F (2003) The lipophilicity of peptide anions – an experimental data set for lipophilicity calculations. J Phys Chem B 107:5650–5657

[B267] Tan WT, Ng GK, Bond AM (2000) Electrochemical oxidation of microcrystalline tetrathiafulvalene (TTF) at an electrode-solid-aqueous (KBr) interface. Malays J Chem 2:34–42

[B268] Tan WT, Chan SY, Lee CK (2002) Electrochemical reduction of microparticles of bismuth vanadate ($Bi_4V_2O_{11}$) mechanically attached to a carbon electrode surface. Malays J Chem 4:34–42

[B269] Bouchard G, Galland A, Carrupt PA, Gulaboski R, Mirčeski V, Scholz F, Girault HH (2003) Standard partition coefficients of anionic drugs in the *n*-octanol/water system determined by voltammetry at three-phase electrodes. Phys Chem Chem Phys 5:3748–3751

[B270] Jayalakshmi M, Mohan Rao M, Scholz F (2003) Electrochemical behaviour of solid lithium manganate ($LiMn_2O_4$) in aqueous neutral electrolyte solution. Langmuir 19:8403–8408

[B271] Scholz F, Gulaboski R, Caban K (2003) The determination of standard Gibbs energies of transfer of cations across the nitrobenzene|water interface with the help of a three-phase electrode. Electrochem Commun 5:929–934

[B272] Inzelt G (2003) Formation and redox behaviour of polycarbazole prepared by electropolymerisation of solid carbazole crystals immobilized on an electrode surface. J Solid State Electrochem 7:503–510

[B273] Wain AJ, Wadhawan JD, Compton RG (2003) Electrochemical studies of vitamin K_1 microdroplets: electrocatalytic hydrogen evolution. Chem Phys Chem 4:974–982

[B274] Grygar T, Dedecek J, Kruiver PP, Dekkers MJ, Bezdička P, Schneeweiss O (2003) Iron oxide mineralogy in late miocene red beds from La Gloria, Spain: rock-magnetic, voltammetric and vis spectroscopy analyses. Catena 53:115–132

[B275] Kovanda F, Grygar T, Dorničák V (2003) Thermal behaviour of Ni-Mn layered double hydroxide and characterization of formed oxides. Solid State Sci 5:1019–1026

[B276] Hasse U, Nießen J, Scholz F (2003) Atomic force microscopy of the electrochemical reductive dissolution of sub-micrometer sized crystals of goethite immobilized on a gold electrode. J Electroanal Chem 556:13–22

[B277] Gulaboski R, Caban K, Stojek Z, Scholz F (2004) The determination of the standard Gibbs energies of ion transfer between water and heavy water by using the three-phase electrode approach. Electrochem Commun 6:215–218

[B278] Hasse U, Wagner K, Scholz F (2004) Nucleation at three-phase junction lines: *in situ* atomic force microscopy of the electrochemical reduction of sub micrometer size silver and mercury(I) halide crystals immobilized on gold electrodes. J Solid State Electrochem 8:842–853

[B279] Hasse U, Scholz F (2004) In situ AFM evidence of the involvement of an oversaturated solution in the course of oxidation of silver nanocrystals to silver iodide crystals. Electrochem Commun 6:409–412

[B280] Komorsky-Lovrić Š, Mirčeski V, Kabbe C, Scholz F (2004) An *in situ* microscopic spectroelectrochemical study of a three-phase electrode where an ion transfer at the water|nitrobenzene interface is coupled to an electron transfer at the interface ITO|nitrobenzene. J Electroanal Chem 566:371–377

[B281] Mirčeski V, Gulaboski R, Scholz F (2004) Square-wave thin-film voltammetry in the presence of uncompensated resistance. Electroanal Chem 566:351–360, Electrochem Commun 12:206–209

[B282] Gulaboski R, Galland A, Bouchard G, Caban K, Kretschmer A, Carrupt PA, Stojek Z, Girault HH, Scholz F (2004) A comparison of the solvation properties of 2-nitrophenyloctyl ether, nitrobenzene, and *n*-octanol as assessed by ion transfer experiments. J Phys Chem B 108:4565

[B283] Bond AM (2002) Illustration of the principles of voltammetry at solid-electrode-solvent (electrolyte) interfaces when redox active microparticles are adhered to an electrode surface. In: Broadening electrochemical horizons, principles and illustration of voltammetric and related techniques. Oxford University Press, Oxford

[B284] Wadhawan JD, Wain AJ, Kirkham AN, Walton DJ, Wood B, France RR, Bull SD, Compton RG (2003) Electrocatalytic reactions mediated by *N,N,N′,N′*-tetraalkyl-1,4-phenylenediamine redox liquid microdroplet-modified electrodes: chemical and photochemical reactions in, and at the surface of femtoliter droplets. J Am Chem Soc 125:11418–11429

[B285] Banks CE, Davies TJ, Evans RG, Hignett G, Wain AJ, Lawrence NS, Wadhawan JD, Marken F, Compton RG (2003) Electrochemistry of immobilised redox droplets: concepts and applications. Phys Chem Chem Phys 5:4053–4069

[B286] Wain AJ, Lawrence NS, Greene PR, Wadhawan JD, Compton RG (2003) Reactive chemistry via the redox switching of microdroplets of 4-nitrophenyl nonyl ether in the presence of aqueous electrolytes. Phys Chem Chem Phys 5:1867–1875

[B287] Wadhawan JD, Wain AJ, Compton RG (2003) Electrochemical probing of photochemical reactions inside femtolitre droplets confined to electrodes. Chem Phys Chem 4:1211–1215

[B288] Davies TJ, Brooks BA, Fisher AC, Yunus K, Wilkins AJ, Greene PR, Wadhawan JD, Compton RG (2003) A computational and experimental study of the cyclic voltammetry response of partially blocked electrodes. Part II: Randomly distributed and overlapping blocking systems. J Phys Chem B 107:6431–6444

[B289] Marken F, Blythe A, Compton RG, Bull SD, Davies SG (1999) Sulfide accumulation and sensing based on electrochemical processes in microdroplets of N^1-[4-(dihexylamino) phenyl]-N^1,N^4,N^4-trihexyl-1,4-phenylenediamine. Chem Commun 1823–1824

[B290] Snook G (2000) Investigation of solid-state reactions by electrochemical and quartz crystal microbalance measurements. Ph.D. thesis, Monash University, Melbourne

[B291] Neufeld A (2003) Investigations using a Kelvin probe instrument and solid state electrochemical techniques: the initiation mechanism of corrosion of zinc and the solid-solid electrochemical transformation of CuTCNQ. Ph.D. thesis, Monash University, Melbourne

[B292] Widmann A (2003) Insertions elektrochemie von festen Metallhexacyano-metallaten. Ph.D. Thesis, Universität Greifswald

[B293] Gulaboski R (2004) The determination of standard Gibbs energies of ion transfer across liquid|liquid interfaces with the help of three-phase electrodes. Ph.D. thesis, Universität Greifswald

[B294] Zhang J, Bond AM (2004) Voltammetric studies with adhered microparticles and the detection of a dependence of organometallic $cis^+ \rightarrow trans^+$ first-order isomerization rate constants on the identity of the ionic liquid. J Phys Chem B 108:7363–7372

[B295] Snook G, Bond AM, Fletcher S (2003) The catalysis of solid state intercalation processes by organic solvents. J Electroanal Chem 554–555:157–165

[B296] Neufeld AK, Madsen I, Bond AM, Hogan CF (2003) Phase, morphology, and particle size changes associated with the solid-solid electrochemical interconversion of TCNQ and semiconducting CuTCNQ (TCNQ = tetracyanoquinodimethane). Chem Mater 15:3573–3585

[B297] Guo S, Zhang J, Elton DM, Bond AM (2004) Fourier transform large-amplitude alternating current cyclic voltammetry of surface-bound azurin. Anal Chem 76:166–177

[B298] Wu P, Cai C (2004) The solid state electrochemistry of samarium(III) hexacyanoferrate. J Solid State Electrochem 8:538–543

[B299] Wooster TJ, Bond AM (2003) Ion selectivity obtained under voltammetric conditions when a TCNQ chemically modified electrode is presented with aqueous solutions containing tetraalkylammonium cations. Analyst 128:1386–1390

[B300] Thompson M, Lawrence NS, Davis J, Jiang L, Jones TGJ, Compton RG (2002) A reagentless renewable *N,N′*-diphenyl-*p*-phenylenediamine loaded sensor for hydrogen sulfide. Sens Actuators B Chem 87:33–40

[B301] Doménech-Carbó A, Alvaro M, Ferrer B, García H (2003) Electrochemistry of mesoporous organosilica of MCM-41 type containing 4,4′-bipyridinium units: voltammetric response and electrocatalytic effect on 1,4-dihydrobenzoquinone oxidation. J Phys Chem B 107:12781–12788

[B302] Grygar T, Kuckova S, Hradil D, Hradilova J (2003) Electrochemical analysis of natural solid organic dyes and pigments. J Solid State Electrochem 7:706–713

[B303] Fay N, Dempsey E, Kennedy A, McCormac T (2003) Synthesis and electrochemical characterisation of $[Ru(bpy)_3]_3[P_2W_{18}O_{62}]$. J Electroanal Chem 556:63–74

[B304] Doménech-Carbó A, Doménech-Carbó MT, Osete-Cortina L, Gimeno-Adelantado JV, Sánchez-Ramos S, Bosch-Reig F (2003) Quantitation of metal ions in archaeological glass by abrasive stripping square-wave voltammetry using graphite/polyester composite electrodes. Electroanalysis 15:1465–1475

[B305] Wildgoose GG, Pandurangappa M, Lawrence NS, Jiang L, Jones TGJ, Compton RG (2003) Anthraquinone-derivatised carbon powder: reagentless voltammetric pH electrodes. Talanta 60:887–893

[B306] Zhang J, Bond AM, Belcher J, Wallace KJ, Steed JW (2003) Electrochemical studies on the modular podand 1,3,5-tris(3-((ferrocenylmethyl)amino)pyridiniumyl)-2,4,6-triethylbenzene hexafluorophosphate in conventional solvents and ionic liquids. J Phys Chem B 107:5777–57

[B307] Galova M, Grygar T, Pikna L, Lux L (2003) Electrochemical study on the reactivity of Fe powders. Chem Anal 48:293–303

[B308] Zhang J, Bond AM (2003) Conditions required to achieve the apparent equivalence of adhered solid- and solution-phase voltammetry for ferrocene and other redox-active solids in ionic liquids. Anal Chem 75:2694–2702

[B309] Pandurangappa M, Lawrence NS, Compton RG (2002) Homogeneous chemical derivatisation of carbon particles: a novel method for functionalising carbon surfaces. Analyst 127:1568–1571

[B310] Pandurangappa M, Lawrence NS, Jiang L, Jones TGJ, Compton RG (2003) Physical adsorption of *N,N′*-diphenyl-*p*-phenylenediamine onto carbon particles: application to the detection of sulfide. Analyst 128:473–479

[B311] Grygar T, Bezdička P, Hradil D, Pikna L (2003) Electrochemical analysis of metal oxides. Solid State Chem Solid State Phenomena 90(91):45–49

[B312] Cepriá G, Usón A, Pérez-Arantegui J, Castillo JR (2003) Identification of iron(III) oxides and hydroxyl-oxides by voltammetry of immobilised microparticle. Anal Chim Acta 477:157–168

[B313] Uzun D, Ozser ME, Yuney K, Icil H, Demuth M (2003) Synthesis and photophysical properties of *N,N′*-bis(4-cyanophenyl)-3,4,9,10-perylene-bis(dicarboximide) and *N,N′*-bis (4-cyanophe-nyl)-1,4,5,8-naphthalenediimide. J Photochem Photobiol A Chem 156:45–54

[B314] Alonso-Sedano AB, Tascón-García L, Vázquez-Barbado D, Sánchez Batanero P (2004) Electrochemical study of copper and iron compounds in the solid state by using voltammetry of immobilized microparticles: application to copper ferrite characterization. J Electroanal Chem 566:433–441

[B315] Wildgoose GG, Leventis HC, Streeter I, Lawrence NS, Wilkins SJ, Jiang L, Jones TGJ, Compton RG (2004) Abrasively immobilised multiwalled carbon nanotube agglomerates: a novel electrode material approach for the analytical sensing of pH. ChemPhysChem 5:669–677

[B316] Rees NV, Wadhawan JD, Klymenko OV, Coles BA, Compton RG (2004) An electrochemical study of the oxidation of 1,3,5-tris[4-[(3-methylphenyl)phenylamino]phenyl] benzene. J Electroanal Chem 563:191–202

[B317] Doménech-Carbó A, Torres FJ, Alarcón J (2004) Electrochemical characterization of cobalt cordierites attached to paraffin-impregnated graphite electrodes. J Solid State Electrochem 8:127–137

[B318] Doménech-Carbó A, Sánchez-Ramos S, Yusá-Marco DJ, Moya-Moreno M, Gimeno-Adelantado J, Bosch-Reig F (2004) Electrochemical determination of boron in minerals and ceramic materials. Anal Chim Acta 501:103–111

[B319] Doménech-Carbó A, Doménech-Carbó MT (2006) Chronoamperometric study of proton transfer/electron transfer in solid state electrochemistry of organic dyes. J Solid State Electrochem 10:949–958

[B320] Doménech-Carbó A, Doménech-Carbó MT (2008) In situ AFM study of proton-assisted electrochemical oxidation/reduction of microparticles of organic dyes. Electrochem Commun 10:1238–1241

[B321] Doménech-Carbó A, Doménech-Carbó MT, Calisti M, Maiolo V (2010) Identification of naphtoquinonic and anthraquinonic dyes via sequential potential steps applied to the voltammetry of microparticles methodology. J Solid State Electrochem 14:465–477

[B322] Doménech-Carbó A, Doménech-Carbó MT, Calisti M, Maiolo V (2010) Sequential identification of organic dyes using the voltammetry of microparticles approach. Talanta 81:404–414

[B323] Doménech-Carbó A, Martini M, de Carvalho LM, Doménech-Carbó MT, Silva M (2013) Screening of pharmacologic adulterant classes in herbal formulations using voltammetry of microparticles. J Pharm Biomed Anal 74:194–204

[B324] Doménech-Carbó A, Martini M, de Carvalho LM, Viana C, Doménech-Carbó MT, Silva M (2013) Standard additions-dilution method for absolute quantification in voltammetry of microparticles. Application for determining psychoactive 1,4-benzodiazepine and antidepressants drugs as adulterants in phytotherapeutic formulations. J Pharm Biomed Anal 80:159–163

[B325] O'Mullane AP, Neufeld AK, Bond AM (2012) Scanning electrochemical microscopy study of the solid-solid interconversion of TCNQ to phase I and phase II CuTCNQ. Electrochem Commun 22:21–24

[B326] Doménech-Carbó A, Montoya N, Alarcón J (2011) Electrochemical characterization of praseodymium centers in $Pr_xZr_{1-x}O_2$ zirconias using electrocatalysis and photoelectrocatalysis. J Solid State Electrochem 16:963–975

[B327] Doménech-Carbó A, Montoya N, Alarcón J (2011) Electrochemical characterization of praseodymia doped zircon. Catalytic effect on the electrochemical reduction of molecular oxygen in polar organic solvents. Electrochim Acta 56:7104–7111

[B328] Karnicka K, Eckhard K, Guschin DA, Stoica L, Kulesza PJ, Schuhmann W (2007) Visualisation of the local bio–electrocatalytic activity in biofuel cell cathodes by means of redox competition scanning electrochemical microscopy (RC-SECM). Electrochem Commun 9:1998–2002

[B329] Szot K, Nogala W, Niedziolka-Jonsson J, Jonsson-Niedziolka M, Marken F, Rogalski J, Kirchner CN, Wittstock G, Opallo M (2010) Hydrophilic carbon nanoparticle-laccase thin film electrode for mediatorless dioxygen reduction: SECM activity mapping and application in zinc-dioxygen battery. Electrochim Acta 54:4620–4625

[B330] Guadagnini L, Malijusch A, Chen X, Neugebauer S, Tonelli D, Schuhmann W (2009) Visualization of electrocatalytic activity of microstructured metal hexacyanoferrates by means of redox competition mode of scanning electrochemical microscopy (RC-SECM). Electrochim Acta 54:3753–3758

[B331] Doménech-Carbó A, Doménech-Carbó MT, Peiró MA (2011) Dating archaeological lead artifacts from measurement of the corrosion content using the voltammetry of microparticles. Anal Chem 83:5639–5644

[B332] Doménech-Carbó A, Doménech-Carbó MT, Peiró MA, Martínez I, Barrio J (2012) Application of the voltammetry of microparticles for dating archaeological lead using polarization curves and electrochemical impedance spectroscopy. J Solid State Electrochem 16:2349–2356

[B333] Doménech-Carbó A, Doménech-Carbó MT, Peiró MA (2011) 'One-touch' voltammetry of microparticles for the identification of corrosion products in archaeological lead. Electroanalysis 23:1391–1400

[B334] Pavlov D, Monakhov B (1987) Effect of Sb on the electrochemical properties of $Pb/PbSO_4/H_2SO_4$ and $Pb/PbO/PbSO_4/H_2SO_4$ electrodes. J Electroanal Chem 218:135–153

[B335] Pavlov D, Monakhov B, Salmi K, Sundholm G (1991) Ring–disk electrode studies of soluble intermediates formed during the polarization of Pb in H_2SO_4. Electrochim Acta 36:953–963
[B336] Cai W-B, Wan Y-Q, Liu H-T, Zhou W-F (1995) A study of the reduction process of anodic PbO_2 film on Pb in sulfuric acid solution. J Electroanal Chem 387:95–100
[B337] Pavlov D (1981) Semiconductor mechanism of the processes during electrochemical oxidation of PbO to PbO_2. J Electroanal Chem 118:167–185
[B338] Bartlett PN, Dunford T, Ghanem MA (2002) Templated electrochemical deposition of nanostructured macroporous PbO_2. J Mater Chem 12:3130–3135
[B339] Pavlov D, Monakhov B (1989) Mechanism of action of Sn on the passivation phenomena in the lead-acid battery positive plate (Sn-free effect). J Electrochem Soc 136:27–33
[B340] Doménech-Carbó A, Doménech-Carbó MT, Peiró M, Osete-Cortina L (2011) Authentication of archaeological lead artifacts using voltammetry of microparticles: the case of the *Tossal de Sant Miquel* Iberian plate. Archaeometry 53:1193–1211
[B341] Doménech-Carbó A, Doménech-Carbó MT, Pasíes T, Bouzas MC (2012) Modeling corrosion of archaeological silver-copper coins using the voltammetry of immobilized particles. Electroanalysis 24:1945–1955
[B342] Rodgers JD, Jedral W, Bunce NJ (1999) Electrochemical oxidation of chlorinated phenols. Environ Sci Technol 33:1453–1457
[B343] Panizza M, Cerisola G (2003) Influence of anode material on the electrochemical oxidation of 2-naphthol: Part 1. Cyclic voltammetry and potential step experiments. Electrochim Acta 48:3491–3497
[B344] Pavlov D, Kirchev A, Monahov B (2005) Mechanism of the oxygen cycle reactions proceeding at the negative plates of VRLA batteries. J Power Sources 144:521–527
[B345] Wu YM, Li WS, Long XM, Wu FH, Chen HY, Yan JH, Zhang CR (2005) Effect of bismuth on hydrogen evolution reaction on lead in sulfuric acid solution. J Power Sources 144:338–345
[B346] Chen P, McCreery RL (1996) Control of electron transfer kinetics at glassy carbon electrodes by specific surface modification. Anal Chem 68:3958–3965
[B347] Barisci JN, Wallace GG, Baughman RH (2000) Electrochemical studies of single-wall carbon nanotubes in aqueous solutions. J Electroanal Chem 488:92–98
[B348] Doménech-Carbó A (2012) Electrochemical analysis: voltammetry of microparticles. In: Dillmann P, Adriaens A, Angelini E, Watkinson D (eds) Green book on corrosion of cultural heritage artifacts, Chap 9. European Federation of Corrosion, Maney, Leeds, pp 165–189
[B349] Doménech-Carbó A (2012) Electrochemical techniques. In: Edwards HGM, Vandenabeele P (eds) Analytical archaeometry, selected topics, Chap 7. The Royal Society of Chemistry, London, pp 210–241
[B350] Doménech-Carbó A, Doménech-Carbó MT, Lee Y, Osete-Cortina L (2012) Potential application of voltammetry of microparticles for dating porcine blood-based binding media used in Taiwanese architectural polychromies. Chem Asian J 7:2268–2273
[B351] Doménech-Carbó A, Doménech-Carbó MT, Costa V (2009) Electrochemical methods for archaeometry, conservation and restoration. In: Scholz F (ed) Monographs in electrochemistry. Springer, Berlin
[B352] Doménech-Carbó A, Doménech-Carbó MT, Edwards HGM (2008) Quantitation from Tafel analysis in solid-state voltammetry. Application to the study of cobalt and copper pigments in severely damaged frescoes. Anal Chem 80:2704–2716
[B353] Doménech-Carbó A, Doménech-Carbó MT, Martínez-Lázaro L (2010) Layer-by-layer identification of copper alteration products in metallic works of art using the voltammetry of microparticles. Anal Chim Acta 680:1–9
[B354] Doménech-Carbó A, Doménech-Carbó MT, Peiró-Ronda MA (2011) 'One-touch' voltammetry of microparticles for the identification of corrosion products in archaeological lead. Electroanalysis 23:1391–1400

[B355] Doménech-Carbó A, Lastras M, Rodríguez F, Osete-Cortina L (2013) Mapping of corrosion products of highly altered archeological iron using voltammetry of microparticles. Microchem J 106:41–50
[B356] Doménech-Carbó A, Lastras M, Rodríguez F, Cano E, Piquero J, Osete-Cortina L (2014) Monitoring stabilizing procedures of archaeological iron using electrochemical impedance spectroscopy. J Solid State Electrochem 18:399–409
[B357] Doménech-Carbó A, Moya-Moreno M, Doménech-Carbó MT (2004) Standard addition method applied to solid–state striping voltammetry. Determination of zirconium in minerals and ceramic materials. Anal Bioanal Chem 380:146–156
[B358] Doménech-Carbó A, Doménech-Carbó MT, Gimeno-Adelantado JV, Bosch-Reig F (2006) H-point standard addition method applied to solid state stripping voltammetry. Application to the determination of lead and tin in archaeological glazes. Anal Bioanal Chem 385:1552–1561
[B359] Bosch-Reig F, Doménech-Carbó A, Doménech-Carbó MT, Gimeno-Adelantado JV (2007) H-Point standard addition method applied to voltammetry of microparticles. Quantitation of dyes in pictorial samples. Electroanalysis 19:1575–1584
[B360] Doménech-Carbó A, Koshevoy IO, Montoya N, Pakkanen TA (2010) Electrochemically assisted anion insertion in Au(I)-Cu(I) heterotrimetallic clusters bearing ferrocenyl groups: application to the flouride/chloride discrimination in aqueous media. Electrochem Commun 12:206–209
[B361] Doménech-Carbó A, Koshevoy IO, Montoya N, Pakkanen TA (2010) Electrochemical anion sensing using electrodes chemically modified with Au(I)-Cu(I) heterotrimetallic alkynyl cluster complexes containing ferrocenyl groups. Anal Bioanal Chem 397:2013–2022
[B362] Barbante GJ, Hogan CF, Hughes AB (2009) Solid state spectroelectrochemistry of microparticles of ruthenium diimine complexes immobilised on optically transparent electrodes. J Solid State Electrochem 13:599–608
[B363] Brown RJC, Kucernak AR (2005) The photoelectrochemistry of platinum phthalocyanine films in aqueous media. J Solid State Electrochem 9:459–468
[B364] Komorsky-Lovrić Š, Quentel F, L'Her M, Scholz F (2008) Voltammetry of microparticles of lutetium bisphthalocyanine. J Solid State Electrochem 12:165–169
[B365] Brown RJC, Brett DJL, Kucernak ARJ (2009) An electrochemical quartz crystal microbalance study of platinum phthalocyanine thin films. J Electroanal Chem 633:339–346
[B366] Doménech-Carbó A, Maciuk A, Figadère B, Poupon E, Cebrián-Torrejón G (2013) Solid state electrochemical assay of heme-binding molecules for screening of drugs with antimalarial potential. Anal Chem 85:4014–4021
[B367] Doménech-Carbó A, García H, Doménech-Carbó MT, Llabrés-i-Xamena F (2006) Electrochemical nanometric patterning of MOF particles: anisotropic metal electrodeposition in Cu/MOF. Electrochem Commun 8:1830–1834
[B368] Doménech-Carbó A, García H, Doménech-Carbó MT, Llabrés-i-Xamena F (2007) Electrochemistry of metal-organic frameworks: a description from the voltammetry of microparticles approach. J Phys Chem C 111:13701–13711
[B369] Babu KF, Kulandainathan MA, Katsounaros I, Burrows AD, Raithby PR, Marken F (2010) Electrocatalytic activity of Basolite™ F300 metal-organic-framework structures. Electrochem Commun 12:632–636
[B370] Cummings CI, Roweth AH, Ching AKZ, Jenkings ATA, Mitchels JM, Shariki S, Liew SY, Thielemans W, Walsh DA, Marken F (2010) Facile cation electro-insertion into layer-by-layer assembled iron phytate films. Electrochem Commun 12:1722–1726
[B371] Halls JE, Ahn SD, Jiang D, Keenan LL, Burrows AD, Marken F (2012) Proton uptake vs. redox driven release from metal-organic-frameworks: alizarin red S reactivity in UMCM–1. J Electroanal Chem 689:168–175

[B372] Halls JE, Hernan-Gomez A, Burrows AD, Marken F (2012) Metal-organic frameworks post-synthetically modified with ferrocenyl groups: framework effects on redox processes and surface conduction. Dalton Trans 41:1475–1480

[B373] Komorsky-Lovrić Š, Vukašinović N, Penovski R (2003) Voltammetric determination of microparticles of some local anesthetics and antithusics immobilized on the graphite electrode. Electroanalysis 15:544–547

[B374] Komorsky–Lovrić Š, Nigović B (2004) Identification of 5–aminosalicylic acid, ciprofloxacin and azithromycin by abrasive stripping voltammetry. J Pharm Biomed Anal 36:81–89

[B375] Komorsky-Lovrić Š, Nigović B (2006) Electrochemical characterization of simvastatin by abrasive stripping and square-wave voltammetry. J Electroanal Chem 593:125–130

[B376] Komorsky-Lovrić Š, Novak I (2009) Estimation of antioxidative properties of tea leaves by abrasive stripping electrochemistry using paraffin-impregnated graphite electrode. Coll Czech Chem Commun 74:1467–1475

[B377] Komorsky-Lovrić Š, Novak I (2011) Abrasive stripping square–wave voltammetry of blackberry, raspberry, strawberry, pomegranate, and sweet and blue Potatoes. J Food Sci 76:C916–C920

[B378] Corduneanu O, Janeiro P, Oliveira-Brett AM (2006) On the electrochemical oxidation of resveratrol. Electroanalysis 18:757–762

[B379] Oliveira Brett AM, da Silva LA, Fujii H, Mataka S, Thienmann T (2003) Detection of the damage caused to DNA by a thiophene-*S*-oxide using an electrochemical DNA-biosensor. J Electroanal Chem 549:91–99

[B380] de Carvalho LM, Hilgemann M, Spengler C, do Nascimento PC, Bohrer D (2010) Voltametria de micropartículas imobilizadas: fundamentos e aplicações analíticas (Voltammetry of immobilized microparticles: fundamentals and analytical applications). Química Nova 33:1765–1772

[B381] de Carvalho LM, Martini M, Moreira APL, de Lima APS, Correia D, Falcao T, Garcia SC, de Bairros AV, do Nascimento PC, Bohrer D (2011) Presence of synthetic pharmaceuticals as adulterants in slimming phytotherapeutic formulations and their analytical determination. For Sci Int 204:6–12

[B382] Cavalheiro ETG, Brett CMA, Oliveira-Brett AM, Fatibello-Filho O (2012) Bioelectroanalysis of pharmaceutical compounds. Bioanal Rev 4:31–53

[B383] Marselli B, Garcia-Gomez J, Michaud P-A, Rodrigo MA, Comninellis C (2003) Electrogeneration of hydroxyl radicals on boron-doped diamond electrodes. J Electrochem Soc 150: D79–D83

[B384] Enache TA, Chiorcea-Paquim A-M, Fatibello-Filho O, Oliveira-Brett AM (2009) Hydroxyl radicals electrochemically generated in situ on a boron–doped diamond electrode. Electrochem Commun 11:1342–1345

[B385] Kapałka A, Foti G, Comninellis C (2009) The importance of electrode material in environmental electrochemistry: formation and reactivity of free hydroxyl radicals on boron–doped diamond electrodes. Electrochim Acta 54:2018–2023

[B386] Hilgemann M, Scholz F, Kahlert H, de Carvalho LM, Barcellos da Rosa M, Lindequist U, Wurster M, do Nascimento PC, Bohrer D (2010) Electrochemical assay to quantify the hydroxyl radical scavenging activity of medicinal plant extracts. Electroanalysis 22:406–412

[B387] Gulaboski R, Lovrić M, Mirčeski V, Bogeski I, Hoth M (2008) Protein-film voltammetry: a theoretical study of the temperature effect using square-wave voltammetry. Biophys Chem 137:49–55

[B388] Bogeski I, Mirčeski V, Hoth M (2008) Probing the redox activity of T-lymphocytes deposited at electrode surfaces with voltammetric methods. Clin Chem Lab Med 46:197–203

[B389] Li G, Liu X, Du L, Wang E (2008) Direct electrochemistry of horseradish peroxidase immobilized in calcium carbonate microsphere doped with phospholipids. Electroanalysis 20:1421–1426

[B390] Hermes M, Scholz F, Walther R, Schild L, Wolke C, Lendeckel U (2011) Electrochemical signals of mitochondria: a new probe of their membrane properties. Angew Chem Int Ed 50:6872–6875

[B391] Doménech-Carbó A, Doménech-Carbó MT, Saurí-Peris MC (2005) Electrochemical identification of flavonoid dyes in work of art samples by abrasive voltammetry at paraffin-impregnated graphite electrodes. Talanta 66:769–782

[B392] Janeiro P, Oliveira Brett AM (2005) Solid state electrochemical oxidation mechanisms of morin in aqueous media. Electroanalysis 17:733–738

[B393] Doménech-Carbó A, Doménech-Carbó MT, Saurí-Peris MC (2005) Identification of curcuma and safflower dyes by voltammetry of microparticles using paraffin-impregnated graphite electrodes. Microchim Acta 152:75–84

[B394] Doménech-Carbó A, Doménech-Carbó MT, Vázquez de Agredos-Pascual ML (2007) Electrochemical monitoring Maya blue preparation from Maya's ancient procedures. J Solid State Electrochem 11:1335–1346

[B395] Hu X-W, He J-B, You Y-H, Zhang Y-M, Zhang S-L (2011) Spectroelectrochemistry of solid indirubin and its sulfonated form. Electrochim Acta 56:1219–1226

[B396] Ngameni E, Tonle IK, Nanseu CP, Wandji R (2000) Voltammetry study of 2-hydroxy-3-isopropenyl-1,4-naphthoquinone using a carbon paste electrode. Electroanalysis 12:847–852

[B397] He J-B, Ma G-H, Chen J-C, Yao Y, Wang Y (2010) Voltammetry and spectroelectrochemistry of solid indigo dispersed in carbon paste. Electrochim Acta 55:4845–4850

[B398] Gómez-Romero P, Sanchez C (2005) Hybrid materials. Functional properties. From Maya blue to 21st century materials. New J Chem 29:57–58

[B399] Doménech-Carbó A (2010) Electrochemistry of porous materials. Taylor & Francis, Boca Raton, FL

[B400] Doménech-Carbó A, Doménech-Carbó MT, Sánchez del Río M, Vázquez de Agredos-Pascual ML, Lima E (2009) Maya blue as a nanostructured polyfunctional hybrid organic-inorganic material: the need to change paradigms. New J Chem 33:2371–2379

[B401] Doménech-Carbó A, Doménech-Carbó MT, Vázquez de Agredos–Pascual ML (2006) Dehydroindigo: a new piece into the Maya blue puzzle from the voltammetry of microparticles approach. J Phys Chem B 110:6027–6039

[B402] Doménech–Carbó A, Doménech–Carbó MT, Vázquez de Agredos–Pascual ML (2007) Indigo/dehydroindigo/palygorskite complex in Maya blue: an electrochemical approach. J Phys Chem C 111:4585–4595

[B403] Doménech-Carbó A, Doménech-Carbó MT, Sánchez del Río M, Goberna S, Lima E (2009) Evidence of topological indigo/dehydroindigo isomers in Maya blue-like complexes prepared from palygorskite and sepiolite. J Phys Chem C 113:12118–12131

[B404] Doménech-Carbó A, Doménech-Carbó MT, Vázquez de Agredos-Pascual ML (2007) Chemometric study of Maya blue from the voltammetry of microparticles approach. Anal Chem 79:2812–2821

[B405] Doménech-Carbó A, Doménech-Carbó MT, Vázquez de Agredos-Pascual ML (2009) Correlation between spectral, SEM/EDX and electrochemical properties of Maya blue. A chemometric study. Archaeometry 51:1015–1034

[B406] Doménech-Carbó A, Doménech-Carbó MT, Vázquez de Agredos-Pascual ML (2011) From Maya blue to 'Maya yellow': a connection between ancient nanostructured materials from the voltammetry of microparticles. Angew Chem Int Ed 50:5741–5744

[B407] Doménech-Carbó A, Doménech-Carbó MT, Vidal-Lorenzo C, Vázquez de Agredos-Pascual ML (2012) Insights into the Maya blue technology: greenish pellets from the ancient city of La Blanca. Angew Chem Int Ed 51:700–703

[B408] Doménech-Carbó A, Doménech-Carbó MT, Sánchez del Río M, Vázquez de Agredos-Pascual ML (2009) Comparative study of different indigo-clay Maya blue-like systems using the voltammetry of microparticles approach. J Solid State Electrochem 13:869–878

[B409] Doménech-Carbó A, Martini M, De Carvalho LM, Doménech–Carbó MT (2012) Square wave voltammetric determination of the redox state of a reversibly oxidized/reduced depolarizer in solution and in solid state. J Electroanal Chem 684:13–19

[B410] Doménech-Carbó A, Doménech-Carbó MT, Osete-Cortina L, Montoya N (2012) Application of solid-state electrochemistry techniques to polyfunctional organic-inorganic hybrid materials: the Maya blue problem. Micropor Mesopor Mater 166:123–130

[B411] Doménech-Carbó A, Montoya N, Alarcón J (2012) Study of solid state kinetics using voltammetry of immobilized particles. Application to tetragonal to monoclinic transition in nanoparticulate zirconia and praseodymia-doped zirconia. Electrochim Acta 67:24–32

[B412] Walcarius A (2003) Implication of zeolite chemistry in electrochemical science and applications of zeolite-modified electrodes. In: Auerbach, SM, Carrado KA, Dutta PK (eds) Handbook of zeolite science and technology, Chap 14. Marcel Dekker, New York, pp 721–783

[B413] Walcarius A (2006) Zeolite-modified paraffin-impregnated graphite electrode. J Solid State Electrochem 10:469–478

[B414] Rolison DR, Nowak RJ, Welsh TA, Murray CG (1991) Analytical implications of zeolites in overlayers at electrodes. Talanta 38:27–35

[B415] Walcarius A (1996) Zeolite-modified electrodes: analytical applications and prospects. Electroanalysis 8:971–986

[B416] Švancara I, Vytřas K, Kalcher K, Walcarius A, Wang J (2009) Carbon paste electrodes in facts, numbers, and notes: a review on the occasion of the 50-years jubilee of carbon paste in electrochemistry and electroanalysis. Electroanalysis 21:7–28

[B417] Bessel DA, Rolison DR (1997) Topological redox isomers: surface chemistry of zeolite-encapsulated Co(salen) and [Fe(bpy)3]2+ complexes. J Phys Chem B 101:1148–1157

[B418] Doménech-Carbó A, Formentín P, García H, Sabater MJ (2002) On the existence of different zeolite-associated topological redox isomers. Electrochemistry of Y zeolite-associated Mn(salen)N_3 complex. J Phys Chem B 106:574–582

[B419] Doménech-Carbó A, Alvaro M, Ferrer B, García H (2003) Electrochemistry of mesoporous organosilica of MCM-41 type containing 4,4′-bipyridinium units: voltammetric response and electrocatalytic effect on 1,4-dihydrobenzoquinone oxidation. J Phys Chem B 107:12781–12788

[B420] Doménech-Carbó A, Koshevoy IO, Montoya N, Pakkanen TA (2011) Estimation of free energies of anion transfer from solid state electrochemistry of alkynyl-based Au (I) dinuclear and Au(I)-Cu(I) cluster complexes containing ferrocenyl groups. Electrochem Commun 13:96–98

[B421] Doménech-Carbó A, Koshevoy IO, Montoya N, Karttunen AJ, Pakkanen TA (2011) Determination of individual Gibbs energies of anion transfer and excess Gibbs energies using an electrochemical method based on insertion electrochemistry of solid compounds. J Chem Eng Data 56:4577–4586

[B422] Doménech-Carbó A, Montoya N, Scholz F (2011) Estimation of individual Gibbs energies of cation transfer employing the insertion electrochemistry of solid Prussian blue. J Electroanal Chem 657:117–122

[B423] Doménech-Carbó A (2012) Solvent-independent electrode potentials of solids undergoing insertion electrochemical reactions: Part I. Theory. J Phys Chem C 116:25977–25983

[B424] Doménech-Carbó A, Koshevoy IO, Montoya N, Pakkanen TA, Doménech-Carbó MT (2012) Solvent-independent electrode potentials of solids undergoing insertion electrochemical reactions: Part II. Experimental data for alkynyl-diphosphine dinuclear Au (I) complexes undergoing electron exchange coupled to anion exchange. J Phys Chem C 116:25984–25992

[B425] Doménech-Carbó A, Scholz F, Montoya N (2012) Solvent-independent electrode potentials of solids undergoing insertion electrochemical reactions: Part III. Experimental data for Prussian blue undergoing electron exchange coupled to cation exchange. J Phys Chem C 116:25993–25999

[B426] Doménech-Carbó A, Cebrián-Torrejón G, De Miguel L, Tordera V, Rodrigues D, Assad-Kahn S, Fournet A, Figadère B, Vázquez R, Poupon E (2014) dsDNA, ssDNA, G-quadruplex DNA, and nucleosomal DNA electrochemical screening using canthin-6-one alkaloid–modified electrodes. Electrochim Acta 115:546–552. doi:10.1016/j.electacta.2013.11.025

[B427] Allanore A, Levalaine H, Valentin G, Birat JP, Delcroix P, Lapicque F (2010) Observation and modeling of the reduction of hematite particles to metal in alkaline solution by electrolysis. Electrochim Acta 55:4007–4013

[B428] Aeschbacher M, Sander M, Schwarzenbach RP (2010) Novel electrochemical approach to assess the redox properties of humic substances. Environ Sci Technol 44:87–93

[B429] Aeschbacher M, Vergari D, Schwarzenbach RP, Sander M (2011) Electrochemical analysis of proton and electron transfer equilibria of the reducible moities in humic acids. Environ Sci Technol 45:8385–8394

[B430] Gorski CA, Aeschbacher M, Soltermann D, Voeglin A, Baeyens B, Fernández MM, Hofstetter TB, Sander M (2012) Redox properties of structural Fe in clay minerals. 1. Electrochemical quantification of electron-donating and -accepting capacities of smectites. Environ Sci Technol 46:9360–9368

[B431] Gorski CA, Khüpfel L, Voeglin A, Sander M, Hofstetter TB (2012) Redox properties of structural Fe in clay minerals. 2. Electrochemical and spectroscopic characterization of electron transfer irreversibility in ferruginous smectite SWa-1. Environ Sci Technol 46:9369–9377

[B432] Doménech-Carbó A, Doménech-Carbó MT, Valle-Algarra FM, Domine ME, Osete-Cortina L (2013) On the dehydroindigo contribution to Maya blue. J Mater Sci 48:7171–7183

[B433] Doménech-Carbó A, Valle-Algarra FM, Doménech-Carbó MT, Domine ME, Osete-Cortina L, Gimeno-Adelantado JV (2013) Redox tuning and species distribution in Maya blue–type materials: a reassessment. App Mater Sci Interfaces 5:8134–8145

[B434] Dracka O (1996) Theory of current elimination in linear scan voltammetry. J Electroanal Chem 402:19–28

[B435] Trnkova L, Dracka O (1996) Elimination voltammetry. Experimental verification and extension of theoretical result. J Electroanal Chem 413:123–129

[B436] Trnkova L (2005) Identification of current nature by elimination voltammetry with linear scan. J Electroanal Chem 582:258–266

[B437] Pikna L, Kovacova Z, Hezelova M, Trnkova L (2013) Anodic dissolution of iron nanoparticles in chloride-containing acetate buffer electrolytes by elimination voltammetry. J Solid State Electrochem 17:2687–2696

[B438] Agrisuelas J, García-Jareño JJ, Giménez-Romero D, Vicente F (2009) Innovative combination of three alternating current relaxation techniques: electrical charge, mass, and color impedance spectroscopy. Part I: The tool. J Phys Chem C 113:8430–8437

[B439] Agrisuelas J, García-Jareño JJ, Giménez-Romero D, Vicente F (2009) Combination of three alternating current relaxation techniques: electrical charge, mass, and color impedance spectroscopy. Part II: Prussian blue ⇆ Everitt's salt process. J Phys Chem C 113:8438–8446

[B440] Agrisuelas J, Gabrielli C, García-Jareño JJ, Perrot H, Vicente F (2012) Kinetic and mechanistic aspects of poly(o-toluidine)-modified electrode. 1. Simultaneous cyclic spectroelectrochemistry and electrogravimetric studies in H_2SO_4 solutions. J Phys Chem C 116:15620–15629

[B441] Agrisuelas J, García-Jareño JJ, Vicente F (2012) Identification of processes associated with different iron sites in Prussian blue structure by in situ electrochemical, gravimetric, and spectroscopic techniques in the dc and ac regimes. J Phys Chem C 116:1935–1947

[B442] Li C, Li Y, Delaunay J-J (2014) A novel method to synthesize highly photoactive Cu_2O microcrystalline films for use in photoelectrochemical cells. Appl Mater Interfaces 6:480–486

[B443] Palacios-Pedrós A, Caballero-Briones F, Díez-Pérez I, Sanz F (2013) Tin passivation in alkaline media: formation of SnO microcrystals as hydroxyl etching product. Electrochim Acta 111:837–845

[B444] Doménech-Carbó A, Labuda J, Scholz F (2013) Electroanalytical chemistry for the analysis of solids: characterization and classification (Technical Report). Pure Appl Chem 85:609–631

[B445] Gulppi MA, Vejar N, Tamayo L, Azocar MI, Vera C, Silva C, Zagal JH, Scholz F, Páez MA (2014) Electrochem Commun 41:8–11

[B446] Roznyatovskaya NV, Vassiliev SYu, Yusipovich AI, Tsirlina GA, Roznyatovskii VV (2005) Aqueous electrochemistry of binuclear copper complex with Robson-type ligand: dissolved versus surface-immobilized reactant. J Solid State Electrochem 9:581–589

[B447] Brahimi R, Trari M, Bouguelia A, Bessekhouand Y (2010) Electrochemical intercalation of O^{2-} in $CuAlO_2$ single crystal and photoelectrochemical properties. J Solid State Electrochem 14:1333–1338

[B448] Ketir W, Saadi S, Trari M (2012) Physical and photoelectrochemical characterization of $CuCrO_2$ single crystal. J Solid State Electrochem 16:213–218

[B449] Eghbalnia M, Dixon DG (2013) In situ electrochemical characterization of natural pyrite as a galvanic catalyst using single-particle microelectrode technique in ferric sulfate solutions. J Solid State Electrochem 17:235–267

[B450] Scholz F, Gulaboski R (2005) Gibbs energies of transfer of chiral anions across the interface water|chiral organic solvent determined with the help of three-phase electrodes. Faraday Discuss 129:169–177

[B451] Doménech-Carbó A, Doménech-Carbó MT, Capelo S, Pasíes T, Martínez-Lázaro I (2014) Dating archaeological copper/bronze artifacts using the voltammetry of microparticles. Angew Chem Int Ed 53:262–9266

[B452] Doménech-Carbó A, Martini M, Valle-Algarra FM (2014) Determination of the depth profile distribution of guest species in microporous materials using the voltammetry of immobilized particles methodology: application to lapachol attachment to palygorskite and kaolinite. PhysChemChemPhys 16:19024–19034

Index

F. Scholz et al., *Electrochemistry of Immobilized Particles and Droplets*,
DOI 10.1007/978-3-319-10843-8

H

I

K

L

M

N

MIX
Papier aus verantwortungsvollen Quellen
Paper from responsible sources
FSC® C105338

If you have any concerns about our products,
you can contact us on
ProductSafety@springernature.com

In case Publisher is established outside the EU,
the EU authorized representative is:
Springer Nature Customer Service Center GmbH
Europaplatz 3, 69115 Heidelberg, Germany

Printed by Libri Plureos GmbH
in Hamburg, Germany